Disaster Risk Reduction

Methods, Approaches and Practices

Series editor
Rajib Shaw, Keio University, Shonan Fujisawa Campus, Fujisawa, Japan

About the Series

Scope of the Series

Disaster risk reduction is a process, which leads to the safety of community and nations. After the 2005 World Conference on Disaster Reduction, held in Kobe, Japan, the Hyogo Framework for Action [HFA] was adopted as a framework of risk reduction. The academic research and higher education in disaster risk reduction has made/is making a gradual shift from pure basic research to applied, implementation-oriented research. More emphasis is given on the multi-stakeholder collaboration and multidisciplinary research. Emerging university networks in Asia, Europe, Africa and the Americas have urged for the process-oriented research in disaster risk reduction field. Keeping this in mind, this new series will promote the outputs of action research on disaster risk reduction, which will be useful for a wider range of stakeholders including academicians, professionals, practitioners and students and researchers in the related field. The series will focus on some of the emerging needs in the risk reduction field, starting from climate change adaptation, urban ecosystem, coastal risk reduction, education for sustainable development, community-based practices, risk communication, human security, etc. Through academic review, this series will encourage young researchers and practitioners to analyse field practices and link it to theory and policies with logic, data and evidences. Thus, the series emphasizes evidence-based risk reduction methods, approaches and practices.

More information about this series at http://www.springer.com/series/11575

Indrajit Pal • Rajib Shaw
Editors

Disaster Risk Governance in India and Cross Cutting Issues

Springer

Editors
Indrajit Pal
Disaster Preparedness, Mitigation and Management, Asian Institute of Technology
Khlong Luang, Pathum Thani, Thailand

Rajib Shaw
Graduate School of Media and Governance, Keio University, Shonan Fujisawa Campus
Fujisawa, Kanagawa, Japan

ISSN 2196-4106 ISSN 2196-4114 (electronic)
Disaster Risk Reduction
ISBN 978-981-10-3309-4 ISBN 978-981-10-3310-0 (eBook)
DOI 10.1007/978-981-10-3310-0

Library of Congress Control Number: 2017947906

Printed on acid-free paper

This Springer imprint is published by Springer Nature
The registered company is Springer Nature Singapore Pte Ltd.
The registered company address is: 152 Beach Road, #21-01/04 Gateway East, Singapore 189721, Singapore

Preface

Natural disasters and human-induced disasters have increased both in frequency and fury over the years. India has suffered enormously in terms of loss of lives and livelihoods and damage to both public and private property due to the recurrence of major natural and human-induced disasters. Effective disaster risk governance strategies not only save lives and minimize damage but also reduce extra burdens on the economy and enhance resilience to respond effectively during emergencies. Spending on risk reduction is always profitable compared to crisis management. Prior to 2005, governmental stress was always on a reactive approach rather than on disaster risk-reduction policies and strategies.

The Government of India recognized the need for a proactive, comprehensive, and sustained approach to disaster management to reduce the detrimental effects of disasters on the overall socioeconomic development of the country. Consequently, they came out with the Disaster Management Act of 2005 and very recently the National Disaster Management Plan of 2016.

There is increasing evidence from recent disasters that well-aware and well-prepared local governments and local communities can minimize the impacts of disasters. It is a well-accepted fact that the community varies from place to place, and its risk perception and ways to respond to disaster also varies. Therefore, it is important to decentralize policy and to customize it based on local needs and priorities. A strong and committed local government system is often found to be effective for successfully implementing disaster risk-reduction plans. There is a significant gap between scientific interventions, their practices, and implementation in society. Disaster risk governance could provide the opportunity to bridge the gap between science and technology interventions and community implementation.

This book, Disaster Risk Governance in India and Cross-Cutting Issues, is a pioneering attempt to provide a balanced approach of theory and practice in disaster risk governance. The book analytically discusses the status of disaster risk governance at national and regional levels, lessons from recent disasters and scientific interventions towards mitigation measures in an Indian context.

The book primarily covers major geophysical and hydrometeorological hazards and related disasters in India. The book is organized into four parts. Part I provides the outline and basics of disaster-risk governance perspectives at the national level with supporting examples from a global review. Part II specifically highlights the detailed perspectives of risk governance at regional and local levels. Part III of the book is devoted to approaches and issues of disaster risk governance and development at various levels, highlighting the practices and certain examples of disaster risk governance, policy options, institutional setup, risk reduction strategies, and key lessons learned. Part IV of this book highlights the risk reduction and cross-cutting issues, focusing more on risk mitigation and scientific intervention for disaster risk reduction. The book contains 19 chapters.

The main purpose and objective of this work is to connect existing data, research, conceptual work, and practical cases on risk resilience and risk reduction from across India under a common umbrella. People from academia, civil administration, the armed forces, and disaster managers describe their deep understanding and field experiences with the efficacy of disaster governance pertaining to Indian crisis situations. Analysis of Indian policy, best practices on disaster risk governance, and response management also are portrayed in this volume. The primary target groups for the book are students and researchers in the fields of disaster risk management, public administration, disaster risk reduction, environmental sciences, geography, geology, earth sciences, and climate change studies. In addition to academicians, the book targets practitioners and policy makers, who will be able to apply the lessons, risk information, and collective wisdom into policy and decision making. This book serves as a comprehensive reference related to disaster risk governance for disaster managers working in India and other countries.

Khlong Luang, Pathum Thani, Thailand Indrajit Pal
Fujisawa, Kanagawa, Japan Rajib Shaw

Contents

Contributors

Rakesh Bhambri Wadia Institute of Himalayan Geology, Dehradun, Uttarakhand, India

Chandrani Bhandery School of Oceanographic Studies, Jadavpur University, Kolkata, West Bengal, India

Mihir R. Bhatt All India Disaster Mitigation Institute (AIDMI), Ahmadabad, India

Shantanoo Bhattacharya Public Works Department (PWD), Guwahati, Assam, India

Abhra Chanda School of Oceanographic Studies, Jadavpur University, Kolkata, West Bengal, India

Ranit Chatterjee Graduate School of Global Environmental Studies, Kyoto University, Kyoto, Japan

Chiranjiv Choudhary Commissioner of Horticulture & Sericulture, Agriculture & Cooperation Department, Government of Andhra Pradesh, Hyderabad, India

Chandan Ghosh GeoHazards Division, National Institute of Disaster Management, Ministry of Home Affairs, New Delhi, India

Tuhin Ghosh School of Oceanographic Studies, Jadavpur University, Kolkata, India

Anil K. Gupta Wadia Institute of Himalayan Geology, Dehradun, Uttarakhand, India

Anil Kumar Gupta National Institute of Disaster Management, New Delhi, India

Sugata Hazra School of Oceanographic Studies, Jadavpur University, Kolkata, West Bengal, India

Narender Kumar Centre for Land Warfare Studies (CLAWS), New Delhi, India

Himadri Maitra District Disaster Management Officer, Govt. of West Bengal, Bengal, India

Manish Mehta Wadia Institute of Himalayan Geology, Dehradun, Uttarakhand, India

Debashis Mitra Indian Institute of Remote Sensing, Dehradun, Uttarakhand, India

Anirban Mukhopadhyay School of Oceanographic Studies, Jadavpur University, Kolkata, West Bengal, India

Srinivasa Rajamani Neeli City Project Coordinator, UNDP-GoI Project, Visakhapatnam, India

Saswati Paik Faculty, School of Education, Azim Premji University, Bengaluru, India

Indrajit Pal Disaster Preparedness, Mitigation and Management (DPMM), Asian Institute of Technology, Khlong Luang, Pathum Thani, Thailand

J. Perumal Wadia Institute of Himalayan Geology, Dehradun, Uttarakhand, India

Sunil Prashar Asian Disaster Preparedness Center, Bangkok, Thailand

Piyoosh Rautela Department of Disaster Management (Government of Uttarakhand), Disaster Mitigation and Management Centre, Dehradun, Uttarakhand, India

Subir Sen Department of Humanities and Social Sciences, Indian Institute of Technology (IIT), Roorkee, India

Rajib Shaw Graduate School of Media and Governance, Keio University, Shonan Fujisawa Campus, Fujisawa, Kanagawa, Japan

Siddharth Singh Public Health Engineering Department, Indian Administrative Service (IAS), Dispur, Assam, India

Pradeep Srivastava Wadia Institute of Himalayan Geology, Dehradun, Uttarakhand, India

Nakul Kumar Tarun Emergency Response Team (ERT), Super Highway Lab, Gurgaon, India

Priya Namrata Topno Jamsetji Tata School for Disaster Studies, Tata Institute of Social Sciences, Mumbai, Maharashtra, India

About the Editors

Indrajit Pal is an assistant professor in Disaster Preparedness, Mitigation and Management, Asian Institute of Technology, Thailand. He served as a faculty member at the Centre for Disaster Management at Lal Bahadur Shastri National Academy of Administration, Mussoorie, India (a premier national institute for training Indian administrative services officers) for more than 8 years. Dr. Pal holds a Ph.D. in Seismotectonic and Earthquake Hazard Assessment, and two master's degrees, one in applied geology and another in sociology with a specialization in urban sociology. He has more than 14 years of experience in teaching, training, research, curriculum development, advocacy, consultancy primarily focused on disaster governance, risk management (DRM), incident command systems, disaster risk education (DRR), hazard and risk assessment, GIS and remote sensing in DRM, climate change adaptation, and DRM. Dr. Pal is working with international organizations on various aspects of DRR and supervising doctoral and master's degree research students in the field of DRM. He is a reviewer for reputed international journals and has published a number of books and research articles in refereed national and international journals. In 2017 Dr. Pal was been recognized as an IRDR Young Scientist by Integrated Research on Disaster Risk, Beijing, China.

Rajib Shaw is a professor of Keio University at its Shonan Fujisawa Campus in Japan. Earlier, he was the executive director of Integrated Research on Disaster Risk (IRDR), a decade-long research program co-sponsored by the International Council for Science (ICSU), the International Social Science Council (ISSC), and the United Nations International Strategy for Disaster Reduction (UNISDR). He is the co-chair of the UNISDR Asia Science Technology Academic Advisory Group (ASTAAG). Dr. Shaw is also a senior fellow of the Institute of Global Environmental Strategies (IGES) Japan and the chairperson of SEEDS Asia, a Japanese NGO. Previously, he was a professor in the Graduate School of Global Environmental Studies of Kyoto University, Japan. His expertise includes community-based disaster risk management, climate change adaptation, urban risk management, and disaster and environmental education. He has published more than 40 books and more than 300 academic papers and book chapters.

Part I
Overview and National Perspectives

Chapter 1
Disaster Governance and Its Relevance

Indrajit Pal and Rajib Shaw

Abstract It is evident that disasters are a true litmus test of governance. Many attributes of governance interplay in disasters, before, during, and after a situation. Governance is the exercise of political, economic, and administrative authority in the management of a country's affairs at all levels. This book is emphasizing the governance of disaster risks that influences the way in which national and subnational actors (including governments, parliamentarians, public servants, the media, the private sector, and civil society organizations) are willing and able to coordinate their actions to manage and reduce disaster-related risk (UNDP. Disaster risk reduction, governance & mainstreaming. UNDP, New York, 2010). This introductory chapter will discuss the risk governance perspectives in the parlance of accepted theoretical base. The present research will make a review of different governance issues related to the disaster risk reduction, from global, regional, national, and local perspectives, and will highlight the needs and relevance of disaster governance at different levels.

Keywords Risk governance • Disaster management • Natural hazards

1.1 Introduction

The Indian subcontinent has diversified geo-climatic conditions and is being vulnerable for multiple natural hazards. Every year India is facing severe level of disasters, which kills thousands of people and affects millions of people especially the underserved communities. Natural hazards continue to pose a major threat to the entire world with prospects of even greater impacts to life and property in the future (Aini

I. Pal (✉)
Disaster Preparedness, Mitigation and Management (DPMM), Asian Institute of Technology, Khlong Nung, Pathum Thani, Thailand
e-mail: indrajit.pal@gmail.com

R. Shaw
Graduate School of Media and Governance, Keio University, Shonan Fujisawa Campus, Fujisawa, Kanagawa, Japan
e-mail: shaw@sfc.keio.ac.jp

I. Pal, R. Shaw (eds.), *Disaster Risk Governance in India and Cross Cutting Issues*, Disaster Risk Reduction, DOI 10.1007/978-981-10-3310-0_1

and Fakhrul-Razi 2010; Hayles 2010). "Natural" disasters have caused more than 780,000 fatalities and destroyed property and infrastructure worth a minimum of $960 billion over the last decade (Guha-Sapir et al. 2010; UNISDR 2010a). It has been projected that should current trends continue, 100,000 lives will be lost each year, while the costs of "natural" disasters will be in excess of $300 billion per year by 2050 (IFRC 2009).

The literature on disaster risk governance often suggests the "governance" in parlance of "common agreement" that originally invented by political and social science in general. The concept of governance is not only complex in nature because of system inbuilt uncertainty but controversial as well. In its most general form of understanding, governance is not government. Governance, as a concept, recognizes the existence of power inside and outside the formal authority and institutions of government. Therefore, government is a major actor, but not the sole actor influencing decisions and how they are implemented. Other entities in the broad spectrum of governance include religious organizations, private enterprise, unions, cooperatives, financial institutions, political parties, and community-based organizations.

The United Nations Development Programme (UNDP) defines governance as "the exercise of economic, political, and administrative authority to manage a country's affairs at all levels. It comprises the mechanisms, processes, and institutions through which citizens and groups articulate their interests, exercise their legal rights, meet their obligations, and mediate their differences. It brings together the actions of state, non-state, and private sector actors" (UNDP 2004). Disaster risk governance also can be categorized into three categories: (a) *economic governance*, influencing decision-making processes that directly and indirectly affect a country's economic activities and their implications for equity, poverty (including risk), and quality of life; (b) *political governance*, the driving process of decision-making to set legislative processes, and formulate laws, regulation, and policies, which is also referred as institutional basis for implementation; (c) *administrative governance*, defined as the system of policy implementation that requires the existence of well-functioning government organizations at the national and local levels, and which play roles as enforcers of regulations related to disaster mitigation, building code enforcement, land use planning, environmental risk, and human vulnerability monitoring and safety standards (UNDP 2004).

Governance is also defined by the process of decision-making and the process by which decisions are implemented (or not implemented) (UNESCAP 2009). The governance also emphasizes the "process" which recognizes the decisions are made based on complex relationships between multi-stakeholders with different priorities. Governance of risk and catastrophe promotes the idea that there is empirical power (such as government, market, civil society, science) that coexists to govern risk and catastrophe (Lassa 2011).

Proponents of risk governance frameworks, such as of the Institute of Risk Governance Council (IRGC), suggest that risk governance encompasses risk assessment, risk management, and risk communication, which requires fundamental understanding of formal and informal institutions, social-economic contexts within which risk is evaluated, and the involvement of actors and stakeholders who represent them in political and policy arenas that range from the local to the global level (Renn and Walker 2007: 334; Renn 2008; Kathleen 2012).

Rayner (2007: 165) suggests that risk may shift responsibility from governmental to private sector and NGO actors, while at the same time as risk can facilitate government control over citizens. He further argues that "control by government is seen to have been supplemented, if not replaced, by a more distributed form of governing usually referred to by the term 'governance' – defined as the management of a system, usually political or organizational, involving mutual adjustment, negotiation, and accommodation between the parties involved rather than direct control."

It is evident that legal frameworks provide a new direction to disaster risk management through the supplementary incorporation of sound disaster risk reduction measures into developmental decisions, policies, and programs. It is noted widely that some governments have successfully adopted and implemented disaster risk reduction (DRR) policies, while others are lagging behind (Williams 2011) especially when it comes to local governments and field level actions.

1.2 Legislations and Policies for Risk Governance

Legislations and guidelines are one of the most important aspects in maintaining good governance in disaster risk. In the low- and middle-income countries, where socioeconomic conditions are inversely impacted by increasing load of population density and urban migrations, risk governance seems to be a challenge for the disaster managers as well as administrators. This edited book will analyze the status and impacts of policies and legislations toward the governance of disasters in India.

At the operational level, governance has been simplified as "good governance characteristics" such as participation, rule of law, transparency, responsiveness, consensus orientation, equity, effectiveness, efficiency, accountability, and strategic vision in regard to disaster risk reduction (UNDP 2004: 75). Good governance is also viewed as the commitment to sharing decision-making power among the stakeholders in disaster risk reduction, where the government remains critical actors in reducing risk as well as in the broader development context, based on its capacity as a mediator between private and public interests and as well as the people at risk, ranging from the local to the international level.

It is a general "catch-all" term used in a wide array of contexts to mean and justify many things, including particular forms of governance (such as democracy). In an attempt to focus on governance issues rather than advocating for a particular type of governance, this chapter and through the book uses a simple definition of governance adopted from internationally relevant norms.

Within the disaster recovery context, governance is the overall process by which affected governments, organizations, and populations (1) determine what is to be done, how it is to be done, and who it is to benefit and (2) apply themselves to implementing these decisions. Within such a dynamic and unpredictable environment, the impacts of these decisions and their implementation can be profound, drastically changing lives, social systems, economies, and the recovery process itself – either for the better or the worse. Summarized at the 2010 International Recovery Forum by the minister of a disaster-prone South Asian nation, "Governance is everything."

Disaster risk reduction (DRR) measures and risk governance include the systematic development and application of policies, strategies, and practices to prevent or prepare for hazards or to mitigate their adverse effects (UNISDR 2010b). The Department for International Development (DFID) (2005) has divided DRR measures into four categories: policy and planning measures, physical coping and/or adaptive measures, physical preventative measures, and community capacity building measures. Four ways of reducing vulnerability, which can be used for DRR, have also been identified by McEntire, Crocker, and Peters (2010). These are engineering methods, focusing on ways to increase resistance through construction practices in the built environment; physical science methods, stressing exposure to hazards and risk reduction in unsafe environments; structural methods, concentrating on socioeconomic factors and demographic characteristics with a focus on cultural and traditional perceptions of vulnerability; and organizational dimensions, which focus on the effectiveness of preparedness, response, recovery, and management operations. DRR takes place under the auspices of governance.

Though local government is expected to play a leading role in disaster risk governance activities, it faces significant challenges. These include inadequate human, material, and financial resources (Manyena 2006; Pearce 2003), including the lack of experience to manage the expectations of disaster-affected people. It has also been observed that empowering local government through decentralization mechanism proved to be beneficial for local population, including marginalized groups, and provide more autonomy to local authorities to ensure the effective implementation of risk reduction measures (Ahrens and Rudolph 2006; Pulido 2008; UNISDR 2010b).

Postindependence India's journey through the 5-year plans points to the understanding of disasters to primarily mitigate challenges of droughts and floods. Schemes such as the Drought Prone Area Program (DPAP), Desert Development Program (DDP), National Watershed Development Project for Rain-fed Areas (NWDPRA), and Integrated Water Development Project (IWDP) are examples of this conventional paradigm. The government has got the whole machinery in place, and the relief work was carried out with the help of the many agencies including the Indian Red Cross Society, Indian Institute of Tropical Meteorology, UNDP India, Tata Energy Research Institute, Housing and Urban Development Corporation, Council for Advancement of People's Action and Rural Technology (CAPART), etc.

Schemes for financing expenditure on relief and rehabilitation in the wake of natural calamities are governed by the recommendations of Finance Commissions appointed by the Government of India after every 5 years. In the 1990s, the period coinciding with UN-IDNDR, the system of financial response underwent a change to reduce the time between occurrence of a calamity and the provision of relief to the victims. Under the scheme in operation for the period 1995–2000, each state had a corpus of funds called Calamity Relief Fund (CRF), administered by a state level committee and headed by the chief secretary of the state government. The size of the corpus is determined having regard to the vulnerability of the state to different natural calamities and the magnitude of expenditure normally incurred by the state on relief operations. The corpus is built by annual contributions of the union government and the state governments concerned. The states are free to draw upon this corpus for providing relief in the event of a natural calamity. In the event of a major disaster warranting intervention at the national level, a provision exists in the form of National Fund for Calamity Relief with a corpus for the union government to supplement the financial resources needed for relief operations. However, disaster prevention and mitigation were not addressed as such but some of the development programs like construction of dams and embankments were executed with reduction of flood hazards as one of the aims. The number of other structural and non-structural actions has been taken throughout the country across various sectors of development which contributed to disaster prevention and mitigation.

Disaster mitigation and prevention were adopted as essential component of the development strategy. India witnessed the passage of Disaster Management Act 2005 by the parliament which installed an era of paradigm shift from "response and relief centric" to "prevention-mitigation, preparedness centric" holistic disaster management approach. Disaster management has emerged as a high priority for the country. The Eleventh Five-Year Plan 2007–2012 aimed at consolidating the process by giving impetus to projects and programs that develop and nurture the culture of safety and the integration of disaster prevention and mitigation into the development process.

Legislation to minimize the disaster risk is fundamental for enhancing human security. It is the first step toward mainstreaming disaster risk reduction into development. Legislation provides the framework around which strategies to build risk reduction into development and reconstruction activities can be empowered. The law can be used to provide penalties and incentives by enforcing standards in construction, land use, and tenant's rights and by defining people's rights during relief and reconstruction (Pelling and Holloway 2006). The goal of sustainable development through disaster reduction needs to be operationalized through a set of policies and an effective legal framework.

Following the Sendai Framework for Disaster Risk Reduction (SFDRR), the first international agreement adopted within the context of post-2015 development agenda, the National Executive Committee prepared the National Disaster Management Plan (NDMP) to place significant emphasis on improving the disaster risk governance at national and subnational level in Indian perspectives.

1.3 Global, Regional, National, and Local Perspectives of Disaster Risk Reduction

Countries in Asia and the Pacific have substantially improved institutions and policies in disaster risk governance. Mega-disasters, such as the Indian Ocean Tsunami, Cyclone Nargis, and the GEJE, became opportunities to strengthen risk governance in affected countries. Various countries have developed risk governance by creating focal point agencies, establishing national platforms, and promoting legislation. These steps have been taken in line with the HFA and subsequently with Sendai Framework for Disaster Risk Reduction (2015–2030). The number of national platforms is increasing globally and rose from 38 in 2007 to 73 in 2011. An increasing number of countries have been adopting or updating existing disaster risk management legislation (UNISDR 2011b). Legislation can establish new agencies or empower existing agencies with new responsibilities as well as create budget lines and policy responsibilities (Pelling and Holloway 2006).

There are various reasons for changing disaster risk management institutions and policies. Changes in governance, such as globalization and devolution, pose substantial challenges for disaster risk management institutions. The Federal Emergency Management Agency in the USA moved from a limited form of direct service delivery to a complex, network-based approach in the 1990s that stretched from the federal government into state and local governments and the private sector (Kettl 2000).

A number of countries have been continuously strengthening national disaster risk management (DRM) systems based on lessons learned from disasters inside and outside the countries (Amini-Hosseini and Hosseinioon 2012; Ikeda 2012; Nishikawa 2010). In Taipei, Taiwan Typhoon Morakot initiated a change in national institutions. The National Fire Agency was transformed to the National Disaster Prevention and Protection Agency through the Disaster Prevention and Protection Act to expand the agencies' functions to include comprehensive countermeasures. Japan enacted the Disaster Countermeasure Basic Act in 1961 after a high tide disaster in Nagoya in 1959, which caused over 5000 deaths. The main driver of the latest version of Japan's DRM plan after the GEJE is the need to account for low-probability, high-impact compound hazards. There is a pressing need for governments to develop mechanisms to collect disaster data to put in place more evidence-based policies in DRM.

Countries in Asia and the Pacific have taken legislative actions to establish focal point agencies within their central government structures and national platforms. Of the 61 countries and areas in the region, 30 have enacted national or central legislation that specifically deals with DRM (UNESCAP and UNISDR 2012). Following the Indian Ocean tsunami in 2004, affected countries have strengthened their focal point agencies. Sri Lanka established the Disaster Management Ministry by newly creating a disaster management center and merging it with the meteorological agency. Thailand has strengthened the coordination roles of the Disaster Management Department in its government. Indonesia has created a national disaster management agency and also local agencies throughout the country. Myanmar is planning to establish a new agency and to enact a national disaster management law reflecting lessons from the Cyclone Nargis disaster in 2008. Vietnam and the Philippines, which are major disaster-prone countries in Asia, have strengthened existing legislation and institutions. Most countries in the Pacific have created national disaster management offices as stand-alone agencies (Hay 2009).

National governments are expected to play a pivotal role in disaster risk management (DRM). The governments in the Asia and the Pacific region have developed a wide range of innovative solutions at the national level. Building DRM into national development strategies, programs, and projects is needed to protect these developments and make certain that new developments do not exacerbate disaster risks. Thus, each developing country should (i) place DRM as a core element within the structure of its government; (ii) legally define its mandate, status, and coordination role with line ministries and a focal point agency; and (iii) establish mechanisms for coordinating with and supporting local governments.

Governance is widely regarded as the key to reducing disaster risks (Ahrens and Rudolph 2006; Castanos and Lomnitz 2008; UNISDR 2011a; Wisner et al. 2004). Many low- and middle-income countries need responsive, accountable, transparent, and efficient governance structures in DRM (Davis 2011 and UNDP 2010). Governance is defined as an exercise of political, economic, and administrative authority in the management of a country's affairs. Governance influences how income and assets are distributed to the people and determines how the people protect themselves from hazards and how they access support in recovery (Turnbull et al. 2013). Since many low- and middle-income countries lack the administrative, organizational, financial, and political capacity to effectively cope with disasters, the poor become particularly vulnerable. Low-income countries have suffered only 9% of worldwide disasters since 1980 but suffered 48% of the fatalities (World Bank 2012c).

The low- and middle-income developing countries need to place DRM as a core element within the structure of the government. Countries with well-established institutions can decrease the number of affected people and economic losses from natural disasters, while mortality is increasing in countries with weak governance capacities (Cannon 2008; Raschky 2008). The Hyogo Framework for Action (HFA), which was adopted at the World Conference on Disaster Reduction at Kobe in 2005, defines "development, and strengthening of institutions, mechanisms and capacities to build resilience to hazards" as one of the strategic goals and emphasizes the action of ensuring DRM as a national and a local priority with a strong institutional basis for implementation (UNISDR 2005). In Japan, the Disaster Countermeasures Basic Act, which was legislated in 1961, stipulates the DRM framework. The framework covers (1) the roles and responsibilities of national and local governments and communities; (2) the details of disaster management plans, institutions, and countermeasures; and (3) the platforms at national and local levels.

Each country should mainstream DRM into policy, planning, and management in all relevant sectors. Mainstreaming DRM has important implications for a country's growth and development agenda, since disasters can pose serious obstacles to socioeconomic development. Governments should develop a range of innovative programs to prevent increasing vulnerability in the course of the economic development process. For example, Bangladesh's Outline Perspective Plan, produced by the Planning Commission, integrates DRM and climate change adaptation into national development strategies (UNESCAP and UNISDR 2012). In Japan, the government is reviewing DRM approaches by learning lessons from the Great East Japan Earthquake (GEJE) in 2011 and will mainstream DRM further in all relevant sectors by assessing risks and vulnerabilities and allocating necessary resources to prepare for disasters (Committee on Promoting Disaster Management of Central Disaster Management Council 2012) (Ishiwatari 2012b).

A wide range of stakeholders must be coordinated, since DRM concerns everyone. DRM requires a multi-sectoral approach, which covers urban development, infrastructure, water, education, health, and many other sectors. Single-sector development planning cannot address the complexity of problems caused by disasters, nor can such planning build resilient societies (World Bank 2012a, b). For example, DRM plans should be linked with urban planning and DRM education at school, which are effective measures to decrease disaster casualties and damage. Since no single organization can have the ultimate responsibility for managing disaster risks, various stakeholders should share the responsibility. In addition to governmental organizations, the private sector and civil society play crucial roles in DRM (IRP 2009). The private sector can contribute to mitigating disaster damage in a wide range of areas, such as the logistics of providing relief goods, payment of insurance claims, restoration of damaged infrastructure, and the continuation of banking services. Civil society organizations (CSOs) can respond to the various needs of affected people at the grassroots level. The local government and CSOs can play strategic roles in creating safety net systems to protect vulnerable groups from disasters.

1.4 Prime Minister's Ten-Point Agenda on Disaster Risk Reduction

Prime Minister Narendra Modi during his address at the seventh Asian Ministerial Conference on Disaster Risk Reduction (AMCDRR) held in New Delhi outlined a ten-point agenda for renewing efforts toward disaster risk reduction. The following are the brief on ten-point agenda on DRR (Gupta et al. 2016):

I. *All development sectors must imbibe the principles of disaster risk management*: This will ensure that all development projects are built to the best standards of disaster safety and contribute to the resilience of communities they seek to serve. Further, the public expenditure must factor in risk considerations. For instance, the "housing for all" program and "smart cities" initiative in India represent such opportunities.

II. *Work toward risk coverage for all*: Currently, in most of the Asian countries, insurance penetration is limited to only middle- and upper-middle-income groups. Thus, nation-states should not only regulate but also encourage risk coverage starting from poor households to small and medium enterprises to multinational corporations to nation-states. The Jan-Dhan Yojana launched in India address this very issue to ensure financial inclusion and risk insurance for the poorest.

III. *Encourage greater participation and leadership of women in disaster risk management*: Women are disproportionately affected by disasters. They also have unique strengths and insights. We must train a large number of women volunteers to support special needs of women affected by disasters.

IV. *Invest in risk mapping globally*: In light of the increasing risk of disasters in the region, it is pertinent that member nations should invest in risk mapping in accordance with widely accepted standards and parameters. This will help us ensure that we have a common understanding of the nature and severity of disaster risks in different parts of the world.

V. *Leverage technology to enhance the efficiency of our disaster risk management efforts*: In this direction an e-platform that brings together organizations and individuals and helps them map and exchange expertise, technology, and resources would go a long way in maximizing our collective impact.

VI. *Develop a network of universities to work on disaster issues*: Recognizing the fact that universities have social responsibilities too, we should strive for developing a global network of universities working together on problems of disaster risk management. This network of universities can be developed in the first 5 years of the Sendai Framework. Further, as part of this network, different universities could specialize in multidisciplinary research on disaster issues most relevant to them. For instance, universities located in coastal areas could specialize in managing risks from coastal hazards, and the ones located in the hill cities could focus on mountain hazards.

VII. *Utilize the opportunities provided by social media and mobile technologies*: Social media is transforming disaster response; thus, we must recognize the potential of social media and develop applications for all aspects of disaster risk management.

VIII. *Build on local capacity and initiative*: This calls for designing specific actions and implementing them locally, that is, we must localize disaster risk reduction with an aim to make the most of traditional best practices and indigenous knowledge. Further, response agencies need to interact with their communities and make them familiar with the essential drill of disaster response.

IX. *Ensure that the opportunity to learn from a disaster is not wasted*: This needs a more vibrant and visual system of learning. In this direction, the United Nations could start an international competition of documentary films on disaster and risk reduction measures. Further, for post-disaster recovery, we also need to put in place systems for quick risk assessments

X. *Bring about greater cohesion in international response to disasters*: In this direction, the United Nations could think of a common logo and branding under which all those who are helping with relief, rehabilitation, and reconstruction operate.

1.5 The Role of Local Government in Disaster Risk Reduction

The phrase "local government" is conceptualized differently in different countries in terms of the duties, structure, composition, size, level, and functional responsibilities between central and subnational governments. These responsibilities can be along fiscal, administrative, and political lines (Col 2007), and the size and levels of the community may be divided into different categories such as regions, provinces, cities, municipality, and townships (The Incheon Declaration 2009). The United

Nations Office for Public Administration considers local government to be a political subdivision of a nation or state that is constituted by law, whose governing body is elected and has substantial control of local affairs (Ola 1984).

Councils constitute local government in Cameroon and have disaster management functions conferred upon them by the legislative and administrative instruments governing local government in Cameroon. Law No 2004/018 of July 22, 2004, stipulates that councils are responsible for such tasks as promoting local development, improving the living conditions of their inhabitants, organizing and coordinating relief operations for needy persons, organizing firefighting activities, and securing the socioeconomic development of their populations. However, in reality, councils generally lack the required resources and are not sufficiently empowered to carry out these functions.

Local governments typically perform the following key roles in implementing DRR initiatives: they play a central role in coordinating and sustaining a multilevel, multi-stakeholder platform to promote DRR; they engage local communities and citizens with DRR activities in an effective manner and link their concerns with government priorities; they strengthen their own institutional capacities and implement practical DRR actions by themselves; and they devise and implement innovative tools and techniques for DRR.

Since natural disasters are fundamentally local in nature, local governments and communities are in a good position to be the first responder and to have the principal responsibility for crisis management along with the disaster risk management. National governments have various roles to support local governments to prepare for and respond to disasters. During normal times, the national government can provide financial and technical support to local governments to promote disaster risk management activities. The national governments should substantially support local governments when the local governments cannot manage large-scale disasters. Decentralization is required for the local governments to promptly respond to disasters on the ground. However, powers and budgets should be gradually devolved to the local governments, taking into account the limits of their capacity.

The national government can support local governments in strengthening DRM during normal times. The national government guides the local governments to establish DRM mechanisms by enacting new laws and budgetary systems. In Japan, prefectural and municipal governments have the primary responsibility for DRM, while the national government has the responsibility for developing large-scale DRM infrastructures, such as dams and embankments for managing floods and droughts in major rivers. The national government in Japan does not have local offices. This is because local governments and communities have developed capacities by coping with disasters as local events through their history. The Indonesian Government has supported local governments to create DRM offices in all 33 states, and in around 400 prefectures and cities, since 2008. The national government is establishing tsunami warning systems with local governments.

1.6 Challenges and Gaps in Risk Governance

The national government can provide financial subsidies to local governments to promote disaster risk management at the local level. It is unlikely for the local governments to put a high priority on DRM among various development areas because of limited financial capacity. A national subsidy mechanism is useful to promote countermeasures as a minimum requirement throughout the country (Ishiwatari 2012a). For example, retrofitting schools and hospitals that are crucial in disaster management operations needs financial support from the national government. Budgets and authorities are often devolved to local governments in the education and health sectors. The local governments tend to put higher priority on allocating budgets for constructing new buildings or purchasing equipment, rather than retrofitting existing buildings in preparation for unpredictable earthquakes.

The national government's support is required when local governments cannot properly respond to large-scale disasters. Various government agencies can mobilize experts, including search and rescue teams, medical teams, and engineers to devastated areas by utilizing national networks. Also, the national government provides meteorological and hydrological services and disaster information, such as monitoring warnings of typhoons, floods, tsunamis, and earthquakes. Based on the disaster information received from the national government, local governments can issue evacuation orders to the public. In the 2011 Japan earthquake incident, many municipalities suffered serious damage to their office buildings and incurred considerable staff losses, which hampered their disaster response timing and effectiveness. The national agencies of the police, fire departments, infrastructures, medical facilities, and self-defense forces had prepared specialized teams by compiling rosters and conducting training during normal times and were able to start deploying these teams on March 11, 2011, the day of the disaster.

In Indian context, powers and budgets in disaster risk governance domain need gradual development from the national government, while taking into account the limited capacity of local governments. Vertical and horizontal integration among the stakeholders and institutions are one of the key elements to focus for prompt response to disasters on the ground. In India, however, limited capacity at the local level is a common problem.

1.7 Analyzing National Disaster Management Plan

The enactment of Disaster Management Act 2005 and its subsequent implementation across the country signifies the efficiency of the political and bureaucratic system at various levels. The National Disaster Management Authority (NDMA) has incorporated Sendai Framework for Disaster Risk Reduction 2015–2030 through its various activities and roadmap and also acquired knowledge and lesions from a rich pool of global good practices. While the DM Act 2005 contributes significantly to the disaster risk governance endeavor in terms of legislations at the national, subnational, and local level, there is a need to provide more structural and coordinated initiatives on proactive risk management. The National Policy on Disaster Management has been adopted in 2009 to provide more structured approach to risk governance mechanism. Recent new global development and institutional agreements of Sendai Framework for Disaster Risk Reduction 2015–2030, Sustainable Development Goals 2015–2030, and Paris Agreement on Climate Change at COP 21 were complemented the landmark steps toward Indian disaster risk governance through the preparation of National Disaster Management Plan in 2016. The National Disaster Management Plan (NDMP) is highly ambitious with short-, medium-, and long-term measures targeted to complete in the time frame of 5, 10, and 15 years. Despite the varied time frames, the scales of implementation as well as institutional complexity involved in implementation also vary depending on the specific measures. The NDMP is a dynamic document in the sense that it will be periodically improved keeping up with the emerging global best practices and knowledge base on disaster management. It is in accordance with the provisions of the Disaster Management Act, 2005; the guidance given in the National Policy on Disaster Management, 2009 (NPDM); and the established national practices.

As mandated by Disaster Management Act, 2005 (DM Act 2005), the Government of India (GoI) created a multitiered institutional system consisting of the National Disaster Management Authority (NDMA) headed by the prime minister, the State Disaster Management Authorities (SDMAs) headed by the respective chief ministers, and the District Disaster Management Authorities (DDMAs) headed by the district collectors and cochaired by chairpersons of the local bodies. The DM Act 2005 lays down institutional and coordination mechanism for effective disaster management (DM) at the national, state, district, and local levels to facilitate a paradigm shift from reactive crisis management to proactive risk management through integrated approach of strengthening disaster preparedness, mitigation, and emergency response. In continuation to the DM 2005 and National Policy on Disaster Management, 2009, the National Disaster Management Plan (NDMP) provides a framework and direction to the government agencies for all phases of disaster management cycle.

NDMP recognizes the need to minimize the ambiguity and confusions in the responsibility of an individual and institution at various stages of managing disasters. The NDMP is designed for implementation in flexible and scalable manner in all phases of disaster management at the national and subnational level. The national government of India developed the NDMP to make a holistic attempt to comply with global approaches following the recommendations in Sendai Framework for Disaster Risk Reduction 2015–2030. The NDMP will help the country to meet the goals set in the SFDRR 2015–2030, aiming to achieve substantial reduction of disaster risk and losses in lives, livelihoods, health, economic, and physical, social, cultural, and environmental assets of persons, business, communities, and countries.

NDMP visioned a disaster-resilient India through the achievement of substantial disaster risk reduction and significant decrease in loss of life, livelihoods, and assets. The national plan also targeted to address the economic, physical, cultural, and environmental vulnerability at various levels.

1.8 Disaster Governance and Risk Communications

The concept of risk governance comprises a broad picture of risk: not only does it include what has been termed "risk management" or "risk analysis," it also looks at how risk-related decision-making unfolds when a range of actors is involved, requiring coordination and possibly reconciliation between a profusion of roles, perspectives, goals, and activities. Indeed, the problem-solving capacities of individual actors, be they government, the scientific community, business players, NGOs, or civil society as a whole, are limited and often unequal to the major challenges facing society today.

The risk governance framework breaks down into three main phases: "pre-assessment," "appraisal," and "management" (Fig. 1.1). The purpose of the *pre-assessment* phase is to capture both the variety of issues that stakeholders and society may associate with a certain risk and existing indicators, routines, and conventions that may prematurely narrow down, or act as a filter for, what is going to be addressed as risk. The first step of pre-assessment, risk framing, therefore places particular importance on the need for all interested parties to share a common understanding of the risk issue(s) being addressed or, otherwise, to raise awareness

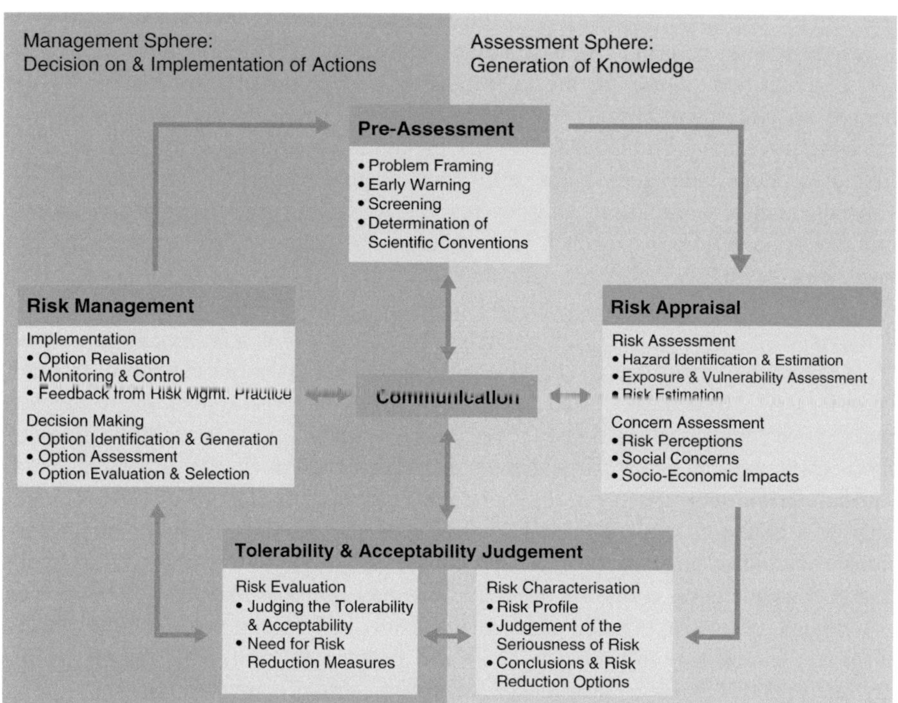

Fig. 1.1 Risk governance framework (Source: Risk Governance – Towards and Integrative Approach, International Risk Governance Council, 2006)

among those parties of the differences in what is perceived as a risk. For a common understanding to be achieved, actors need both to agree with the underlying goal of the activity or event generating the risk and be willing to accept the risk's foreseeable implications on that very goal. The second step of the pre-assessment phase, early warning and monitoring, establishes whether signals of the risk exist that would indicate its realization. This step also investigates the institutional means in place for monitoring the environment for such early warning signals. The third step, prescreening, takes up and looks into the widespread practice of conducting preliminary probes into hazards or risks and, based on prioritization schemes and existing models for dealing with risk, of assigning a risk to predefined assessment and management "routes." The fourth and final step of pre-assessment selects major assumptions, conventions, and procedural rules for assessing the risk as well as the emotions associated with it.

The objective of the *risk appraisal* phase is to provide the knowledge base for the societal decision on whether or not a risk should be taken and, if so, how the risk can possibly be reduced or contained. Risk appraisal thus comprises a scientific assessment of both the risk and of questions that stakeholders may have concerning its social and economic implications. The first component of risk appraisal, risk assessment, seeks to link a potential source of harm, a hazard, with likely consequences, specifying probabilities of occurrence for the latter. Depending on the source of a risk and the organizational culture of the community dealing with it, many different ways exist for structuring risk assessment. Despite such diversity, three core steps can be identified. These are the identification and, if possible, estimation of the hazard, an assessment of related exposure and/or vulnerability, and an estimation of the consequent risk. The latter step – risk estimation – aggregates the results of the first two steps and states, for each conceivable degree of severity of the consequence(s), a probability of occurrence. Confirming the results of risk assessments can be extremely difficult, in particular when cause-effect relationships are hard to establish, when they are instable due to variations in both causes and effects, and when effects are both scarce and difficult to understand. Depending on the achievable state and quality of knowledge, risk assessment is thus confronted with three major challenges that can best be summarized using the risk categories through "complexity," "uncertainty," and "ambiguity." For a successful outcome to the risk process and, indeed, overall risk governance, it is crucial that the implications of these challenges are made transparent at the conclusion of risk assessment and throughout all subsequent phases. It is also important to understand the physical attributes of the risk on detailed knowledge of stakeholders' concerns and questions – emotions, hopes, fears, and apprehensions – about the risk as well as likely social consequences, economic implications, and political responses. The second component of risk appraisal, concern assessment, thus complements the results from risk assessment with insights from risk perception studies and interdisciplinary analyses of the risk's (secondary) social and economic implications.

The *risk management* phase designs and implements the actions and remedies required to tackle risks with an aim to avoid, reduce, transfer, or retain them. Risk management thereby relies on a sequence of six steps which facilitates systematic decision-making. To start with, and based on a reconsideration of the knowledge gained in the risk appraisal phase and while judging the acceptability and/or tolerability of a given risk, a range of potential risk management options is identified. The options are then assessed with regard to such criteria such as effectiveness, efficiency, minimization of external side effects, sustainability, etc. These assessment results are next complemented by a value judgment on the relative weight of each of the assessment criteria, allowing an evaluation of the risk management options. This evaluation supports the next step in which one (or more) of the risk management options is selected, normally after consideration of possible trade-offs that need to be made between a number of second-best options. The final two steps include the implementation of the selected options and the periodic monitoring and review of their performance.

The most controversial phase of handling risk, risk characterization and evaluation, aims at judging a risk's acceptability and/or tolerability. The risk process has "communication" as a companion to all phases of addressing and handling risk and is itself of a cyclical nature. However, the clear sequence of phases and steps offered by this process is primarily a logical and functional one and will not always correspond to reality (Fig. 1.1). A risk deemed "acceptable" is usually limited in terms of negative consequences so that it is taken on without risk reduction or mitigation measures being envisaged. A risk deemed "tolerable" links undertaking an activity – which is considered worthwhile for the value-added or benefit it provides – with specific measures to diminish and limit the likely adverse consequences. This judgment is informed by two distinct but closely related efforts to gather and compile the necessary knowledge which, in the case of tolerability, must additionally support an initial understanding of required risk reduction and mitigation measures. While risk characterization compiles scientific evidence based on the results from the risk appraisal phase, risk evaluation assesses broader value-based issues that also influence the judgment. Such issues, which include questions such as the choice of technology, societal needs requiring a given risk agent to be present, and the potential for substitution as well as for compensation, reach beyond the risk itself and into the realm of policy making and societal balancing of risks and benefits.

The *risk communication* is of major importance throughout the entire risk handling chain. Not only should risk communication enable stakeholders and civil society to understand the rationale of the results and decisions from the risk appraisal and risk management phases when they are not formally part of the process, but it should also help them to make informed choices about risk, balancing factual knowledge about risk with personal interests, concerns, beliefs, and resources, when they are themselves involved in risk-related decision-making. Effective risk communication consequently fosters tolerance for conflicting viewpoints and provides the basis for their resolution and creates trust in the institutional means for assessing and managing risk and related concerns. Eventually, risk communication can have a major impact on how well society is prepared to cope with risk and react to crises and disasters. Risk communication has to perform these functions both for the experts involved in the overall risk process – requiring the exchange of information between risk assessors and managers, between scientists and policy makers, and between academic disciplines and across institutional barriers – and for the "outside world" of those affected by the process.

1.9 Challenges in Risk Assessment and Governance

Risk assessment is confronted with three major challenges that can be best described using the terms "complexity," "uncertainty," and "ambiguity." These three challenges are not related to the intrinsic characteristics of hazards or risks themselves but to the *state and quality of knowledge available* about both hazards and risks. Since risks are mental constructs, the quality of their explanatory power depends on the accuracy and validity of their predictions and forecast information. Unlike some other scientific constructs, validating the results of risk assessments is particularly difficult because, in theory, one would need to wait indefinitely to prove that the probabilities assigned to a specific outcome were correctly assessed. If the number of predicted events is frequent and the causal chain is obvious (as is the case with car accidents), validation is relatively simple and straightforward. If, however, the assessment focuses on risks where cause-effect relationships are difficult to discern, effects are rare and difficult to interpret, and variations in both causes and effects are obscuring the results, the validation of the assessment results becomes a major problem. In such instances, assessment procedures are needed to characterize the existing knowledge with respect to complexity and remaining uncertainties and ambiguities (Klinke and Renn 2002). Complexity in the risk assessment for governance possesses a difficulty of identifying and quantifying causal links between a multitude of potential causal agents and specific observed effects.

It is useful to differentiate between *horizontal and vertical governance* (Benz and Eberlein 1999; Lyall and Tait 2004). The horizontal level includes the relevant actors in decision-making processes within a defined geographical or functional segment (such as all relevant actors within a community, region, nation, or continent); the vertical level describes the links between these segments (such as the institutional relationships between the local, regional, and state levels).

Risk governance involves the "translation" of the substance and core principles of governance to the context of risk and risk-related decision-making. In IRGC's understanding, risk governance includes the totality of actors, rules, conventions, processes, and mechanisms concerned with how relevant risk information is collected, analyzed, and communicated, and management decisions are taken. Encompassing the combined risk-relevant decisions and actions of both governmental and private actors, risk governance is of particular importance in, but not restricted to, situations where there is no single authority to take a binding risk management decision but where, instead, the nature of the risk requires the collaboration of, and coordination between, a range of different stakeholders. Risk governance however not only includes a multifaceted, multi-actor risk process but also calls for the consideration of contextual factors such as institutional arrangements (e.g., the regulatory and legal framework that determines the relationship, roles, and responsibilities of the actors and coordination mechanisms such as markets, incentives, or self-imposed norms) and political culture, including different perceptions of risk.

1.10 Conclusion

While devolution to local governments is needed to effectively respond to the needs of people affected by disasters on the ground, local governments typically face difficulties in taking on powers and budgets devolved from the national government because of their limited capacity. In some cases, national governments have established local offices or seconded staff to the local governments to promote disaster risk management at the grassroots level. National and local disaster risk governance means governing and managing disaster risk beyond government's hierarchy at different levels of jurisdiction and category of actors. The success of a disaster risk governance process is partly measured by legitimacy principles. Legitimacy principle in DRR policy making means "rightfulness" of decision process or the extent to which the policy making process is accepted by the participants (Lassa 2008) as fair or unfair. Its synonymous is "technology of participation" which is often ignored by top down disaster risk governance. Legitimacy may also mean how one DRR policy stakeholder perceives the policy making process as unbiased and should be politically and procedurally correct and fair. It "involves the belief that a policy making process is "fair" and consider appropriate values, interests, concerns, and specific circumstances from multiple perspectives."

The countries across the Asia Pacific region have significantly developed the disaster risk governance by creating focal point agencies, establishing national platforms, promoting legislation in line with the HFA, and developing mechanisms for cooperation of national and local governments. The second World Conference on Disaster Risk Reduction in 2015 also emphasized and reiterated the significance of risk governance through the Sendai Framework for Disaster Risk Reduction (SFDRR) as one of the priorities for the international framework from 2015 to 2030. The mega-disasters, such as the Indian Ocean Tsunami, Cyclone Nargis, and the GEJE, became opportunities for making these improvements. A designated coordinating agency with high authority in the government is theoretically required to neutrally coordinate and to lead organizations concerned with disaster risk reduction process, but implementation of the desired and planned risk reduction mechanism is challenging for various reasons. The Indian national government seems to be proactive to comply with global priorities and agreements to make India disaster resilient but many-fold and complex challenges are hindering the affirmative actions achieving these goals in designated time frame. This publication is an academic effort to support the national goals to comply the SFDRR priorities in reality. The 4 parts and 19 chapters on various perspectives analyzed the science and technology perspectives as well as social dimension of disaster risk governance in Indian context.

References

Ahrens J, Rudolph PM (2006) The importance of governance in risk reduction and disaster management. J Conting Crisis Manag 14(4):207–220

Aini MS, Fakhrul-Razi A (2010) Development of socio-technical disaster model. Saf Sci 48(10):1286–1295

Amini-Hosseini K, Hosseinioon S (2012) Evaluation of recent developments in laws regulations for earthquake risk mitigation and management in Iran. Risk Hazards Crisis Public Policy 3(1):1–20

Benz A, Eberlein B (1999) The Europeanization of regional policies: patterns of multi-level governance. J Eur Publ Policy 6:2

Cannon T (2008) Reducing people's vulnerability to natural hazards: communities and resilience, Research Paper 2008/34. United Nations University- World Institute for Development Economic Research, Helsinki

Castanos H, Lomnitz C (2008) Ortwin Renn, risk governance: coping with uncertainty in a complex world. Nat Hazards 48(2):313–314

Col JM (2007) Managing disasters: the role of local government. Publ Admin Rev (Special Issue):114–124.

Committee on Promoting Disaster Management of Central Disaster Management Council (2012) Final report: reestablishing resilient Japan (Yuruginai Nihonno Saikochikuwo Mezashite [in Japanese]). Committee on promoting Disaster Management of Central Disaster Management Council, Tokyo

Davis I (2011) Reducing disaster risks 1980–2010: some reflections and speculations. Environ Hazards 10(1):80–92

Department for International Development (DFID). (2005). Natural disaster and disaster risk reduction measures – a desk review of costs and benefits

Development Programme (UNDP). http://www.unisdr.org/we/inform/publications/18869

Guha-Sapir D, Vos F, Below R, with Ponserre S (2010) Annual disaster statistical review 2010: the numbers and trends. CRED, Brussels; 2011

Gupta A, Chopde S, Singh S, Wajih SA, Katyal S (2016) Prime minister's agenda 10: India's disaster risk management roadmap to climate resilient and sustainable development, CDKN.

Hayles CS (2010) An examination of decision making in post disaster housing reconstruction. Int J Dis Resil Built Environ 1(1):103–122

Hay JE (2009) Institutional and policy analysis of disaster risk reduction and climate change adaptation in Pacific Island countries. United Nations International System for Disaster Reduction (UNISDR) and the United Nations, Suva

Ikeda M (2012) Knowledge note 2-2 disaster management plans. In: Learning from megadisasters. http://wbi.worldbank.org/wbi/document/disaster-management-plans

Incheon Declaration (2009) Education 2030 Incheon declaration towards inclusive and equitable quality education and lifelong learning for all, The Economic and Social Commission for Asia and the Pacific.

International Recovery Platform (IRP) (2009) Why governance issues are important in recovery? Knowledge for recovery series info kit governance 1. IRP Secretariat, Kobe

Ishiwatari M (2012a) Government roles in community-based disaster risk reduction. In: Shaw R (ed) Community-based disaster risk reduction: community, environment and disaster risk management. Emerald Group Publishing Limited, Bingley

Ishiwatari M (2012b) Review of countermeasures in the East Japan Earthquake and tsunami. In: Shaw R, Takeuchi Y (eds) East Japan Earthquake and tsunami: evaluation, communication, education and volunteerism. Research Publishing, Singapore

Kathleen T (2012) Disaster governance: social, political, and economic dimensions. Annu Rev Environ Resour 37:341–363

Kettl DF (2000) The transformation of governance: globalization, devolution, and the role of government. Public Adm Rev 60(6):488–497

Klinke K, Renn O (2002) A new approach to risk evaluation and management: risk-based, precautionbased, and discourse-based strategies. Risk Anal 22(6):1071–1094
Lassa J (2008) Measuring the sustainability of tsunami early warning systems: an interdisciplinary research agenda. Georisk Assess Manag Risk Eng Syst Gehazards 2(4):185–192
Lassa J (2011) Institutional vulnerability and governance of disaster risk reduction: Macro, meso and micro scale analysis. PhD dissertation, University of Bonn
Lyall C, Tait J (2004) Shifting policy debates and implications for governance. New modes of governance: developing an integrated policy approach to science, technology, risk and the environment. pp 3–17
McEntire D, Crockers CG, Peters E (2010) Addressing vulnerability through an integrated approach. Int J Dis Resil Built Environ 1(1):50–64. doi:10.1108/17595901011026472
Ministry of Finance, Japan. 2012. Joint statement by the finance minister of Japan and the World Bank President at the Sendai Dialogue. http://www.mof.go.jp/english/international_policy/mdbs/wb/press_release/20121010sendai_statement.htm
Nishikawa S (2010) From Yokohama strategy to Hyogo framework: sharing the Japanese experience of disaster risk management. Asian J Environ Disaster Manag 2(3):249–262
Ola R (1984) Local administration in Nigeria. Kegan Paul International, London
Pelling M, Holloway A (2006) Legislation for mainstreaming disaster risk reduction. Tearfund, Teddington
Pearce, L. (2003). Disaster management and community planning, and public participation: how to achieve sustainable hazard mitigation, Nat Hazards, 28: 211. doi:10.1023/A:1022917721797
Pulido L (2008) Colombia: integrating disaster risk reduction at the local level Colombia. DPAD, viewed 29 Feb 2010
Raschky PA (2008) Institutions and the losses from natural disasters. Nat Hazards Earth Syst Sci 8:627–634
Rayner S (2007) The rise of risk and the decline of politics. Environ Hazard 7:165–172
Renn O (2008) Risk governance. coping with uncertainty in a complex world. Earthscan, London
Renn O, Walker K (2007) Lessons learned: a re-assessment of the IRGC framework on risk governance. In: Renn O, Walker K (eds) The IRGC risk governance framework: concepts and practice. Springer, Heidelberg/New York, pp 331–367
Turnbull M, Sterrett LC, Hilleboe A (2013) Toward resilience: a guide to disaster risk reduction and climate change adaptation. Practical Action, Rugby
UNDP (2004) Reducing disaster risk: a challenge for development a global report. Bureau of Crisis Prevention and Recovery, United Nations Development Program, New York
UNESCAP (2009) UNESCAP working paper macroeconomic policy and development division from human development to human security: a prototype human security index prepared by David A. Hastings
UNESCAP, UNISDR (2012) Reducing vulnerability and exposure to disasters: the Asia-Pacific disaster Report 2012. UNISDR, Bangkok
UNISDR (2005). Hyogo framework for action 2005–2015: building the resilience of nations and communities to disasters
UNISDR (2010a) Information kit for ISDR 2010–2011 world disaster reduction campaign. Geneva, Switzerland. Retrieved from http://www.unisdr.org/english/campaings/campaigns2010-2011/documents/campaign-kit.pdf
UNISDR (2010b) Local governments and disaster risk reduction: good practices and lessons learned. Geneva, Switzerland. Retrieved from http://ww.unisdr.org/files/13627_LocalGovernmentsandDisasterRiskRedu.pdf
UNISDR (2011a) Global assessment report on disaster reduction. UNISDR, Geneva
UNISDR (2011b) Mid-term review, Hyogo framework for action 2005–2015: building the resilience of nations and communities to disasters. UNISDR, Geneva
United Nations Development Programme (UNDP) (2010) Disaster risk reduction, governance & mainstreaming. UNDP, New York

Williams JC (2008) State of emergency preparedness of Kentucky's rural public health workforce: assessing its ability to identify community health problems. Public Health Rep 123:178–188
Wisner B et al (2004) At risk: natural hazards, people's vulnerability and disasters, 2nd edn. Routledge, London
World Bank (2012a) Acting today for tomorrow: a policy and practice note for climate- and disaster-resilient development in the Pacific Islands region. World Bank, Washington, DC
World Bank (2012b) Executive summary of knowledge note: learning from megadisasters. World Bank, Washington, DC
World Bank (2012c) Sendai report: managing disaster for a resilient future. World Bank, Washington, DC

Chapter 2
Disaster Risk Reduction: The Indian Landscape

Mihir R. Bhatt

Abstract India has registered a remarkable progress in reducing disaster risk since the Hyogo Framework for Action (HFA) went into effect in 2005; still it remains as one of the most disaster-prone countries of the world. Even though the HFA Monitor shows that nations across the world are making good progress towards the goal of reduced disaster losses and impacts, national and global disaster loss databases and global risk models indicate rising economic and livelihood losses (FLACSO, UNISDR. The future of disaster risk reduction. A Scoping Meeting for GAR 2015. San Jose, Costa Rica. Available at: http://www.unisdr.org/files/35715_thefutureofdisasterriskmanagement.pdf, 2013). Also, there is a little evidence that current risk governance arrangements are fit or suitable to effectively reduce underlying risk factors (UNISDR. Chair's summary fourth session of the global platform for disaster risk reduction, Geneva, 21–23 May 2013. Available at: http://www.preventionweb.net/files/33306_chairssummarypostdraft1.4.pdf, 2013). The institutional and legislative arrangements have largely taken the form of disaster-focused organisations and systems, which had little real influence on development processes (UNISDR. Synthesis report: consultations on a post-2015 framework on disaster risk reduction (HFA2) Geneva: Switzerland. Available at: http://www.unisdr.org/files/32535_hfasynthesisreportfinal.pdf, 2013). Thus, the call for appropriate risk governance, defined as system of norms, institutions and interactions that determine how decisions are made and enforced, is clearly identified in the Sendai Framework. This paper discusses disaster risk, governance structures, stakeholders and opportunities for effectively managing disaster risk in the Indian disaster risk reduction policy landscape. This paper tries to trace the progress made in disaster risk reduction efforts from national to local levels from the perspectives of DRR.

Keywords Disaster risk reduction • Policy • Risk governance • Sendai Framework

M.R. Bhatt (✉)
All India Disaster Mitigation Institute (AIDMI), Ahmedabad, India
e-mail: mihir@aidmi.org

I. Pal, R. Shaw (eds.), *Disaster Risk Governance in India and Cross Cutting Issues*, Disaster Risk Reduction, DOI 10.1007/978-981-10-3310-0_2

2.1 Natural Disasters and Climate Extremes in India

2.1.1 Vulnerability

India is highly vulnerable to natural disasters and climate extremes, including conflict and displacement. These hazards pose significant risk to India's economy and to its citizens every year. According to the Ministry of Home Affairs, Government of India, India is one of the ten worst disaster-prone countries of the world. Out of 35 states and union territories in the country, 27 of them are disaster-prone. Almost 58.6% of the landmass is prone to earthquakes of moderate to very high intensity; over 40 million hectares (12% of land) are prone to floods and river erosion; of the 7516-km-long coastline, close to 5700 km is prone to cyclones and tsunamis; 68% of the cultivable area is vulnerable to drought and hilly areas are at risk from landslides and avalanches (Ministry of Home Affairs (MoHA) 2011). 'Disaster risks in India are further compounded by increasing vulnerabilities related to changing demographics and socio-economic conditions, unplanned urbanization, development within high-risk zones, environmental degradation, climate change, geological hazards, epidemics and pandemics' (NDMA, 2016).

2.1.2 Disaster Risks

In India, risk in terms of both human and economic exposure to disasters is extremely high. As per the recently released World Risk Report 2016, India *ranks 77 in disaster risk index of the world* (The Hindu 2016). And, with the impact of global warming, it is likely that both human and economic exposure to disasters risk will only increase. The SREX report has concluded that the economic losses from weather- and climate-related disasters have increased over time and that these losses as a proportion of GDP are higher in developed countries, whereas deaths from natural disasters occur at a much higher rates in developing countries (IPCC 2012). Even vulnerability to disasters or emergencies of Chemical, Biological, Radiological and Nuclear (CBRN) origin in India has increased on account of socio-economic development (Disaster Management in India 2011).

2.1.3 Impact

In the decade 1990–2000, an average of about 4344 people lost their lives and about 30 million people were affected by disasters every year (Ministry of Home Affairs, GoI 2005). According to UNICEF, every year, between 2000 and 2009, 65 million people on average in India were affected by disasters (UNICEF). Data from the

(EM-DAT) OFDA/CRED International Disaster Database for the period of 1990–2014 (PreventionWeb) (CRED EM-DAT: The OFDA/CRED – International Disaster Database) shows that while earthquakes, floods and storms are the biggest killers in India, droughts and floods affect most people in India. Further analysis of data from EM-DAT shows that both, in terms of frequency and economic damage, floods top the list with highest economic damages in India followed by storms and earthquakes (Table 2.1). As per the recent UN report, India's average annual economic loss due to disasters is estimated to be $9.8 billion; these includes more than $7 billion loss on account of floods (Thakur 2015).

Table 2.1 People affected, lives lost and economic damage due to disasters in India between 1990 and 2015

Disaster type	Disaster subtype	Events count	Total deaths	Total affected	Total damage ('000 US$)
Drought	Drought	5	20	351,175,000	2,041,122
Earthquake	Ground movement	12	32,911	7,832,486	4,122,000
	Tsunami	1	16,389	654,512	1,022,800
Epidemic	Bacterial disease	10	1155	55,032	0
	Parasitic disease	4	121	30,135	0
	Viral disease	19	1867	171,453	0
Extreme temperature	Cold wave	19	3186	25	0
	Heat wave	14	9522	25	400,000
	Severe winter conditions	2	320	0	0
Flood	Riverine flood	132	24,367	297,272,962	39,504,729
	Flash flood	21	2530	15,943,526	322,000
	Coastal flood	4	569	11,500,000	275,000
	–	35	8834	230,943,112	7,923,600
Landslide	Landslide	24	1542	1,332,748	4500
	Avalanche	6	532	10,256	50,000
Storm	Convective storm	30	1637	699,639	2,387,000
	Tropical cyclone	42	17,940	52,613,655	14,419,012
Wildfire	Forest fire	1	6	0	0

Source: EM-DAT: The OFDA/CRED International Disaster Database

2.2 Institutional Framework for Disaster Risk Reduction in India

2.2.1 Historical Perspective

According to the Ministry of Home Affairs, Government of India, a permanent and institutionalised setup for DRR began in the decade of 1990s with setup of a disaster management cell under the Ministry of Agriculture, following the declaration of the decade of 1990s as the 'International Decade for Natural Disaster Reduction' (IDNDR) by the UN General Assembly. Following a series of disasters such as Latur earthquake (1993), Malpa landslide (1994), Orissa super cyclone (1999) and Bhuj earthquake (2001), a high-powered committee under the Chairmanship of Mr. J.C. Pant, Secretary, Ministry of Agriculture, was constituted for drawing up a systematic, comprehensive and holistic approach towards disasters (Ministry of Home Affairs (MoHA) 2011). There was a shift in policy from an approach of relief through financial aid to a holistic one for addressing disaster management. Consequently, the disaster management division was shifted under the Ministry of Home Affairs in 2002, and a hierarchical structure for disaster management evolved in India.

2.2.2 The Foundation Stone

Recognising the critical need for having stronger legislation to promote and protect human rights during disasters, the government of India enacted the Disaster Management Act in 2005. The Act deals with the provisions regarding protection of human rights by virtue of providing specific guidelines with reference to minimum standards of relief and number of other administrative measures for reducing disaster risks in India. These legal provisions are considered to be the sacrosanct legislative measures as the intent of the legislature in enacting this Act is to provide for the effective management of disasters and to uphold the human rights of the disaster-affected people (Pratap 2012). The Act has allowed the country to evolve a more organised disaster risk governance system and shift its approach from relief and rehabilitation to preparedness and mitigation by formation of the national-, state- and district-level authorities with clear plans and guidelines.

In this structure, the National Disaster Management Authority is the authority for formulation of policy and guidelines for all disaster management work in the country. The state authorities further lay down the guidelines for departments of the state and the districts falling in their respective jurisdictions. Similarly, district authorities direct the civil administration, departments and local authorities such as the municipalities, police department and civil administration. The state executive committees are responsible for execution of the tasks envisaged by the authorities.

2.2.3 DRR Progress

Being a signatory to the Hyogo Framework of Action 2005–2015, the country incorporated global knowledge and know-how on DRR to implement international strategy for DRR at national, state and local levels in India. The HFA reporting has improved monitoring and reporting of DRR actions in India. India has registered a remarkable progress on several fronts in reducing disaster risks since the HFA went into effect in 2005. This included legislation on disaster management and setting up of national, state and district level authorities with clear plans and guidelines. The HFA has also helped India improve accountability and coordination with international and global disaster risk reduction sector, including regional cooperation, especially in the South Asian Association for Regional Cooperation (SAARC) region. Since the enforcement of the Disaster Management Act in 2005, DRR landscape in India has become much wider and more inclusive.

In 2009, the government of India released India's first ever national policy on disaster management to guide DRR actors in the country. Similarly, a comprehensive report on 'disaster management in India' by the Ministry of Home Affairs, Government of India, was released in 2011 to capture India's initiatives on DRR and guide DRR stakeholders across India (Fig. 2.1). Established since 2005, the NDMA has produced numbers of guidelines to support and strengthen various aspects of disaster management in India and has produced a number of guidelines to facilitate

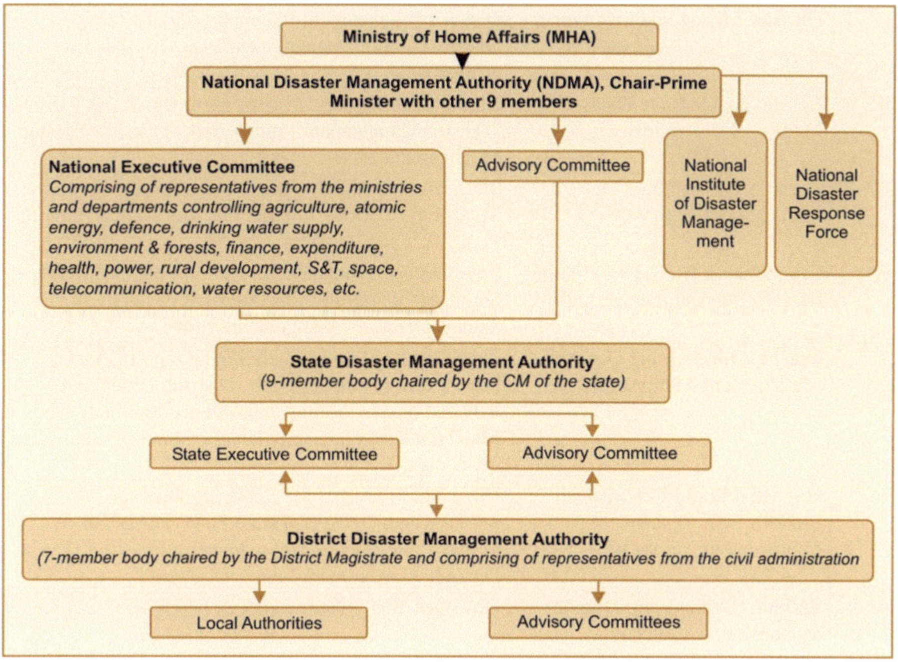

Fig. 2.1 Disaster management structure in India (Disaster Management in India 2011)

disaster recovery and preparedness planning. India has not only set up disaster relief funds at all levels but also launched the National Disaster Relief Fund, which is financed through the levy of a cess (The Ministry of Environment & Forests (MoEF) 2015). Likewise, State Disaster Management Authorities in India have developed their own plans and policies on DRR, including District Disaster Management Plan (DDMP) for each district (Table 2.2). But the National Disaster Management Authority has not reviewed DDMPs that are made by the districts with help from National Institute of Disaster Management guidelines and UNDP support (AIDMI) (2012). With an overarching goal of making local disaster management planning

Table 2.2 Laws, policies and plans to guide DRR actions of state and non-state actors at state levels

No.	State/authority	Key initiatives/projects/interventions
1.	Assam: Assam State Disaster Management Authority (ASDMA)	Assam State Disaster Management: Rules and Policy, State Disaster Management Plan, 27 District Disaster Management Plans and three City Disaster Management Plans
2.	Andhra Pradesh: Andhra Pradesh Revenue Disaster Management Department/ SDMA	The Andhra Pradesh Disaster Management Rules 2007, Andhra Pradesh State Disaster Management Plan (APSDMP), Preparing City level DM plan for all four cities (i.e. Srikakulam, Vijayawada, Khammama and Kurnool)
3.	Jharkhand: Jharkhand State Disaster Management Authority	State Disaster Management Plan, Departmental Disaster Management Planning Guidelines and District Disaster Management Plans for Palamu and Ramgarh
4.	Kerala: Kerala State Disaster Management Authority	Kerala State Disaster Management Rules, 2007, State Disaster Management Policy, 2010, District Disaster Management Plans
5.	Tripura: Tripura State Disaster Management Authority	State Disaster Management Policy, State Disaster Management Rules, District Disaster Management Plans
6.	Uttarakhand: Uttarakhand State Disaster Management Authority	State Disaster Management Action Plan (SDMAP), Standard Operating Procedures for key departments, Village Disaster Management Plans
7.	Odisha: Odisha State Disaster Management Authority	State Disaster Management Plan, Departmental Disaster Management Plans, District, Block and Village Disaster Management Plans, State Disaster Management Policy, City Disaster Management Plans in six cities
8.	Himachal Pradesh: Himachal Pradesh State Disaster Management Authority	Himachal Pradesh State Policy on Disaster Management 2011, DM Rule 2011, Himachal Pradesh State Disaster Management Plan and District Disaster Management Plans
9.	Sikkim: Sikkim State Disaster Management Authority	DM rules, SM policies, Sikkim DM Plan
10.	Maharashtra: Maharashtra State Disaster Management Authority	State Disaster Management Plan, District Disaster Management Plans

inclusive, AIDMI has launched a national campaign on making DDMPs inclusive. The initiative has covered select districts from the states of Assam, Bihar, Jammu and Kashmir and Odisha.

With a new government coming to power in May 2014, one of the key pending requirements of the Disaster Management Act, National Disaster Management Plan (NDMP), is now fulfilled. The NDMP is an opportunity to integrate climate action and India's NDCs as it not only lays down guidelines for preparation of state-level disaster management plans and plans by each central ministry and department but also provides for horizontal and vertical integration of government agencies and departments at various levels (Prime Minister's Office 2016). However, 'the national plan needs to be supplemented by national roadmaps for disaster resilience with clear goals, targets, timeframe, and ideas about how resources shall be mobilised for its implementation' (Dhar Chakrabarti 2016).

India's contribution and commitment continued during the third UN World Conference for Disaster Risk Reduction (WCDRR) in March 2015 in Sendai, Japan, where the new international framework for disaster risk reduction – Sendai Framework for Disaster Risk Reduction 2015–2030 (SDFRR) – was adopted by 187 member states. At the third UN WCDRR, the government of India assumed responsibility of hosting the seventh AMCDRR in 2016. As a follow up from the sixth Asian Ministerial Conference outcome and as a requirement of the Sendai Framework implementation, the intended outcome of the first Ministerial Conference in India will be to develop a 'Regional Action plan for Implementation of the Sendai Framework'.

2.3 Key Stakeholder Mapping in the Country

2.3.1 Governance

Governance can be simply defined as 'the process of decision-making and the process by which decisions are implemented (or not implemented)' (UNESCAP 2009). Although the government is a major actor, issues of governance are not limited to government as government is not the sole actor influencing decisions and how they are implemented (IRP) (Guidance Note on Recovery: Governance 2010). Risk governance is far more complex and comprehensive for a single institution or ministry to fully comprehend or cover. In a broader context, risk governance refers to the way in which the authorities, public servants, media, private sector and civil society coordinate in communities and on regional and national levels in order to manage and reduce disaster and climate-related risks (UNDP 2012). Thus, apart from the government, DRR landscape includes many other players such as UN agencies, national and international civil society organisations, financial institutions, private sector entities and so on.

2.3.2 Stakeholders

The DRR landscape in India includes the central government, state governments, private sector and civil society actors, including local communities, all playing significant roles in DRR at various levels. Well-defined policies, establishment of new authorities and proper guidelines, triggered by the Disaster Management Act, 2005, have stimulated DRR-related finance from a variety of public-private and international sources in India at various levels. Because many natural disasters are linked to climate change, a significant overlap in approaches, policies and measures is seen. India's INDCs have recognised planning and implementation of actions to enhance climate resilience and reduce vulnerability to climate change as one of the key priorities for achieving its commitments (The Ministry of Environment & Forests (MoEF) 2015).

This overlap, in many ways, is useful and beneficial. Since SFDRR 2015–2030 specifically addresses climate change and climate action, providing measures, guiding principles and means of implementation (UNISDR 2015a), integration of SFDRR implementation strategies and plans with NDCs is needed. However, the DRR landscape mostly remains isolated from the climate change landscape and vice versa. Although importance of better integration between DRR and CCA is recognised in India, current capacities in the government setup are limited to organise and prioritise such integration systematically. AIDMI's Risk and Resilience Programme (March 2008 to July 2016) has made systematic efforts to integrate DRR and CCA efforts at various levels through trainings, research and policy advocacy (AIDMI 2016). AIDMI works on integrating climate change and disaster risk reduction plans in over 50 districts in India under the National Disaster Management Plan of Government of India.

The recently release NDMP of India states that 'from time to time, the central government notifies hazard-specific nodal ministries to function as the lead agency in managing particular types of disasters (Tables 2.3 and 2.4).

2.4 Key Issues, Challenges and Opportunities for Strengthening Disaster Risk Governance

The World Conference has adopted the Sendai Framework for Disaster Risk Reduction 2015–2030 in Sendai, Japan, on March 18, 2015. The four priorities for action agreed by the World Conference focuses on (a) a better understanding of risk, (b) strengthened disaster risk governance, (c) more investment and (d) more effective disaster preparedness and embedding the 'build back better' principle into recovery, rehabilitation and reconstruction. To support the assessment of global progress in achieving the outcome and goal of this framework, seven global targets have been agreed.

Table 2.3 Current list of disaster-specific nodal ministries notified by GoI (NDMA, Government of India 2016)

	Disaster	Nodal ministry/department
1	Biological	Min. of Health and Family Welfare (MoHFW)
2	Chemical and industrial	Min. of Environment, Forest and Climate Change (MoEFCC)
3	Civil aviation accidents	Min. of Civil Aviation (MoCA)
4	Cyclone/tornado	Min. of Earth Sciences (MoES)
5	Tsunami	Min. of Earth Sciences (MoES)
6	Drought/hailstorm/cold wave and frost/ pest attack	Min. of Agriculture and Farmers Welfare (MoAFW)
7	Earthquake	Min. of Earth Sciences (MoES)
8	Flood	Min. of Water Resources (MoWR)
9	Forest fire	Min. of Environment, Forests and Climate Change (MoEFCC)
10	Landslides	Min. of Mines (MoM)
11	Avalanche	Min. of Defence (MoD)
12	Nuclear and radiological emergencies	Dept. of Atomic Energy (DAE)
13	Rail accidents	Min. of Railways (MoR)
14	Road accidents	Min. of Road Transport and Highways (MoRTH)
15	Urban floods	Min. of Urban Development (MoUD)

In the federal structure of our country with multiple agencies and stakeholders working on similar issues, entry points for improving disaster risk reduction governance can be many. 'Strengthening disaster risk governance for prevention, mitigation, preparedness, response, recovery and rehabilitation is therefore necessary and fosters collaboration and partnership across mechanisms and institutions for the implementation of instruments relevant to disaster risk reduction and sustainable development' (UNISDR 2015b) (Table 2.5). Although issues of governance cut across all the priorities of the Sendai Framework, we focus specifically on the governance priority of the framework, i.e. SFDRR priority 2: Strengthen disaster risk governance to manage disaster risk in this paper.

2.4.1 Mainstreaming

Mainstreaming DRR approaches and measures into development plans and setting achievable targets to measure impact are key to improving disaster risk governance in India. The government has established a policy framework for disaster management with a focus on mainstreaming disaster risk reduction into development planning and programmes. This framework has been successful in strengthening and implementing various provisions of the Disaster Management Act 2005 but has

Table 2.4 The list of entities play an important role in Indian landscape of DRR. This list by no means is exhaustive

Government	Intergovernmental bodies	Donors	CSO: NGOs	CSO: community groups	Other stakeholders
Ministry of Home Affairs, National Disaster Management Authority	UN agencies	Office of Direct Assistance	NGO networks/coalition	CBOs	Academic and research think tanks
State Disaster Management Authorities	Asia Disaster Preparedness Center	Multilateral and bilateral donors/financing institutions	INGOs	Volunteer groups	Private businesses
District Disaster Management Authorities	SAARC Center for Disaster Management, ASEAN disaster management bodies		National NGOs	Philanthropic groups	Media
National Institute of Disaster Management Military	Foreign embassies		Local NGOs		

Table 2.5 Governance issues and advocacy

Key governance issues in India	Entry points for strengthening disaster risk governance	Advocacy vehicles
Mainstreaming disaster risk reduction	Disaster Management Act 2005	Networks, alliances and working groups
Financing disaster risk reduction	National Policy for Disaster Management 2009	Workshops, conferences and forums
Participation of non-state actors	National Disaster Management Plan 2016	Funding
Accountability and performance	State Disaster Management Policies	Lobbying
	State Disaster Management Plans	Research
	District Disaster Management Plans	Pilot/demonstration projects
	Smart Cities Mission and City Disaster Management Plans	Policy briefs and publications
	Five-year plans and annual budgets	Media tools
	National and State Action Plan on Climate Change	Web influence
	Nationally determined contributions (NDCs)	
	Sectoral polices (water, energy, tourism, transport, livelihoods, transport, etc.)	

achieved limited impact in terms of mainstreaming DRR into development planning across authorities, stakeholders and sectors. 'Most SDMPs contain a high degree of detail on sectoral and departmental responsibilities for disaster responses, but fall short of outlining pathways for mainstreaming risk reduction' (CDKN and ODI 2016). Not enough attention is given to strengthen lowest governing structures such as *panchayats* in villages and urban local bodies (ULBs) in towns and cities of India. Although these structure shares roles and responsibilities laid by the 2005 Disaster Management Act of India, they are often found ill-equipped to deal with disaster in India both in terms of power and resources at their disposal. Effective decentralisation of disaster management responsibilities and mainstreaming cannot be done without employing these most essential structures at the local levels. Overall, results of mainstreaming efforts of key stakeholders in the DRR landscape of India have been mixed. For example, AIDMI experience of working with UNICEF and Save the Children on child-centred disaster risk reduction in India shows that DRR mainstreaming goal in the education sector has been quite successful but similar efforts in promoting development of new financial tools such as risk insurance or risk transfer in the finance sector have tested limited success.

The role of the private sector, such as insurance companies for 'risk informed investments in recovery efforts', is highlighted in the NDMP. However, much needs to be done especially at the grassroots level to make micro-insurance products available to the poor and vulnerable. AIDMI is currently piloting a disaster micro-insurance innovation supported by the Humanitarian Innovation Fund (HIF) in three cities of India (Puri, Cuddalore and Guwahati), which has a focus on small and microbusiness enterprises. The project is creating a platform to promote risk transfer as a tool for urban disaster risk reduction, showcasing risk transfer products and their impacts and importance of inclusion of women for a successful risk transfer model (HIF) (n.d.a).

Similarly, results of integrating DRR and CCA approaches in India have been encouraging but widespread coverage of such integration is lacking. National missions such as smart cities development and national scheme such as MGNREGS opportunities for DRR mainstreaming are plenty. However, limited capacities of authorities to explore such opportunities in planning are limiting India's efforts to reduce risk through DRR mainstreaming. A lot has been done to mainstream DRR through the Disaster Risk Management Project of UNDP India. But more efforts are required, especially at state and district levels. One such effort is recently initiated by the ADPC and Government of Bihar with support from Bill and Melinda Gates Foundation to provide technical support to the state government for the implementation of the Bihar Roadmap for Disaster Risk Reduction (2015–2030). Under this partnership, ADPC will build the capacity of Bihar's government officials on mainstreaming disaster risk reduction into development planning with a special focus on the agricultural and health sectors (ADPC 2016).

2.4.2 Financing DRR

An analysis of National Progress Reports on HFA by *Duryog Nivaran* (2014) shows that, except Bangladesh and Maldives, all other countries in the region, including India, have made available legislations that mandate local governments for DRR functions. However, when it comes to regular provision of financial resources to local governments for undertaking these functions, except Bangladesh, no country in the region provides the local governments the required financial resources for DRR. All the countries in the region (including India) have highlighted the limited capacity of the local government functionaries and elected representative on account of DRR as one of the key challenges they are facing. Both the SREX and AR5 reports indicate that the cost of both mitigation and adaptation is expected to rise substantially, and financing these two areas effectively would not only require more money but also changes in investment patterns and governance structure. However, opportunities to create synergies in international finance for disaster risk management and adaptation to climate change and its convergence with national climate finance resources are yet to be realised fully (IPCC 2012). Resource allocation beyond conventional DRR areas will be key to the successful implementation of Sendai Framework and India's nationally determined contribution. 'In the absence of reliable financing support, it would be unrealistic to expect many developing countries to take full advantage of the post-2015 DRR framework (Ministry of

External Affairs, Government of India 2014). Countries such as India will require to secure and sustain substantial amounts of financial resources to overcome the impacts of natural disasters and climate extremes in the coming years. This will require regular budgetary allocations, especially at state and district levels. The recently concluded review of state disaster management plans in India has revealed that although 'equal legal importance is given to financing for disaster response and risk reduction at the national level, but there are limited funds for risk reduction across states, despite legal and legislative mandates for this' (CDKN and ODI 2016).

2.4.3 Private Sector Engagement

The private sector's engagement in DRR landscape of India so far has been limited. The private sector is key to filling the financing gap in DRR and climate change adaptation, including meeting emission targets. But private sector bodies are hardly represented in ongoing DRR activities. The country needs to find ways to engage the private sector more to share the burden of reducing disaster risk. Public policy needs to create an enabling environment that supports and encourages micro-, small and medium enterprises (MSMEs) to integrate disaster resilience into their business through processes such as business continuity planning (6th AMCDRR 2014). The role of private sector in risk-proofing business investments is well understood, but most governments have not been able to capitalise on this experience. Many SDMPs in India encourage risk transfer mechanisms such as insurance, where the private sector has a significant role to play, but the uptake of insurance has not been widespread (CDKN and ODI 2016). AIDMI project supported by the Humanitarian Innovation Fund (HIF) is demonstrating how this could be achieved (n.d.b). Governments need to maximise the potential for private investment as fully as possible through enabling policy measures, regulations, taxations and other incentives. 'Improve existing governance systems, techno-legal regimes and policy frameworks to motivate and facilitate the Industry towards a viable and comprehensive Business Continuity Planning structure which incorporates DRR in their policies, plans and investment decisions' (SAARC 2014). India has become 'the first country to mandate a minimum spending on corporate social responsibility (CSR), Indian and foreign companies based in India that meet a certain turnover threshold are obliged to contribute two percent of their average net profits towards social development measures' (Chandrasekhar 2014). This additional resource has a lot of potential but has to be invested strategically in areas such as creating access to renewable energy, resilient crops, risk transfer and smart cities development.

2.4.4 Accountability and Performance

Issues of participation, decentralisation, accountability and transparency, including awareness of citizens on the rights of risk information, are critical to effective risk governance (UNISDR 2014). Having policies and legal framework in place will not be enough; policies and law supporting DRR must be supported with indicators in planning and actions against non-compliance, including incentives for responsible actions. Accountability and performance are often related to what is reported and to whom. The HFA Monitor has been used as an important tool for collecting information to self-assess and to provide means of verification to capture the progress and challenges of HFA implementation by nations. India has been most consistent in submitting these reports. But mere reporting is not enough. Accountability cannot exist without clearly defined monitoring, reporting and review systems, which underscore performance; clear targets and indicators in the Sendai Framework which drive and hold actors accountable for disaster and climate risk management actions. To this end, coherence of the targets and indicators with the Sustainable Development Goals and harmonisation of monitoring and review mechanisms, especially at the national level, are of significance (6th AMCDRR 2014). The 2005 Disaster Management Act of India has made pioneering steps in this direction; Chapter X on offences and penalties (The Disaster Management Act 2005; Ministry of Law and Justice, Government of India 2005) of the Act are specific and relevant for effective implementation of various provisions of the Act. Evidence shows that the use of empowering provisions of the Act such as peaceful but forced evacuations has saved many lives during the Cyclone *Phailin* in Odisha. How to avoid misuse of such provisions is also an area of debate and discussion as most states in India have reported significant progress on strengthening their laws, policies and regulation in the last decade or so.

2.5 Conclusion

The disaster risk reduction landscape in India is rapidly changing and like any other landscape is influenced by national and international policy changes. Although a lot of progress has been made in advancing DRR policies and practices, India faces several challenges posed by rising risks and economic losses from natural hazards and extreme events. Overcoming these challenges and turning them into opportunity not only require strengthening government systems and processes but also how non-state actors are governed and act. Achieving a more effective DRR governance system in India requires better integration of DRR and CCA approaches, improved representation of private sector players, clear and well-defined flow of finance for DRR, robust monitoring and accountability mechanics and participation of the poor and at-risk communities in decisions that impact them.

References

6th AMCDRR (2014) Asia-Pacific input document for the post-2015 framework for disaster risk reduction (HFA2). Available at: http://www.preventionweb.net/files/33306_chairssummarypostdraft1.4.pdf

ADPC (2016) ADPC and Government of Bihar aiming to enhance capacity for disaster risk reduction – see more at: http://www.adpc.net/igo/contents/Media/media-news.asp?pid=1111#sthash.QV6ST4uW.dpuf

AIDMI (2012) District disaster management plans. http://www.aidmi.org/DistrictDisasterManagementPlans

AIDMI (2016) Using SREX and AR5 for better implementation of India's NDC

CDKN, ODI (2016) Strengthening disaster risk management in India: a review of five state disaster management plans. http://cdkn.org/wp-content/uploads/2016/07/India-disaster-management-web.pdf

Chandrasekhar A. (2014) Swissinfo.ch, http://www.swissinfo.ch/eng/swiss-know-how_helping-india-cope-with-climate-change/40587432

CRED EM-DAT: The OFDA/CRED – International Disaster Database. www.emdat.be Université catholique de Louvain Brussels - Belgium, http://www.preventionweb.net/countries/ind/data/

Dhar Chakrabarti, P. G., (2016) The new National Disaster Management Plan has several flaws. http://www.hindustantimes.com/analysis/the-new-national-disaster-management-plan-has-several-flaws/story-LrHOFHXg9gwrtOgYjXJDrK.html

Disaster Management in India (2011) Ministry of Home Affairs (MoHA). New Delhi, Government of India (GoI)

FLACSO, UNISDR (2013) The future of disaster risk reduction. A Scoping Meeting for GAR 2015. San Jose, Costa Rica. Available at: http://www.unisdr.org/files/35715_thefutureofdisasterriskmanagement.pdf

Guidance Note on Recovery: Governance, (2010) International recovery platform. http://www.unisdr.org/files/16774_16774guidancenoteonrecoverygovernan.pdf

HIF (n.d.-a) Risk transfer and insurance: implementing SFDRR – 2015–2030. http://www.elrha.org/hif-blog/risk-transfer-insurance-implementing-sfdrr-2015-2030/

HIF (n.d.-b) Innovating disaster micro-insurance for local market recovery. http://www.elrha.org/map-location/innovating-disaster-micro-insurance-local-market-recovery/

IPCC (2012) Summary for policymakers. In: Field CB, Barros V, Stocker TF, Qin D, Dokken DJ, Ebi KL, Mastrandrea MD, Mach KJ, Plattner G-K, Allen SK, Tignor M, Midgley PM (eds) Managing the risks of extreme events and disasters to advance climate change adaptation, A special report of Working Groups I and II of the Intergovernmental Panel on Climate Change. Cambridge University Press, Cambridge, UK, pp 1–19. Available at: http://ipcc-wg2.gov/SREX/images/uploads/SREX-SPMbrochure_FINAL.pdf

Ministry of External Affairs, Government of India, (2014), Excerpts of statement by the leader of the Indian Delegation during Second Preparatory Committee Meeting for the Third UN World Conference on Disaster Risk Reduction (17–18 November 2014, Geneva), Statement by the Leader of the Indian Delegation at Second Preparatory Committee Meeting for the Third UN World Conference on Disaster Risk Reduction [17–18 November 2014, Geneva]. https://www.google.co.in/url?sa=t&rct=j&q=&esrc=s&source=web&cd=18&cad=rja&uact=8&ved=0CEUQFjAHOAo&url=http%3A%2F%2Fmea.gov.in%2FPortal%2FCountryNews%2F3243_Statement_by_Secy__BM__2nd_PrepCom_WC_DRR_17_Nov_2014.doc&ei=4cy6VNTBPM-dugSC14GIAQ&usg=AFQjCNGWq7166k5Hh0yKEJbegWdck_e2ZA&bvm=bv.83829542,d.c2E. Accessed on 17 Jan 2015

Ministry of Home Affairs (MoHA) (2011) Government of India (GoI), disaster management in India. New Delhi, MoHA

Ministry of Home Affairs, GoI (2005) Disaster management in India, http://www.unisdr.org/2005/mdgs-drr/national-reports/India-report.pdf

The Disaster Management Act 2005 Ministry of Law and Justice, Government of India (2005). http://www.ndma.gov.in/images/ndma-pdf/DM_act2005.pdf
NDMA (2016) Vulnerability profile, http://www.ndma.gov.in/en/vulnerability-profile.html
NDMA, Government of India (2016) National disaster management plan. https://www.google.co.in/url?sa=t&rct=j&q=&esrc=s&source=web&cd=3&cad=rja&uact=8&ved=0ahUKEwjFxqHEgqbOAhWJRo8KHYAPB5UQFggyMAI&url=http%3A%2F%2Fwww.ndma.gov.in%2Fimages%2Fpolicyplan%2Fdmplan%2FNational%2520Disaster%2520Management%2520Plan%2520May%25202016.pdf&usg=AFQjCNHFAFoJBby__T7aAf6Uq3LsvCs8uA&bvm=bv.128987424,d.c2I
Nivaran, D (2014) Recommendations for post-2015 framework for DRR: strengthening governance for risk reduction. Available at: http://www.duryognivaran.org/images/duryognivaran_inputs_to_hfa2strengthening_riskgovernance.pdf
Pratap CE (2012) Disaster management and law – a human rights perspective, http://www.lawyersclubindia.com/articles/Disaster-Management-and-Law-A-Human-Rights-Perspective-4497.asp#.UeZuPvWqnps
Prime Minister's Office (2016) The plan aims to make India disaster resilient and reduce loss of lives, Press Release. http://pib.nic.in/newsite/PrintRelease.aspx?relid=145840
SAARC (2014) Conference report: SAARC regional conference on business continuity planning & disaster risk reduction in South Asia, 15–16 December 2014, New Delhi, India, http://saarc-sdmc.nic.in/pdf/bcp-2014.pdf
Thakur P. (2015), Disasters cost India $10bn per year: UN report, http://timesofindia.indiatimes.com/india/Disasters-cost-India-10bn-per-year-UN-report/articleshow/46522526.cms
The Hindu (2016). India ranked 77 in disaster risk index of the world, http://www.thehindu.com/sci-tech/energy-and-environment/india-ranked-77-in-disaster-risk-index-of-the-world/article9036125.ece
The Ministry of Environment & Forests (MoEF) (2015) India's intended nationally determined contributions. http://www4.unfccc.int/submissions/INDC/Published%20Documents/India/1/INDIA%20INDC%20TO%20UNFCCC.pdf
UNDP (2012) Issue brief: disaster risk governance. Available at: http://www.alnap.org/resource/11701
UNESCAP (2009) What is good governance? http://www.unescap.org/pdd/prs/ProjectActivities/Ongoing/gg/governance.asp
UNICEF. UNICEF in emergencies. http://www.unicef.org/india/emergencies.html
UNISDR (2013a) Chair's summary fourth session of the global platform for disaster risk reduction, Geneva, 21–23 May 2013. Available at: http://www.preventionweb.net/files/33306_chairssummarypostdraft1.4.pdf
UNISDR (2013b) Synthesis report: consultations on a post-2015 Framework on Disaster Risk Reduction (HFA2) Geneva: Switzerland. Available at: http://www.unisdr.org/files/32535_hfasynthesisreportfinal.pdf.
UNISDR (2014) Asia-pacific input document for HFA2, building the resilience of nations and communities to disasters. http://6thamcdrr-thailand.net/6thamcdrr/Portals/0/Downloadable/PrevAMCDRR/HFA_input_document_Asia_Pacific-ver_16April.pdf
UNISDR (2015a) Briefing note on the 21st conference of the parties to the UNFCCC (COP21) in Paris. http://www.preventionweb.net/files/46311_unisdrbriefingnotedrrcop21.pdf
UNISDR (2015b) Sendai framework for disaster risk reduction 2015–2030. Switzerland, Geneva. http://www.preventionweb.net/files/43291_sendaiframeworkfordrren.pdf

Chapter 3
Disaster Governance and Legal Systems in India

Anil Kumar Gupta

Abstract India is a multi-hazard-prone country with diversity of eco-geological and socio-economic settings. As witnessed by historical time disasters and their management, the disaster governance in India was primarily of contingency management approach. Codes, practices and guidelines for preparedness and particularly on relief existed even before Independence. However, with the growing understanding on risk causes, the paradigm shift to prevention-mitigation and now on mainstreaming disaster risk reduction and climate change adaptation into development forms a key agenda in development governance. Legal and policy framework has given rise and mandates to key institutions for policy guidelines, capacity development and emergency response at national, subnational and local levels. Laws directly and indirectly addressing disaster-related hazards and factors of vulnerability and capacity have been reviewed besides opportunities, resources, and responsibilities of organizations/agencies stated herein. Mechanisms of financial strategies, international cooperation and recent advances are discussed, besides the journey of disaster management governance in India. A detailed account on national authority (NDMA), institute (NIDM), emergency response and subnational and district framework for disaster governance is also enumerated.

Keywords Disaster governance • India • Law • Institutions • Roles • Resources

3.1 Introduction

Disasters are events of environmental extremes as inevitable entities of this living world. Disaster management highlights the interdependence of the economy, environment and inclusive development. The United Nation Report titled ‘Living with risk’ claims that though there has been decline in the loss of human lives from disaster, the occurrence of disaster is on rise (UNISDR 2004). The world’s biggest industrial disaster – gas tragedy (1984) that took place in Bhopal a central city of

A.K. Gupta (✉)
National Institute of Disaster Management, New Delhi, India
e-mail: envirosafe2007@gmail.com

I. Pal, R. Shaw (eds.), *Disaster Risk Governance in India and Cross Cutting Issues*, Disaster Risk Reduction, DOI 10.1007/978-981-10-3310-0_3

India – shaken the world and triggered global awakening towards human environment and safety and led to a journey of environmental jurisprudence. However, this was followed by several environmental, technical and natural disasters in India and this part of the world. Disasters, especially the ones originating from natural environmental process – geological, hydrological, meteorological or biophysical – were dealt mostly in a conventional ad hoc mode of response and relief-centric approach. Most international environmental strategies starting from Stockholm Conference of 1972 referred to the natural hazards as key impediment to the process of sustainable development. Agenda-21 adapted at the Rio Summit 1992 is still believed to be relevant as a milestone strategy towards sustainable development by adopting integrated approach of environment, disaster reduction, safety and sustainable economic growth. India's commitment and activities following the Rio Summit, which coincided with the UN-IDNDR (International Decade for Natural Disaster Reduction 1991–1999), helped naturally recognize the need of integrated and holistic approach to disaster management.

Major disasters, for example, earthquakes of Uttarkashi (1991), Latur (1993) and Chamoli (1999), Assam floods (1998) and Orissa super cyclone (1999), created an environment of serious brainstorm on the state of disaster management in India and on the steps needed to improve the situation. India was a party to Yokohama Declaration on disaster reduction for a safer world. The constitution of High Power Committee (HPC), chaired by Mr. J. C. Pant, former secretary of the Indian government, was a key step in this direction. HPC resulted in a detailed study report with a set of fundamental and practical recommendations. The Gujarat Bhuj earthquake 2001 experience triggered the proposing of the Disaster Management Bill, which, however, was in dormancy until another major disaster – Indian Ocean tsunami 2004 – severely struck coastal states of India, along many other nations in the region (UNISDR 2015). The year 2005 yielded with passage of the Disaster Management Act which legally established a paradigm shift to 'prevention-mitigation-based holistic disaster management'. Interestingly, the year 2005 also witnessed India's participation in Kobe World Conference on disaster reduction that resulted into Hyogo Framework for Action (2005–2015) (UNISDR 2007).

Though India's pioneering law which enabled for systemic planning and preparedness for disaster emergencies and tiered approach of authorities was the 'Emergency Planning, Preparedness and Response Rules 1996' under the Environmental Protection Act 1996 of India, the mechanism of holistic planning for disaster management and authorities at the national, state, district and local level has been brought by the 2005 Disaster Management Act. In the backdrop of current awakening on climate change, environmental extremes and ecological degradation aggravating disasters and vulnerabilities, a systemic thought process of integrating environment management, disaster risk reduction and human development as imperative facets of sustainable development has emerged following the Intergovernmental Panel on Climate Change (IPCC) Fourth Assessment Report in 2007. This is suggested as the second paradigm shift in disaster management (Gupta et al. 2010), which now coincides with the new epoch of the Sendai Framework for DRR, along Paris Agreement on Climate Change contributing towards the Sustainable Development Goals (2015). Social dimensions, thus, particularly those related with livelihood and food security, shelter, water and sanitation, environmental health, gender sensitivity, empowerment, growth opportunities, ecological security and sustainable habitat have adequate and multiple references in the current strategies and plans of disaster risk reduction at almost all levels.

The disaster governance in India can be understood in four major segments:

(a) Authority/decisions and inter-agency co-ordination: National Disaster Management Authority covering supervision of National Disaster Response Force and National Emergency Operation Centre, National Executive Committee, Cabinet Committee, Disaster Management Division of Government of India in the Ministry of Home Affairs, authorities at state, district and local levels
(b) Capacity building, policy analysis, advocacy, research and training/education, human resource development: National Institute(s) of Disaster Management, policy think tanks and planning agencies (National Institute of Transforming India, NITI Aayog), state institutes/centres, state planning boards, etc.
(c) Financial strategies: Disaster Response Fund, Mitigation Fund, Adaptation Fund, Relief Fund, Insurance, etc. at national, state, district and local levels
(d) Field missions – emergency response and relief: Armed forces, paramilitary, National Disaster Response Force, State Disaster Response Force, institutional network of district/local authority with NGOs, iNGOs, community organizations, etc.

However, with the paradigm shift from 'response and relief-centric' to 'prevention-mitigation'-centric approach and further to the second paradigm of 'environment based integrated approach with adaptation and socio-economic sustainability' along the evolution of institutional/governance mechanisms over the two decades have led to certain 'naturally possible' but serious drawbacks as well, indicated as following:

1. Overlap of authority and activity types: Central disaster response coordination, training and research coordination, policy making vs policy analysis and advocacy, etc.
2. Issues in hierarchy and autonomy/reporting among authorities in different segments of disaster governance at the same level (e.g. NDMA, NIDM, NDRF, NEC, etc.).
3. Staffing: Certain agencies are having huge but unsustainable professional or unprofessional staff, whereas other agencies have less or the least and the strengths and hierarchy incompatible with the mandates.
4. Lack of clarity on certain funds and financial mechanisms, for example, in the case of disaster mitigation fund, implications of State Action Plan on Climate Change implementation for disaster risk reduction, etc.

The current regime of planned and documentation-based approach of disaster management at various levels of government and functions/departments has tried to address these aforementioned issues. India has released its first national disaster management plan in 2016, the final draft of which was developed by the National Executive Committee on Disaster Management, chaired by the Union Home Secretary, with professional assistance of the National Institute of Disaster Management faculty (Gupta et al. 2013). All the states and the districts therein, and many of the local bodies (examples also include village level), have developed their disaster management plan and are in the stage of updating/revising to integrate the aspects of climate resilience and sustainable development. 'Capacity building' being the most inadequately addressed aspect is a prerequisite to the HFA priority 'addressing underlying risk factors' and the Sendai DRR Framework priority 'understanding disaster risk'. The components of capacity building – human resource development – teaching/training and research now attribute a topmost agenda at the highest level of public governance in the country for necessary expansion and strengthening in the form of multi-institutional autonomous framework of the national institute(s) of disaster management, in close coordination with other institutes of professional excellence (in technology, environment, medicine, science, planning, humanities, etc.), central ministries (environment, human resource/education, science and technology, rural development, NITI Aayog, etc.) and international/intergovernmental agencies (SAARC, UNESCO, ICIMOD, GIZ, UNESCAP, etc.).

3.2 Disaster Governance: Principles and Researching

Not incidentally or accidently but thoughtfully, disasters have been referred to as outcome of hazardous environmental process or events, technical or operational failures, faulty developmental practices and poor civil administration or governance. Figure 3.1 summarizes good governance and capacity building as a central component regarding the process and implementation of disaster risk management

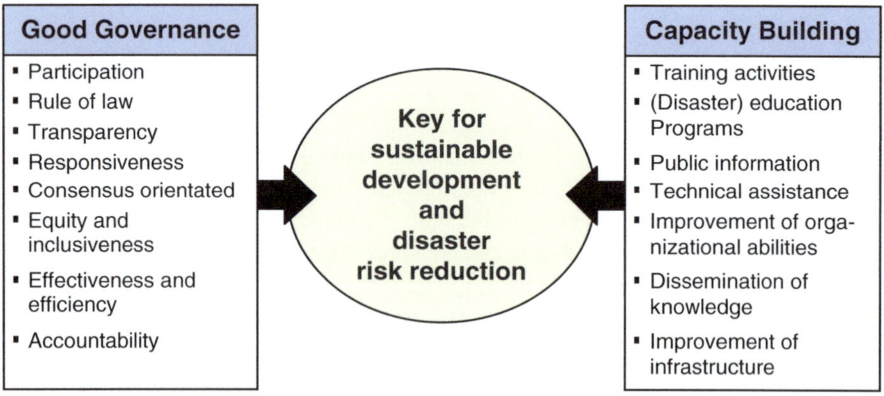

Fig. 3.1 Governance and capacity building for risk reduction (Source: IFS 2006)

and sustainable development (IFS 2006). Governance represent a set of tools and administrative environment wherein decision-makers and community leaders help them appreciate the importance and benefits of good policies, appropriate institutional and legislative systems at the national level as frameworks for the effective recovery plans and programmes (UNISDR 2004).

'Governance' is a function of actors, structures and processes by which societies share power and make collectively binding decisions (Lebel et al. 2006; van Asselt and Renn 2011). Adopting a 'governance perspective entails giving attention to the distribution of political power both internal and external to the state (Goodwin 1998). The distribution of power between state and non-state actors has changed significantly over the last few decades, as a consequence of neoliberal economic and political restructuring. This is often referred to as the shift from 'government to governance'. This shift of thoughts and practice at policy, planning and field actions is crucial for the decentralization and diversification of disaster management in the country, as the basis for mainstreaming disaster risk reduction and climate change adaptation together into developmental planning and practices of government and communities. Nongovernmental organizations (NGOs) have found a new place within the neoliberal global order in terms of the 'outwards' redistribution of state functions (Frewer 2013). NGOs are an extremely diverse group of independent organizations that are neither run by government nor driven by profit (Lewis and Kanji 2009). The disaster risk reduction approach requires redefining the role of the governments as decentralization, capacity building and involvement of other stakeholders towards self-reliance at the local level. Nevertheless, the role of the state/civil administration is always critical as governments and local institutions to be the most important set of actors in disaster management (Christoplos et al. 2001; Bulkeley and Jordan 2012).

The modern paradigm of disaster management – disaster risk reduction (DRR) – and convergence with environmental management, climate change adaptation and socio-economic security systems towards achieving the Sustainable Development Goals (SDGs) calls for a holistic vision of 'good governance'. In the neo-liberal environment of reduced responsibility for the state, alongside global platforms established to implement the Hyogo Framework for Action, a new arena has been opened for a multitude of stakeholders to engage in disaster risk reduction (Jones et al. 2014).It is noted widely that some governments have successfully adopted and implemented disaster risk reduction (DRR) policies, while others lag behind (Williams 2011) especially when it comes to local governments and field level actions. The shortcomings in DRR are increasingly being regarded as a consequence of poor governance and lack of political will (Williams 2011). However, very little attention has been given to researching the processes of governance of DRR, such as the formulation of policy and the roles of different stakeholders in actual field implementation. In addition, there is a lack of evidence on the effectiveness of different governance systems (UNISDR 2011).

Efficacy of governance is particularly hindered by the complex interplay of power and knowledge among diverse groups of actors with irrational command over resources, roles and responsibilities (Ojha et al. 2009). In DRR, this complex interplay of power and knowledge among diverse stakeholder groups gives rise to different governance approaches. For long-term recovery operations, it is particularly important to advocate for 'good governance' at the local government level and also in the policy programming. As the immediate public service provider and the interface with citizens, local governments are naturally situated in the best position to raise awareness of the citizens on resilient recovery (i.e. building back better approaches) and listen to their concerns. Recovery operations may fail, if the communities are not properly informed and engaged (NHRAIA 2001).

3.3 Disaster Management Evolution in India

India is a large country and has had more than its share of major natural hazards like drought, floods, earthquakes and cyclones throughout its history of civilization. With the industrial growth and technical revolution, the risk of industrial-chemical accidents also grew manifolds. Besides natural, industrial/chemical, technological, fire and transport disasters, there is a list of other mishaps, resulting into disasters, totally of civil origin, those related with law and order, or homeland security, for example, stampede, sabotage, hijacking, riots, terrorist attack, serial bomb blasts, etc. Naturally, the country developed its own practices and strategies for dealing with various natural calamities and human-made disasters. However, the mainframe of disaster management in India has been concerned mostly with the natural, environmental, industrial and technological disasters. Management of socio-civil management is adhered primarily to the law and order enforcement agencies.

3.3.1 Pre-IDNDR Disaster Management in India

Post-independence (1947) and until the United National International Decade for Natural Disaster Reduction (IDNDR 1991–1999) disaster management in India has been more of a contingency management oriented. A contingency action plan for dealing with contingencies arising in the wake of natural calamities was developed at the national level, with provision of periodic updating. The major emphasis in the plan was on providing relief in response to the occurrence of a major calamity due to any of the natural hazards beyond the coping capacity of the states or union territories through their own resources. The preparedness for providing relief was ensured through contingency plans formulated for operations at the state levels as well as at the district levels. The primary relief functions of the central government included the following aspects:

1. Forecasting and operation of warning systems
2. Maintenance of uninterrupted communication
3. Wide publicity to warnings of impending calamity, disaster preparedness and relief measures through television, radio and newspapers

4. Transport with particular reference to evacuation and movement of essential commodities and petroleum products
5. Ensuring availability of essential commodities at reasonable prices particularly the commodities through the public distribution system
6. Ensuring availability of medicines, vaccine and drugs
7. Preservation and restoration of physical communication links
8. Investments in infrastructure
9. Mobilization of financial resources

3.3.2 UN-IDNDR, High Powered Committee and HFA Period

India has developed a nationwide relief administration where a lead role of the state governments has been envisaged. The 10-year period of International Decade for Natural Disaster Reduction (UN-IDNDR 1991–1999), therefore, has been taken as a good opportunity for the country to look back at what had been done in the past, take new initiatives during the decade and plan ahead for reducing the impact of the natural hazards on its people, settlements and economic development. The decade also witnessed an international policy paradigm in the form of the Yokohama Strategy and Plan of Action for a Safer World (guidelines for natural disaster prevention, preparedness and mitigation) as outcome of the World Conference on Natural Disaster Reduction, held in Yokohama, Japan, in May 1994.

Following a series of disasters such as the Latur earthquake (1993), Malpa landslide (1994) and Orissa super cyclone (1999), a welcome step in the strategic move was the setting up of a High Powered Committee (HPC) on Disaster Management in 1999, under the chairmanship of Mr. J.C. Pant, former Secretary to the Government of India, for drawing up a systematic, comprehensive and holistic approach towards disasters. The High Powered Committee report in 2001 recommended that at least 10% of 'plan funds' at the national, state and district levels be earmarked and apportioned for schemes that specifically address areas such as prevention, reduction, preparedness and mitigation of disasters. It was also for the first time in the planning history of India that planners devoted a separate chapter titled 'Disaster Management: The Development Perspective' in the Tenth Five-Year Plan document (Planning Commission 2002). In India, capacity building in disaster management has been relatively less organized and was in the form of certain schemes of the central government. The central government, in the Ministry of Agriculture, instituted a central sector scheme, of supporting establishment of the Disaster Management Faculty in the country and the National Centre for Disaster Management (NCDM) was established as a project centre at the Indian Institute of Public Administration (IIPA) in 1995. The Ministry of Environment also implemented certain World Bank-aided projects with disaster prevention, industrial safety and water-related issues as thematic areas. Later the NCDM was carved out of the IIPA and instituted by upgrading to a National Institute of Disaster Management (NIDM) in 2003, with its own management structure.

The High Powered Committee (HPC) on Disaster Management in its report submitted in 2001 gave clear recommendations on the preparation of disaster management plan(s) and suggestions for effective mitigation mechanisms. Its recommendation included the draft of the Disaster Management Act, a national response plan, strategy to move from disaster response to disaster preparedness, and establishment of National Disaster Management Authority. The 10-year period of post-HPC paradigm shift in India was parallel and also in compliance with the Hyogo Framework of Action (2005–2015). The Eleventh Five-Year Plan 2007–2012 (Planning Commission 2008) aimed at consolidating the process by giving impetus to projects and programmes that develop and nurture the culture of safety and the integration of disaster prevention and mitigation into the development process.

3.4 Disaster Management Governance and Institutions

The India Disaster Report (Parsuraman and Unikrishnan 2000) reviewed on the nature of disaster response by the Government of India. It identified key issues with respect to the availability of and access to disaster-related information and its quality, the absence of coherent disaster preparedness and response policy and urgent actions and interventions needed. It expressed that significant advances in health and social and economic development have been repeatedly interrupted and reversed by disasters.

During the HFA period, India made big strides in the area of disaster management by enacting the dedicated national law on disaster management (DM Act 2005) which was followed by state legislations as well. A national policy on disaster management was notified in 2009. The National Act and Policy on Disaster Management gives rise to an institutional mechanism in the country. The tiered approach at the national level (National Disaster Management Authority which also supervises National Disaster Response Force; Emergency Operation Centre, besides Cabinet Committee; National Executive Committee; Disaster Management Division of the Central Government in the Ministry of Home Affairs), state level (State Disaster Management Authority covering State Disaster Response Force, State Disaster Management Department/Secretariat), district authority and local authority has come into existence following the legislature. An institutional framework of capacity building in the form of National Institute of Disaster Management (NIDM) is under plan of multiplication with Vijaywada (Andhra Pradesh) as the second NIDM after New Delhi. Various legal provisions and institutional settings of disaster governance in India are discussed in the proceeding sections.

India is a sovereign socialist secular democratic republic with a parliamentary system of government at the federal level. For the administrative purpose, India has been divided into 36 jurisdictions known as states and union territories. The union territories consist of six jurisdictions that are centrally or federally administered. Disaster management is the responsibility of local administration, under the supervision of the state government, facilitated by the Government of India. The 36 states and union territories are divided into 687 districts. Each district is administrated by a collector and district magistrate (same person performs both the duties). As mandated under the Disaster Management Act, 2005, each district is supposed to have a disaster management plan, district disaster management committee and district emergency operation centre (EOC), conduct mock drills and carry disaster prevention, preparedness and mitigation functions.

3.4.1 Apex Coordination

A National Disaster Management Authority (NDMA) has been established at the centre, and the State Disaster Management Authority (SDMA) at state and district authorities at the district level is gradually being formalized. In addition to this, the National Crisis Management Committee, part of the earlier setup, also functions at the centre. For different disaster types/disaster management function the nodal ministries are identified in the national disaster management plan, under the overall coordination of the Ministry of Home Affairs (nodal ministry for disaster management). This makes the stakeholders interact at different levels within the disaster management framework. Apex coordination agencies of disaster management at the central government level are the following:

National Disaster Management Authority As the nodal apex agency in disaster management, NDMA is chaired by the prime minister of India. It has certain members at the level of secretary to the Government of India and a vice-chairman of the equivalent status of the cabinet secretary. NDMA has been mandated for the following responsibilities:

1. Lay down policies on disaster management.
2. Approve the national plan.
3. Approve plans prepared by the ministries or departments of the Government of India in accordance with the national plan.
4. Lay down guidelines to be followed by the state authorities in drawing up the state plan.
5. Lay down guidelines to be followed by the different ministries or departments of the Government of India for the purpose of integrating the measures for prevention of disaster or the mitigation of its effects in their development plans and projects.
6. Coordinate the enforcement and implementation of the policy and plan for disaster management.
7. Recommend provision of funds for the purpose of mitigation.
8. Provide such support to other countries affected by major disasters as may be determined by the central government.
9. Take such other measures for the prevention of disaster or the mitigation or preparedness and capacity building for dealing with the threatening disaster situation or disaster as it may consider necessary.
10. Lay down broad policies and guidelines for the functioning of the National Institute of Disaster Management.

National Executive Committee NEC is constituted, under the DM Act 2005, to assist the National Authority in the performance of its functions. NEC consists of home secretary as its chairperson, ex officio, with other secretaries to the Government of India in the ministries or departments having administrative control of the agriculture, atomic energy, defence, drinking water supply, environment and forest, finance (expenditure), health, power, rural development science and technology, space, telecommunication, urban development and water resources. The chief of integrated defence staff of the chiefs of staff committee, ex officio, is also its member. NEC may as and when it considers necessary constitute one or more subcommittees for the efficient discharge of its functions. NEC has been given the responsibility to act as the coordinating and monitoring body for disaster management, to prepare a national plan, monitor the implementation of national policy, etc.

Cabinet Committees The Cabinet Committee on Management of Natural Calamities (CCMNC) oversees all aspects relating to the management of natural calamities including assessment of the situation and identification of measures and programmes considered necessary to reduce its impact, monitor and suggest long-term measures for prevention of such calamities and formulate and recommend programmes for public awareness for building up society's resilience to them. The Cabinet Committee on Security (CCS) deals with the matters relating to nuclear, biological and chemical emergencies. The National Crisis Management Committee (NCMC) under the cabinet secretary oversees the command, control and coordination of the disaster response.

The Disaster Management Act, 2005, has created new institutions at the national, state, district and local levels. The new apex institutional framework at different level for disaster management in the country is shown in Fig. 3.2.

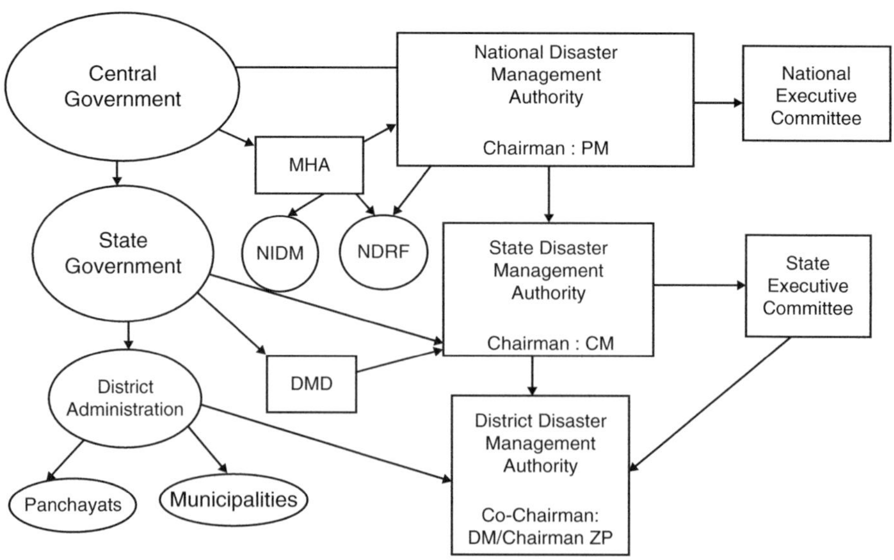

Fig. 3.2 Apex institutions of disaster management at different levels

The Ministry of Home Affairs (MHA) in the central government has the overall responsibility for disaster management in the country. A disaster management division has been established which is headed by the joint secretary (disaster management). For a few specific types of disaster issues, the concerned ministries have the nodal responsibilities for its management, as the following:

Drought:	Ministry of Agriculture and Farmers Welfare
Epidemics and biological disasters:	Ministry of Health and Family Welfare
Chemical disasters:	Ministry of Environment, Forests and Climate Change
Nuclear disasters:	Ministry of Atomic Energy
Air accidents:	Ministry of Civil Aviation
Railway accidents:	Ministry of Railways
Forest fire:	Ministry of Environment, Forests and Climate Change
Floods:	Ministry of Water Resources
Road accidents:	Ministry of Surface Transport

However, certain functions and aspects of forecasting/warning and monitoring are looked by various departments/ministries of the Government of India as nodal ministries, for example:

Climate change, desertification:	Ministry of Environment, Forests and Climate Change
Rainfall, cyclone, earthquake:	Indian Meteorological Department (IMD)
Flood forecasting:	Central Water Commission (CWC)
Tsunami:	Indian Centre for Ocean Information System (INCOIS)
Landslides:	Geological Survey of India
All natural disasters:	Indian Space Research Organization

The Government of India has constituted a multi-stakeholder National Platform for Disaster Risk Reduction (NPDRR) in 2013 with aims of bringing together the whole range of India's disaster risk community from government, parliamentarians, mayors, media, international organizations, NGOs, local community representatives, scientific and academic institutions, corporate businesses, etc. The NPDRR is chaired by the prime minister. The main functions of NPDRR are:

- To review the progress made in the field of disaster management from time to time
- To appreciate the extent and manner in which the disaster management policy has been implemented by the central and state governments, and other concerned agencies, and to give appropriate advice on the matter
- To advise in coordination between the central and state governments/union territory (UT) administrations, local self-governments and civil society organizations for disaster risk reduction
- To advise *suo-moto* or on a reference made by the central government or any other state government or union territory administration on any question pertaining to disaster management
- To review the National Disaster Management Policy

3.4.2 Policies and Legal Framework

India's disaster management policy is geared to make a paradigm change from response and calamity relief to disaster prevention, preparation and mitigation. Another significant vision is to move from disaster management largely from government-centric to public-private partnership and community-based disaster management. As far as law is concerned, India seldom lags behind in enacting a law. The real challenge is realized in implementation and appropriation of the legal provisions to befit the circumstances. Laws relating to disaster management in India, in fact, date back to the period of Emperor Chandragupta Maurya, as described nicely in the book *Kautilya's Arthashastra* (Economics – the first ever book on political science). The Great Famine of 1876–1878 lead to the constitution of the Famine Commission of 1880 and eventual adoption of the Famine Relief Code. India probably has the world's oldest disaster relief code which started in 1880. This relief code provided details of the relief to be given by the government to the affected people.

The Government of India approved the national disaster management policy in 2009. The themes underpinning this policy are the following:

1. Community-based disaster management, including integration of the policy and plans and execution at the grassroots level
2. Capacity development in all related areas
3. Consolidation of past initiatives and best practices
4. Cooperation with agencies at national, regional and international levels
5. Compliance and coordination to generate a multisectoral synergy

The objectives guiding the policy formulation have evolved to include:

- Promoting a culture of prevention and preparedness – by centre-staging disaster management (DM) as an overriding priority at all levels and at all times
- Encouraging mitigation measures based on state-of-the-art technology and environmental sustainability
- Mainstreaming DM concerns into the development planning process
- Putting in place a streamlined institutional techno-legal framework in order to create and preserve the integrity of an enabling regulatory environment and a compliance regime
- Developing contemporary forecasting and early warning systems backed by responsive and fail-safe communications and information technology (IT) support
- Promoting a productive partnership with the media, NGOs and the corporate sector in the areas of awareness generation and capacity development
- Ensuring efficient response and relief with a caring humane approach towards the vulnerable sections of the society
- Making reconstruction an opportunity to build back better and construct disaster-resilient structures and habitats

Legislation for DRR is fundamental to the enhancement of human security. It is the first step towards mainstreaming disaster risk reduction into development. Legislation provides the framework around which strategies to build risk reduction into development and reconstruction activities can be empowered. The law can be used to provide penalties and incentives by enforcing standards in construction, land use and tenant's rights and by defining people's rights during relief and reconstruction (Pelling and Holloway 2006). The goal of sustainable development through disaster reduction needs to be operationalized through a set of policies and an effective legal framework. Legal measures and provisions can play significant roles in following specific ways.

Following the Stockholm Conference 1972, constitutional sanction was given in 1976 to environmental concerns through the 42nd Amendment, which were incorporated into the Directive Principles of State Policy (Article 48 A) and Fundamental Rights and Duties (Article 51A(g)). The Indian constitution is unique in having specific provisions related to environment and human rights. Article 21 (Right to Life) of the Indian Constitution states 'No person shall be deprived of his life or personal liberty except according to procedure established by law'. The right to life has been employed in a diversified manner in India. Besides the mere right to survive as a species, quality of life, the right to live with dignity, the right to livelihood, etc. also come within the purview of Article 21. The Constitution of India provides that all are equal before the law and shall be accorded equal protection of the law. Article 14 (Right to Equality) states that 'The State shall not deny to any person equality before the law or the equal protection of the laws within the territory of India'. Article 14 can be used to challenge government sanctions for mining and other activities with high stakes on human rights and environmental impact, where the permissions are arbitrarily granted without adequate consideration of environmental impacts. The Constitution Act of 1976 (42nd Amendment) explicitly incorporated environmental protection and improvement as a part of state policy. Article 48 A, which is Directive Principles of State Policy, provides that the state shall endeavour to protect and improve the environment and safeguard the forests and wildlife of the country. Article 51A (g) imposes a similar responsibility on every citizen to protect and improve the natural environment including forests, lakes, rivers and wildlife and to have compassion for living creatures.

The Government of India, accepting the risk of climate change implications of different dimensions of environment and development, including on disasters, and foreseeing the needs of adaptation strategies, established a Prime Minister's Council on Climate Change. On June 30, 2008, the prime minister released India's first National Action Plan on Climate Change (NAPCC), outlining existing future policies and programmes and addressing climate mitigation and adaptation. The plan identifies eight core 'national missions' running through 2017 and directs ministries to submit detailed implementation plans to the Prime Minister's Council on Climate Change by December 2008. Emphasizing the overriding priority of maintaining high economic growth rates to raise living standards, the plan 'identifies measures that promote the development objectives while also yielding co-benefits for addressing climate change effectively'. The Ministry of Environment, Forests and Climate

Change is the nodal ministry on climate change-related issues. The National Action Plan on Climate Change identified eight missions:

- National Solar Mission
- National Mission on Sustainable Habitat
- National Mission for Enhanced Energy Efficiency
- National Mission for Sustaining the Himalayan Ecosystem
- National Water Mission
- National Mission for Green India
- National Mission for Sustainable Agriculture
- National Mission for Strategic Knowledge on Climate Change

The Disaster Management Act 2005 lays down strong institutional mechanisms at the national, state and district level that will work together in close harmony. Section 2(d) of the act defines disaster as 'a catastrophe, mishap, calamity or grave occurrence in any area arising due to natural or man-made causes or by accident or negligence which results in substantial loss to life or human suffering or damage to, and destruction of property or damage to, or degradation to environment and is of such a nature or magnitude which is beyond the coping capacity of the community of the affected area'. The DM Act 2005 defines disaster management as a continuous cycle and integrated process of planning, organizing, coordinating and implementing measures which are necessary or expedient for:

1. Prevention of danger or threat of any disaster
2. Mitigation or reduction of risk of any disaster or its severity or consequences
3. Capacity building
4. Preparedness to deal with any disaster
5. Prompt response to any threatening disaster situation or disaster
6. Assessing the severity or magnitude of effects of any disaster
7. Evacuation, rescue and relief
8. Rehabilitation and reconstruction

From time to time, the central government issued notifications under the Environmental Protection Act 1986 for the protection of ecologically sensitive areas or issues relevant to hazards and vulnerabilities associated with disaster risks. Some of such notifications are the following:

- Doon Valley Notification (1989)
- Coastal Regulation Zone Notification (2011)
- Dhanu Taluka Notification (1991)
- Revdanda Creek Notification (1989)
- The Environmental Impact Assessment of Development Projects Notification (2006)
- Ash Content Notification (1997)
- Taj Trapezium Notification (1998)
- Disposal of Fly Ash Notification (1999)

Other environmental and other legislation in India which are relevant in disaster risk reduction aspects are the following:

- Manufacture, Storage and Import of Hazardous Chemical Rules, 1989 (Amended 1994, 2000)
- Bio-Medical Waste (Management and Handling) Rules, 1998
- Recycled Plastics Manufacture and Usage Rules, 1999
- Ozone Depleting Substance (Regulation and Control) Rules, 2000
- Municipal Solid Wastes (Management and Handling) Rules, 2000
- Batteries (Management and Handling) Rules, 2001
- E-waste (Management and Handling) Rules, 2011
- Chemical Accidents (Emergency Planning Preparedness and Response) Rules, 1996
- Public Liability Insurance Act, 1991
- Shore Nuisance (Bombay and Kolaba) Act, 1853
- Obstruction in Fairways Act, 1881
- Indian Fisheries Act, 1897
- River Boards Act, 1956
- Merchant Shipping Act, 1958
- Water (Prevention and Control of Pollution) Cess Act, 1977
- Atomic Energy Act of 1982
- Motor Vehicles Act, 1988
- Indian Forest Act, 1927
- Wildlife (Protection) Act, 1972, Amendment 1991
- Forest (Conservation) Act, 1980.
- Biological Diversity Act, 2002.
- Forest Right Act, 2006.
- Panchayats (Extension to the Scheduled areas) Act, 1996
- Factories Act, 1948 and its Amendment in 1987
- National Green Tribunal Act, 2010
- National Highways Act, 1956
- Inland Waterways Authority of India Act, 1985
- Railway (Notices of and Inquiries into Accidents) Rules, 1998
- Statutory Investigation into Railway Accidents Rules, 1998
- Carriage by Road Act, 2007, Rules 2011
- Epidemic Diseases Act, 1897

3.4.3 Capacity Building: Training and Research

The National Institute of Disaster Management (NIDM) is the apex organization of disaster management capacity building in the country, which also coordinates the India Disaster Resource Network (an inventory mechanism on disaster response-related resources including material and human resources at the district level). Besides New Delhi, a NIDM is recently established at Vijayawada under the Andhra

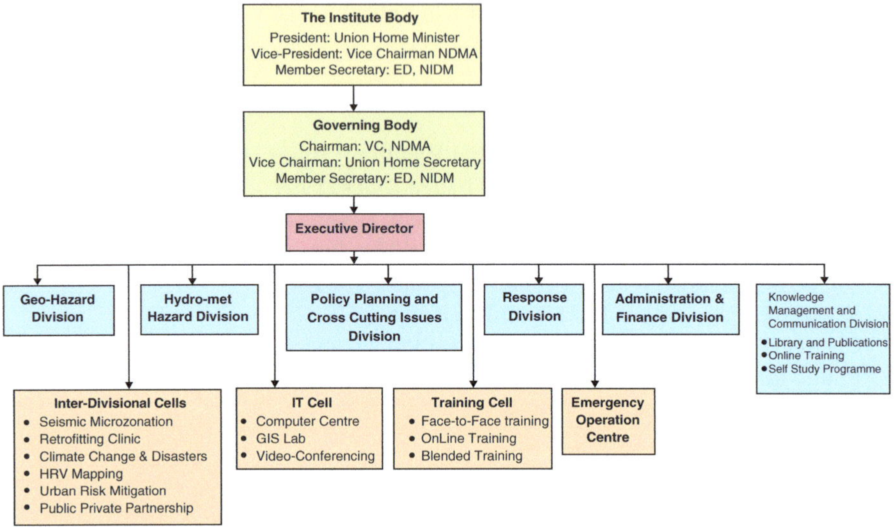

Fig. 3.3 Structure and function of NIDM New Delhi

Pradesh Reorganization Act. Besides NIDM, there are a number of Indian Institutes on Management (IIMs), Technology (IITs), Science (IISC, IISERs), Medicine (AIIMS) and sector institutes like Rural Development (NIRD), Health (NIHFW), Public Administration (IIPA), Environment (DMI, MoEFCC notified it to be a national institute of disaster resilience), Water (NWA Pune), Agriculture (NAARM), Forestry (ICFRE), Urban (NIUA), etc. undertaking activities in the area of disaster management through specialized faculty and engaging in cooperation activities. Most central universities in India has either established or provisioned a Centre for Disaster Management, aimed at research, training and education, under the environmental and sustainable development faculty. Figure 3.3 gives an outline of the structure/function of NIDM New Delhi. NIDM New Delhi, as per the mandate given in the DM Act 2005, developed India's national human resource plan for disaster management in 2012 and developed the expansion plan of the NIDM to cater to the needs of implementing the envisaged human resource plan, in active coordination with institutes of national and international eminence and across sectors.

3.4.4 Subnational Framework

At the state level, the State Disaster Management Authority (SDMA), headed by the chief minister, lays down policies and plans for disaster management in the state. The State Disaster Management Department (DMD) which is mostly positioned in the Revenue and Relief Department is the nodal authority. In the district level, the District Disaster Management Authority (DDMA) is headed by the district magistrate, with the elected representative of the local authority as the co-chairperson. DDMA is the planning, coordinating and implementing body for disaster management at the district level. It is expected to inter alia prepare the District Disaster Management Plan and monitor the implementation of the national and state policies and the national, state and the district plans.

The local authorities both the rural local self-governing institutions (Panchayati Raj Institutions) and urban local bodies (municipalities, cantonment boards and town planning authorities) are mandated to ensure capacity building of their officers and employees for managing disasters and carry out relief, rehabilitation and reconstruction activities in the affected areas. The local authorities are expected to prepare disaster management plans in consonance with guidelines of the national, state and district authorities.

3.4.5 *Emergency Response*

The National Emergency Operation Centre (NEOC) in the Ministry of Home Affairs functions 24×7 to monitor the disaster or disaster-like situation. Based on the feedback received from national forecasting agencies, viz. the Indian Meteorological Department, Central Water Commission, Snow and Avalanche Study Establishment, etc., advisories to the concerned states/UTs are issued from time to time for keeping watch on the developing situation and take necessary measures such as evacuation of the vulnerable persons, operation of relief camps, pre-positioning of essential commodities, etc. During the southwest monsoon, daily situation reports (sitreps) are prepared based on the feedback received from the affected states and concerned central ministries and organizations and are sent to all concerned. During the calamities of severe nature, special situation reports are also prepared and issued to all concerned. NEOC also issue SMS alerts to the concerned officers at the central and state government level.

The DM Act has mandated the NDMA to recommend guidelines for the minimum standards of relief to be provided to persons affected by disasters, which shall include minimum requirements to be provided in the relief camps in relation to shelter, food, drinking water, medical cover and sanitation, special provisions for widows and orphans, ex gratia assistance on account of loss of life as also assistance on account of damage to houses and for restoration of means of livelihood and such other relief as may be necessary. A multidisciplinary, multiskilled, high-tech National Disaster Response Force (NDRF) of 12 battalions has been set up for dealing with all types of disasters capable of insertion by air, sea and land. This is a military related response force. All the battalions are to be equipped and trained for all natural disasters including four battalions in combating nuclear, biological and chemical disasters. Each battalion will provide self-contained search and rescue specialist teams of personnel including engineers, technicians, electricians, dog squads and medical/paramedics. These NDRF battalions are located at nine different locations in the country based on the vulnerability profile to cut down the response time for their deployment. During the preparedness period/in a threatening disaster situation, proactive deployment of these forces is carried out by the NDMA in consultation with state authorities.

The civil defence (CD) organization has the vital role of guarding the hinterland, supporting the armed forces, mobilizing the citizens and helping civil administration for saving life and property, minimizing damage, maintaining continuity in production centres and raising public morale. Fire services are mandated to the municipal bodies as per the provision of the Constitution of India. Presently, fire prevention and firefighting services are organized by the concerned states and UTs. The Ministry of Home Affairs, Government of India, renders technical advice to the states and UTs and central ministries.

3.4.6 Financial Strategies for DRR

Good policies and plans may be in place, but without the necessary resources for implementation, these will remain hollow commitments. Governments need to allocate financial, human and material resources to disaster risk management structures. Resource allocation poses a real challenge in situations where so many demands compete for limited resources. Insufficient capacity and weak governance structure, corruption and a weak national resource base may undermine development of innovative mechanisms for resource mobilization and the providing of task incentives.

Financial assistance in the wake of natural calamities is provided in accordance with the schemes of relief funds. These schemes are based on the recommendations of the successive Finance Commissions. While the budgetary provision of these relief funds is dealt with by the Ministry of Finance, the processing of request of the state government for these funds is done by the Ministry of Home Affairs (DM Division). The present scheme of State Disaster Response Fund (SDRF) and National Disaster Response Fund (NDRF) is based on the recommendations of the 13th Finance Commission, operative from April 1, 2010 to March 31, 2015. Over and above the provisions of the SDRF, funding is provided from the NDRF in the wake of calamities of severe nature. On receipt of the memorandum from the affected states, an Inter-Ministerial Central Team comprising of representatives of the central ministries/departments is constituted, and its report after examination by the Inter-Ministerial Group (IMG) headed by the home secretary is placed before the High Level Committee (HLC) for their consideration and approval of funds from NDRF. In exercise of the powers conferred by Section 7(A) of the Public Liability Insurance Act, 1991, the central government has established the Environment Relief Fund Scheme on November 4, 2008. Besides the fund which are available through plan and non-plan schemes, efforts have also been made by the centre to mobilize the resources from external funding agencies for vulnerability assessment, capacity development, institutional strengthening of response mechanism and mitigation measures, etc.

3.4.7 Mainstreaming DRR

Mainstreaming of disaster risk reduction into different aspects of developmental planning and projects are a key concern at the international and therefore at national and subnational level in India as well. NITI Aayog (previously Planning Commission) at the national level and the State Planning Commission or State Planning Board and the Land-Use Board/Department at subnational level play a key role in mainstreaming disaster risk into developmental planning process across sectors and departments of the government. Key facets of mainstreaming disaster risk reduction integration with climate change adaptation (CCA), and the objectives of Sustainable Development Goals (SDGs) can be categorized into three major clusters:

(a) Infrastructure and commerce/trade sectors
(b) Environment and natural resources sectors
(c) Social welfare and humanitarian sectors

However, as of now, India's subnational risk reduction mainstreaming activities have been largely project-based. Major projects related to risk reduction in India include:

- National Cyclone Risk Mitigation Project
- National Earthquake Risk Mitigation Project
- National Flood Mitigation Project
- National Landslide Mitigation Project
- National Disaster Communication Network.
- School Safety Programme
- Training of Architects and Engineers in Earthquake Risk Management

This project-based approach brings external funding and expertise to support disaster risk reduction for recipient states, which is essential for building capacity and investing into infrastructural needs. However, there are disadvantageous to this project-based approach to reducing disaster risks. For example, reducing people's vulnerability to multiple hazards rarely fits neatly into the confines of a project cycle. Risks are dynamic and change with population growth, urbanisation, land-use change, climate change and economic and political institutions. Incorporating disaster risk reduction into development planning ultimately should be led by local, state and national authorities that are committed to reducing risk in the long term and adapt their approaches according to their constituents' changing circumstances.

Now there is a major shift in approach to mainstreaming DRR in India. Planning mechanisms at the subnational and district (and local) level and also in the ministries/departments are being reviewed and customized to accommodate the mainstreaming of DRR along with CCA. The Gorakhpur model of district level departmental planning for mainstreaming CCA-DRR integration has been up-scaled in the studies recently undertaken in Uttarakhand (Almora district) and Odisha (Puri district) and at state-level planning process. Interventions of applying environmental knowledge and tools in enabling DRR environment with the use of existing tools like EIA, SEA and other decision tools have been recommended through the practical pilots undertaken in technical assistance of GIZ, Germany; CDKN, UK; Potsdam University; UNU Institute of Environment and Human Security, Bonn (Germany); UNEP DM Branch, Geneva; and UNESCAP, Bangkok. UNDP, through its intense project activities jointly with the Government of India, mainly starting with tsunami recovery programme had a major stake in contribution towards promoting mainstreaming DRR into projects and schemes of the government. Guidelines and tools have been prescribed for mainstreaming DRR across sectors based on the NIDM-UNDP studies on key sectors.

3.5 Journey to 2030: Road Ahead in DRR Leadership

'While disaster management and response coordination requires centralized command, there is a need to decentralize disaster risk reduction. Along with the decentralization of power and devolution of government authority, disaster risk reduction at the local level needs to be encouraged and supported' (Living with Risk, ISDR 2004). There is paradigm shift in India from reactive approach of responding and calamity relief after the disaster to proactive approach of disaster prevention, preparedness and mitigation. The enactment of the Disaster Management Act, 2005, establishment of the National Disaster Management Authority with the prime minister as its chairperson and disaster management training by the National Institute of Disaster Management along with the Disaster Management Cells of the State Administrative Training Institutes will help India in becoming disaster resilient. India has also taken significant steps in starting disaster management education from middle and high school.

Although disaster-led displacement seems to be a post-disaster issue of concern, its effective management has to be taken into consideration within the overall disaster management planning framework in the pre-disaster and preparedness phase itself. The objective has to minimize the impact of development or a disaster in the form of displacement – temporary or permanent – and to adapt measures for early rehabilitation and recovery in case when displacement is imperative. The concept of 'risk sensitive land-use planning' for urban and rural areas is being promoted in India as well to facilitate preventive approach for managing the challenges of disaster-led displacements by reducing vulnerability and exposure. The Indian disaster management policy is geared to make a paradigm change from response and calamity relief to disaster prevention, preparation and mitigation. Another significant change is to move from disaster management largely from government to public-private partnership and community disaster management. The efforts have been made to move from disaster management to disaster risk management and finally disaster risk reduction. In this regard, significant changes have been made, but still miles remains to go.

A major audit of institutional setup, laws and policy implementation tools is the need of the hour, in the backdrop of success-failure yardsticks of the major disasters the country faced during the past years. A new set of SDG alongside the Paris Agreement on Climate Change calls for an integrated mechanism for compliance to the Sendai Framework of Action. Capacity building is still the weakest area in India in DRR framework of effectiveness. Strengthening network of institutes and expansion/replication of disaster management centres/institutes to address specific needs of the sectors and geographical environments such as coastal and hilly areas, industrial sectors, etc. and promotion of fundamental research towards innovations and hypothesis testing are critically needed. Foreseeable improvement in effectiveness also calls for greater clarity in roles, authority and resources of institutions, with greater engagement of private sector, community-based organizations, voluntary organizations and voluntary/territorial corps, local experts and institutions in disaster management planning and operations.

Acknowledgements A range of reports and documents, statements, records and data, published and unpublished, and opinions have been referred in developing this chapter, and their sources are gratefully acknowledged, and are cited to the extent possible, as these have been interpreted to bring a perspective towards proper understanding of the subject. Authors are thankful to the respective sources. Views expressed are not necessarily of any government organization/institutes.

References

Bulkeley H, Jordan A (2012) Trans-national environmental governance new findings and emerging research agendas. Environ Plan 30(4):556–570

Christoplos I, Farrington J, Kidd A D (2001) Extension, poverty and vulnerability: inception report of a study for the Neuchâtel initiative. ODI Working Paper No. 144. Overseas Development Institute, London. https://www.odi.org/sites/odi.org.uk/files/odi-assets/publications-opinion-files/2995.pdf

Frewer T (2013) Doing NGO work: the politics of being 'civil society' and promoting 'good governance' in Cambodia. Aust Geogr 44(1):97–114

Goodwin H (1998) Sustainable tourism and poverty elimination. A Discussion Paper for the Department for the Environment, Transport and the Regions and the Department for International Development

Gupta A K, Nair S S, Chopde S, Singh PK (2010) Policies and strategic options for disaster risk management in India. In: Risk to resilience – strategic tools for disaster risk management. Proceedings of the Intl. Conf. Risk to Resilience, New Delhi, Feb 2009, NIDM, ISET, WII, NOAA, DFID. http://i-s-e-t.org/resources/major-program-reports/risk-to-resilience-workshop.html

Gupta AK, Nair SS, Mahajan R (2013) National Disaster Management Plan in developed and developing nations – drawing lessons for India. People and Planet 1(1):27–33

International Federation of Surveyors (2006) The Contribution of the Surveying Profession to Disaster Risk Management. FIG Publication No 38. FIG Working Group 8.4. Published by the International Federation of Surveyors (FIG), October 2006, Frederiksberg, Denmark http://www.fig.net/resources/publications/figpub/pub38/pub38_screen.pdf

International Strategy for Disaster Reduction (2004) Living with risk: a global review of disaster reduction initiatives (Vol. 1). United Nations Publications

Jones S, Oven KJ, Manyena B, Aryal K (2014) Governance struggles and policy processes in disaster risk reduction: a case study from Nepal. Geoforum 57:78–90

Lebel L, Anderies JM, Cambell B, Folke C, Hatfield-Dodds S, Hughes TP, Wilson J (2006) Governance and the capacity to manage resilience in regional social-ecological systems. Ecol Soc 11(1):19. http://www.ecologyandsociety.org/vol11/iss1/art19/

Lewis D, Kanji N (2009) Non-governmental Organisations and development. Routledge, London. 0-415-45430-1.
Natural Hazards Research and Applications Information Center (2001) Holistic disaster recovery: ideas for building local sustainability after a disaster. http://www.oregonshowcase.org/downloads/pdf/resources/holistic_recovery_guide.pdf. Accessed 15 Oct 2009
Ojha HR, Cameron J, Kumar C (2009) Deliberation or symbolic violence? The governance of community forestry in Nepal. Forest Policy Econ 11(5):365–374
Parasuraman S, Unnikrishnan PV (2000) India disasters report: towards a policy initiative. Oxford University Press, New Delhi. 384 p
Pelling M, Holloway A (2006) Legislation for mainstreaming disaster risk reduction. Tearfund, Teddington
Planning Commission (2002) Disaster management: the development perspective. Chapter 7, Tenth Five Year Plan: 2002–07. Government of India
UNISDR (2004) Living with risk: a global review of disaster reduction initiatives, Inter-Agency Secretariat of UNISDR, accessible at www.unisdr.org/eng/about_isdr/bd-lwr-2004-eng.htm
UNISDR (2007) Hyogo framework for Action 2005–2015. Building the resilience of nations and communities to disasters. Extract from the final report of the World Conference on Disaster Reduction (A/CONF.206/6). United Nations Office for Disaster Reduction (UNISDR). http://www.unisdr.org/files/1037_hyogoframeworkforactionenglish.pdf. Accessed 12.07.13
UNISDR (2011) Global assessment report on disaster risk reduction. Revealing risk, redefining development. Geneva, Switzerland: United Nations international strategy for disaster reduction. http://www.preventionweb.net/english/hyogo/gar/2011/en/home/download.html. Accessed 12.07.13
UNISDR (2015) The human cost of weather-related disasters 1995–2015. UNISDR Geneva and CRED Belgium. https://www.unisdr.org/2015/docs/climatechange/COP21_WeatherDisastersReport_2015_FINAL.pdf
Williams G (2011) Study on disaster risk reduction, decentralization and political economy: the political economy of disaster risk reduction. Background report for the Global Assessment Report on Disaster Risk Reduction. UN/ISDR, Geneva

Chapter 4
National-Level Disaster Risk Governance for Rapid Response

Indrajit Pal and Nakul Kumar Tarun

Abstract India has a robust institutional system in place with a wide range of national-level institutions related to early warning, meteorology, remote sensing, information and communication technology, satellite technology and disaster response management, which have substantially contributed to high level of preparedness, in terms of effective response to disaster.

The National Disaster Response Force (NDRF), a unique force of its kind globally, with 15,000 members dedicated to disaster response has proved its importance as a highly skilled organisation in rescue and relief operations. With its vision, the most visible, vibrant, multidisciplinary, multiskilled, highly technical and standalone force is capable of dealing with all types of natural and man-made disasters and of mitigating their effect.

In early warning system, the Indian meteorological department, which was set up more than 200 years ago, has a network of ten Global Atmosphere Watch (GAW) stations, as per WMO protocols and standards. IMD also acts as a Regional Specialized Meteorological Centre for providing cyclone advisory services to the world.

Enactment of the National Disaster Management Act 2005 provided technical institutional mechanism in the form of National Disaster Management Authority (NDMA), supported by the National Institute of Disaster Management (NIDM), and a similar setup at state and district levels for proactive disaster risk reduction measures along with other technical organisations, network of professional and experts in various sectors of disaster/risk management.

Keywords Governance • Risk management • Disaster response • Natural disaster

I. Pal (✉)
Disaster Preparedness, Mitigation and Management (DPMM), Asian Institute of Technology, Khlong Nung, Pathum Thani, Thailand
e-mail: indrajit.pal@gmail.com

N.K. Tarun
Emergency Response Team (ERT), Super Highway Lab, Gurgaon, India
e-mail: nakultarun.dgcd@gmail.com

I. Pal, R. Shaw (eds.), *Disaster Risk Governance in India and Cross Cutting Issues*, Disaster Risk Reduction, DOI 10.1007/978-981-10-3310-0_4

4.1 Introduction

4.1.1 Hazard and Vulnerability Scenario

India is vulnerable to natural disasters in varying degrees. More than 58.6% of the landmass is prone to earthquakes of moderate to very high intensity; over 40 million hectares (12%) of its land is prone to floods and river erosion; close to 5700 km, out of the 7516-km-long coastline is prone to cyclones and tsunamis; 68% of its cultivable area is vulnerable to droughts; and, its hilly areas are at risk from landslides and avalanches. Moreover, India is also vulnerable to chemical, biological, radiological and nuclear (CBRN) emergencies and other man-made disasters (Vulnerability Atlas of India 1999).

Disaster risks in India are further compounded by increasing vulnerabilities related to changing demographics and socio-economic conditions, unplanned urbanisation, development within high-risk zones, environmental degradation, climate change, geological hazards, epidemics and pandemics (UNDG 2003). Clearly, all these contribute to a situation where disasters seriously threaten India's economy, its population and sustainable development.

4.1.2 Response to National and Major International Disasters

The Disaster Management Act has statutory provisions for constitution of the National Disaster Response Force (NDRF) for the purpose of specialised response to natural and man-made disasters. Accordingly, in 2006, NDRF was constituted with eight battalions. At present, NDRF has a strength of 12 battalions with each battalion consisting of 1149 personnel. NDRF aims to emerge as the most visible and vibrant multidisciplinary, multiskilled, high-tech force capable to deal with all types of natural as well as man-made disasters and to mitigate the effects of disasters. Its vision is to be the most efficient and effective response force for any disaster in the country and region. NDRF's mission is to achieve excellence in providing effective and timely rescue and relief to the disaster victims. NDRF is a distinguished, unique force across the country functioning under the Ministry of Home Affairs, Government of India, within the overall command, control and leadership of the NDRF director general (NDRF 2016). Apart from national-level responses, NDRF also responded to major international disasters in Japan (2011) and Nepal (2015) for relief and rescue operations.

The Indian National Disaster Response Force (NRDF) efficiently responded to the 2011 Japan catastrophe during their deployment for about 2 weeks. The team managed to extricate 07 bodies from the rubble more than 2 weeks after the disaster struck the area. The team recovered and handed over cash worth 50 million yen to the authorities apart from the valuables. The NDRF team gains appreciation by the local authorities, media and the public at large in recognition of efforts by the Indian

NDRF team. Commendation of work was conveyed by the Govt. of Japan to the embassy of India on April 5, 2011. People who lived in rehabilitation camp in Onagawa town hall gave vote of thanks to the Indian team and shared their experience with the contingent commander. The apparent dedication and honesty of the team were highly appreciated by the mayor of Onagawa. The mayor of Rifu Cho also thanked and expressed his gratitude to the Indian team for the help and support provided to the people of the Miyagi prefecture in general and in particular Onagawa. It is pertinent to mention here that the team was reached there after 16 days.

The Himalayan nation Nepal was hit by a massive earthquake with a magnitude of 7.8 on the Richter scale, as recorded by Nepal's National Seismological Centre (NSC) on April 25, 2015. The epicentre of this earthquake was the historic district of Gorkha situated at about 76 km northwest of Kathmandu. Indian disaster response teams reached to Nepal to support response in 6 h and rescued 11 lives from the rubble. This is again the matter of applaud disaster governance of India, at how international governance time reduced from 16 days to 6 h.

In 2015 Gorkha, Nepal, earthquake, the death toll was of more than 8000 people and several thousands are injured. More than 7 lakh houses have collapsed/damaged badly leaving about 35 lakh population (especially in the rural areas) homeless. A total of 28 districts have been affected out of which emergency has been declared in the 14 most affected districts where the damage of more than 70% households has been reported. Rural areas especially hill districts have been affected badly where as high as 90% houses have collapsed. Affected areas include the epicentre of the earthquake: Lamjung District, northwest of Kathmandu, and the areas south and east of the city including Kathmandu Valley districts. In the central region, the most affected districts are reported to be Sindulpal chowk, Kavre, Nuwakot, Rasuwa and Dolakha and, in the western region, Kaski, Gorkha and Lamjung. Affected areas include densely populated cities and rural, mountainous regions. In response to this huge regional crisis, 134 international SAR teams including teams from SAARC member states, the United Nations, international NGOs, etc. from 34 countries responded to Nepal's request for help. Just 17 days after a 7.8 earthquake struck Nepal again, the country has been hit by two powerful aftershocks: a magnitude 7.3 earthquake at 07:05:19 UTC and a magnitude 6.3 earthquake at 07:36:53 UTC, both on May 12, 2015. The 7.3 magnitude quake was centred 18 km SE of Kodari. The 6.3 magnitude quake was centred 33 km NNE of Ramechhap. The 7.3 magnitude earthquake was one of the 11 aftershocks that struck within a 34 km radius of Kodari, Nepal. Three other earthquakes struck within 33 km of Ramechhap, Nepal, and one struck 5 km SSW of Zuobude, China (67 km E of Banepa, Nepal). More buildings have collapsed, and new landslides are blocking roads and highways. The number of deaths and injuries varies. According to Nepalese government spokesman Minendra Rijal, at least 48 people have died and another 1261 have been injured in Nepal and 32 districts have been affected. At least 17 people in India have died according to Home Ministry spokesman Kuldeep Dhatwalia. Sixteen of those deaths were in Bihar state, with the other in Uttar Pradesh. The earthquake also affected China, killing a woman in Tibet, according to China's state-run Xinhua news agency. In the same report, local police officer Wu Aijun said that landslides occurred in the Gyirong area in Tibet. The idea is that disaster was widespread and had a regional impact. Hence a regional response is the best answer to support the efficient response in the aftermath of a disaster.

4.1.3 National Legislations in the Global Context

Towards effective resilience building, the Government of India has put into place legal and institutional framework for disaster management. Enacting Disaster Management Act in 2005 and Disaster Management Policy in 2009 for prevention and mitigation effects of disasters and for undertaking a holistic coordinated and prompt response to any disaster situation is an important landmark effort for creating resilient community in India. It also indicates coordination of action of the ministries or departments of the Government of India, state government constitution of national authority and state authorities in relation to disaster management. The Act also recommends the integration of measures for prevention of disaster and mitigation by ministries or departments of the Government of India into their development plan and projects; establishing national/state disaster response forces; setting up a national institute for training and capacity building, standardised national building codes and BIS codes; and setting up financial mechanism for disaster response (World Bank 2010).

Recently, the government has also earmarked 10% funds in all Centrally Sponsored Schemes for disaster mitigation, restoration and innovation activities. Also, any programme proposal, requiring financial approval of the central government, must specifically incorporate whether the project is secured against natural/man-made disasters like floods, cyclones, earthquake, tsunamis, etc.

India has already incorporated climate change adaptation through DRR while laying down the structure for early warning, assessment of hazard impact, piloting of last mile connectivity (early warning) and development of multi-hazard preparedness and response plans (Govt. of India 2008, 2009a). Designing a hazard-specific mitigation plan and dovetailing the preparedness components in development activities at the state and district level followed by DRR bilateral programmes to strengthen and fill the gaps that emerged with specific focus on urban risk reduction are the highlights. Mainstreaming of DRR and CCA into social protection schemes for livelihood promotion, food security, health and hygiene promotion, housing, drinking water, etc. has already been initiated for poorer section of society which is not only contributing to risk reduction but also enhancing the inbuilt capacity of community towards sustainable development and social equity (Govt. of India 2009b).

The Central Government is empowered to take measures as it deems necessary or expedient for the purpose of disaster management like deployment of naval, military and air forces and other armed forces of the union or any other civilian personnel as may be required for the purposes of this Act; coordination with the United Nations agencies, international organisations and governments of foreign countries for the purposes of this Act; and establishment of institutions for research, training and developmental programmes in the field of disaster management (Govt. of India 2009c, d). It is also empowered to deal with all such other matters as it deems necessary or expedient for the purpose of securing effective implementation of the provisions of the Act. The following are the excerpt from the national Disaster Management Act 2005:

Chapter VI – Local Authorities: Subject to the directions of the District Authority, the local authorities shall ensure that the officers and employees are trained, resources are so maintained as to be readily available, carry out relief rehabilitation and reconstruction activities in the affected areas and may take such other measures as may be necessary for the disaster management.

Chapter VII – National Institute of Disaster Management: The Central Government is empowered to constitute an institute to be called the National Institute of Disaster Management. The institute functions within the broad policies and guidelines laid down by the National Authority and is responsible for planning and promoting training and research in the area of disaster management, documentation and development of national level information base relating to disaster management policies, prevention mechanisms and mitigation measures.

Chapter VIII – National Disaster Response Force: A National Disaster Response Force for the purpose of specialist response to a threatening disaster situation or disaster is to be constituted. The general superintendence, direction and control of the Force shall be vested and exercised by the National Authority and the command and supervision of the Force shall vest in an officer to be appointed by the Central Government as the Director General of the National Disaster Response Force.

Chapter IX– Finance, Accounts and Audits: The Central Government is empowered to constitute a fund to be called as the National Disaster Response fund for meeting any threatening disaster situation or disaster and there shall be credited thereto an amount which Central Government may, after due appropriation made by parliament by law in this behalf provide any grants that may be made by any person or institution for the purpose of disaster management.

Chapter X – Offences and Penalties: The Act imposes punishments to persons/companies for contravening the provisions of this Act, 2005 such as obstructing or abandoning, refusing to comply with any of the provisions of this Act, making false claims, misappropriation of money or materials or false warning, etc. The punishment in such cases could be imprisonment or fine or both.

Chapter XI – Miscellaneous: The National Authority, the State Authority, or a District Authority is empowered to recommend the Government to give direction to any authority or person in control of any audio or audiovisual media or such other means of communication as may be available to carry any warning or advisories regarding any threatening disaster situation or disaster, and the said means of communication and media as designated shall comply with such direction.

In Indian context, HFA has been implemented at the national level through enactment of DM Act 2005 and other emergency-related Acts and guidelines, to enhance the national and state level response mechanism and roles and responsibility of the institutions. In context of Sendai framework, all seven targets and four priorities have been taken care in above said acts and policies.

4.2 Early Warning System Disaster Response and Crisis Management

India has developed a very robust early warning system which showed result also in recent past. The Indian Meteorological Department (IMD) is mandated to monitor and give warnings regarding a tropical cyclone (TC). Monitoring process has been revolutionised by the advent of remote sensing techniques. A TC intensity analysis and forecast scheme have been worked out using satellite image interpretation techniques which facilitate forecasting of storm surges. Data resources are crucial to early forecasting of cyclones. Satellite-based observations are being extensively utilised. Satellite-integrated-automated weather stations have been installed on islands, oilrigs and exposed coastal sites. Buoys for supplementing the surface data network in the tropical ocean have been deployed. The government has also started a National Data Buoy Programme. A set of 12 moored buoys have been deployed in the northern Indian Ocean to provide meteorological and oceanographic data.

Dynamic forecasting of TCs requires knowledge of the vertical structure of both the cyclone and the surrounding environment. The rawinsonde remains the principal equipment for sounding. The Doppler radar wind profiler provides hourly soundings. A mesosphere, stratosphere and troposphere (MST) radar has also been installed at Tirupati. Another profiler is being developed and will be deployed at IMD Pune. Another important source of upper-level data is the aircraft reports. An increasing number of commercial jet aircraft are equipped with the Aircraft Meteorological Data Relay system. This data that is being made available is also being used by the IMD for analysis and predictions. Radars have been used to observe TCs since long. Surveillance of the spiral rain bands and the eye of the TC is an important function of the coastal radars. Ten cyclone detection radars have already been installed. These radars are providing useful estimates of storm centres up to a range 300–400 km. Doppler radars provide direct measurements of wind fields in TCs. Due to range limitation, Doppler wind estimates are usually within a range of about 100 km. IMD has deployed Doppler radars at three sites on the east coast. Another set of three Doppler radars are being deployed in Andhra Pradesh in the near future. The meteorological satellite has made a tremendous impact on the analysis of cyclones. All developing cloud clusters are routinely observed through satellite cloud imagery, and those showing signs of organisation are closely monitored for signs of intensification. TC forecasters everywhere use the Dvorak technique to estimate storm location and intensity. It has been found to provide realistic estimates for TCs in the Bay of Bengal as well as the Arabian Sea. INSAT data has also been used to study the structures of different TCs in the Bay of Bengal. IMD is also producing cloud motion vectors (CMVs).

INSAT–2E satellite has been improved with Very high-resolution radiometer (VHRR) sensor to provide water vapor channel data in addition to VIS, IR and charge-coupled device (CCD).

'Cyclone warning' is issued 24 h in advance. Landfall point is forecasted in this stage of cyclone warning. In addition to the forecasts for heavy rains and strong winds, the storm surge forecast is also issued. Since the storm surge is the biggest killer so far as the devastating attributes of a storm are concerned, information in this regard is the most critical for taking follow-up action for evacuation from the low-lying areas likely to be affected by the storm.

In India a beginning in scientific flood forecasting was made in November, 1958, by the Central Water Commission (then known as the Central Water and Power Commission) when a flood forecasting centre was set up at its headquarters, at Delhi, for giving timely forecasts and warnings of the incoming floods to the villages located in the river areas around the national capital, Delhi. The network has been expanding, and by now the flood forecasting network of the Central Water Commission (CWC) covers all the major flood-prone interstate river basins in the country. At present there are more than 166 flood forecasting stations on various rivers in the country which include 134 level forecasting and 32 inflow forecasting stations across India.

4.3 National Governance Mechanism and Response

4.3.1 Social, Political, Environmental and Economic Perspective of Governance in the Context of Risk Management

After frequent strikes by disaster including a super cyclone in 1999, Gujrat Earthquake in 2001 and Indian Tsunami in 2004, the government decided to have a very structured and specific disaster governance and response system in India. Hence a team at national-level NDMA headed by the prime minister of India, a team at state-level SDMA headed by CM and a team at district-level DDMA headed by DM were constituted. The National Disaster Response Force, a specialised force capable to respond to natural as well as man-made disaster, is in place. There are numbers of emergency support functions that are also to respond in case of any emergency including fire and emergency services, police and civil defence. There are a lot of volunteer organisations like NCC, NSS and so many non-governmental organisations preparing themselves for responding to a disaster, if any.

In India the Disaster Management Act 2005 came into the statute book on December 26, 2005, by a Gazette notification, exactly on the first anniversary of the devastating tsunami of 2004, which killed nearly 13,000 people in India alone and affected 18 million people. The Act provides a legal and institutional framework for 'the effective management of disasters and for matters connected therewith or incidental thereto'. It provides for the establishment of the National Disaster Management Authority (NDMA), State Disaster Management Authority (SDMA) and District Disaster Management Authorities (DDMA) at the national, state and district levels with adequate financial and administrative powers and creation of the National Institute of Disaster Management (NIDM) with the mandate of undertaking training and capacity building, developing training modules on various aspects of disaster management, undertaking research and documentation, formulating and implementing comprehensive HRD plan covering all aspects of DM, providing assistance in national-level policy formulation and providing assistance to state governments and state training institutions. The Act also provides guidelines for the creation of the National Disaster Response Fund and National Mitigation Fund, establishment of funds by the state government and allocation of funds by ministries and departments for emergency procurement. The Act also provides for the establishment of the National Disaster Response Force (NDRF) (World Bank 2003).

4.3.2 National Policy on Disaster Management 2009

The National Policy on Disaster Management was approved by the government in November 2009. This comprehensive policy document lays down policies on every aspect of holistic management of disasters in the country. The policy has 13 chapters as under:

1. Preamble
2. Approach and Objectives
3. Institutional and Legal Arrangements
4. Financial Arrangements
5. Disaster Prevention, Mitigation and Preparedness
6. Techno-Legal Regime
7. Response
8. Relief and Rehabilitation
9. Reconstruction and Recovery
10. Capacity Development
11. Knowledge Management
12. Research and development
13. Road Ahead

Salient features of India's National Policy on Disaster Management: India's National Policy on Disaster Management was approved by the Union Cabinet of India on October 22, 2009 with the aim to minimise the losses of lives, livelihoods and property, caused by natural or man-made disasters with a vision to build a safe and disaster-resilient India by developing a holistic, proactive, integrated, multi-disaster-oriented and technology-driven strategy. With this national policy in place in India, a holistic and integrated approach will be evolved towards disaster management with emphasis on building strategic partnerships at various levels. The themes underpinning the policy include community-based disaster management, capacity development in all spheres, consolidation of past initiatives and best practices and cooperation with agencies at national and international levels with multisectoral synergy.

The policy is also intended to promote a culture of prevention, preparedness and resilience at all levels through knowledge, innovation and education. It encourages mitigation measures based on environmental sustainability. It seeks to mainstream disaster management into the developmental planning process and provides for institutional and financial arrangements at national, state and district levels for disaster prevention, mitigation, preparedness and response as it ensures adequate budgeting for disaster mitigation activities in all ministries and departments.

4.3.3 The High-Powered Committee (HPC) Prepared a Model District Disaster Management Plan with the Following Salient Sections

High-Powered Committee (HPC): Guidelines for District Plans 2002

- Assigning responsibility to organisations and individuals for carrying out specific actions of projected items. This includes setting forth lines of authority and organisational relationships. It also includes activities regarding identification of personnel, equipment, facilities, supplies and other resources.
- Identifying and recording basic information on the district, viz. topography, communication links, etc.
- Procedures and organisational arrangements for hazard identification and vulnerability analysis at district level.
- Identifying mitigation measures for long-term management of risk to reduce the adverse fallout of the disaster events on the physical and social infrastructure.
- Identification of functions that might be required during the response phase to protect life and assets of the community.
- Arrangements for organising short-term and long-term recovery and rehabilitation of the affected communities.
- Developing interface with the media, NGOs, relief and donor agencies and other stakeholders.

The HPC underlined the need for a district plan structured in a way that permits easy and quick retrieval of relevant information on which the authority/individual may have to act upon (HPC 2001).

4.3.4 National Disaster Response Force (NDRF)

4.3.4.1 Constitution and Role of NDRF

The National Disaster Response Force (NDRF) has been constituted under Section 44 of the DM Act 2005 by upgradation/conversion of eight standard battalions of the central paramilitary forces, i.e. two battalions each from the Border Security Force (BSF), Indo-Tibetan Border Police (ITBP), Central Industrial Security Force (CISF) and central reserve. The National Disaster Response Force (NDRF) has been constituted under Section 44 of the DM Act 2005 by upgradation/conversion of eight standard battalions of the central paramilitary forces, i.e. two battalions each from the Border Security Force (BSF), Indo-Tibetan Border Police (ITBP), Central Industrial Security Force (CISF) and Central Reserve Police Force (CRPF) to build them up as a specialist force to respond to disaster or disaster like situations. At present, the National Disaster Response Force consists of 12 battalions, three each from the BSF and CRPF and two each from CISF, ITBP and SSB. Each battalion has 18 self-contained specialist search and rescue teams of 45 personnel each including engineers, technicians, electricians, dog squads and medical/paramedics. The total strength of each battalion is 1149. All the 12 battalions have been equipped and trained to respond to natural as well as man-made disasters. Battalions are also trained and equipped for response during chemical, biological, radiological and nuclear (CBRN) emergencies.

Based on vulnerability profile of different regions of the country, these specialist battalions have been presently stationed at the following ten places as may be seen from the map below (Fig. 4.1).

4.3.4.2 State Disaster Response Force (SDRF)

The states/UTs have also been advised to set up their own specialist response force for responding to disasters on the lines of the National Disaster Response Force vide Ministry of Home Affairs letter dated July 26, 2007, and March 8, 2011. The National government is providing assistance for training of trainers. The state governments have been also advised to utilise 10% of their State Disaster Response Fund and capacity building grant for the procurement of search and rescue equipment and for training purposes of the response force.

4.3.4.3 Civil Defence

Aims and Objectives of the Civil Defence Act The Civil Defence Policy of the GoI until 1962 was confined to make the states and UTs conscious of the need of civil protection measures and to keep in readiness civil protection plans for major cities and towns under the Emergency Relief Organisation (ERO) scheme. The

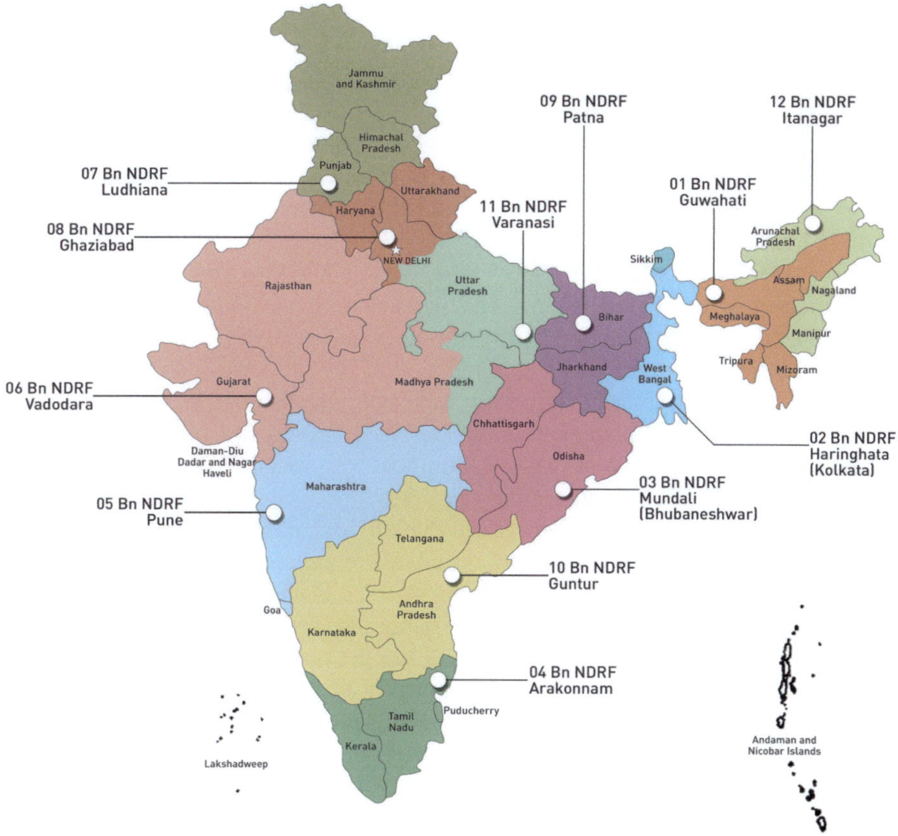

Fig. 4.1 NDRF battalion deployment in strategic locations (Modified After NDRF 2016)

legislation on civil defence (CD) known as the Civil Defence Act was enacted in 1968 which is enforced throughout the country. The Act defines CD and provides for the powers of the Central Government to make rules for CD, spelling out various actions to be taken for CD measures. It further stipulates for constitution of CD corps, appointment of members and officers, functions of members, etc. The Act has since been amended in 2010 to cater to the needs of disaster management so as to utilise the services of civil defence volunteers effectively for enhancement of public participation in disaster management-related activities in the country. The CD organisation is raised only in such areas and zones which are considered vulnerable to enemy attacks. The revision and renewal of categorised CD towns is done at regular intervals, with the level of perceived threat or external aggression or hostile attacks by anti-national elements or terrorists to vital installations.

4.3.4.4 Fire Service

Fire services are mandate of the municipal bodies as estimated in item 7 of Schedule 12 under Article 243 W of the constitution. The structure across is not uniform. Presently fire prevention and firefighting services are organised by the concerned states and UTs. The Ministry of Home Affairs, Govt. of India, renders technical advice to the States and UTs and central ministries on fire protection, fire prevention and fire legislation.

The Government of India in 1956 formed a 'Standing Fire Advisory Committee' under the Ministry of Home Affairs. The mandate of the committee was to examine the technical problems relating to fire services and to advise the Government of India for speedy development and upgradation of fire services all over the country. This committee had representation from each state fire services, as well as the representation from the Ministry of Home, Defence, Transport, Communication and Bureau of Indian Standards. This committee was renamed as 'Standing Fire Advisory Council' (SFAC) during the year 1980.

Fire services in Gujarat, Chhattisgarh, Punjab, Maharashtra, Himachal Pradesh, Haryana and Madhya Pradesh excluding Indore are under the respective concerned municipal corporations. In other remaining states, it is under the home department. While some states have enacted their own fire act, some others have not. As of today, there is no standardisation with regard to the scaling of equipment, the type of equipment or the training of their manpower. In each state it has grown according to the initiatives taken by the states and the funds provided for the fire services.

Presently the only basic lifeline of fire and emergency services which is fully committed to the common public is the municipal in some states and state fire services. The airport authority, big industrial establishments, CISF and armed forces, however, also have their own fire services and many a times in case of need rush in aid to the local fire services. Apart from the lack of being a proper government department with a complete developmental plan, state fire services have their own organisational structure, administrative setup, funding mechanism, training facilities and equipment.

4.4 Risk Assessment and Early Warning of Disasters

A number of scientific, technical and research institutions in the country under various ministries and departments are engaged in fundamental and applied research on different aspects of assessment and analysis of natural hazards and risks and provided early warning and disaster management support services. The foremost amongst them are the institutions under the Ministry of Earth Sciences which include the India Meteorological Department (IMD), Indian Institute of Tropical Meteorology (IITM), National Centre for Medium Range Weather Forecasting (NCMRWF), Indian National Centre for Ocean Information Sciences (INCOIS), etc. IMD is responsible for early warning of cyclones in India and South Asia, while

INCOIS has developed a system for early warning of tsunami. NCMRFW is engaged in medium-range weather forecasting which is useful for agricultural activities particularly in the contexts of drought or drought-like conditions. IITM is engaged in fundamental research on long-term climate change and its implications on extreme weather events such as cyclones, flood, drought, etc. The Central Water Commission under the Ministry of Water Resources has the nodal responsibility for early warning of flood, while the Geological Survey of India under the Ministry of Mines has the responsibility for research, analysis and documentation of landslides and earthquakes. A network of seismic observatories functioning under IMD, GSI, DST, etc. provide valuable services in decoding the earthquakes based on their epicentre, intensities, focal depths, etc. and developing earthquake zonation and microzonation maps (UNISDR 2009 and 2011).

Scientific and technological establishments under the Ministry of Science and Technology such as the Survey of India and the laboratories of the Council of Scientific and Industrial Research such as the Structural Engineering Research Centre, Central Building Research Institute, Central Road Research Institute, etc. and the Indian Institutes of Technologies and the Building Materials and Technology Promotion Council are doing pioneering works that have important bearing for reducing the risks of built environment. Similarly, the Indian Space Research Organisation (ISRO) and its units such as National Remote Sensing Centre are providing valuable support services for assessing the hazards and risks before, during and after disasters.

India has, over the past years, produced rich base maps through systematic topographic surveys, geological surveys, soil surveys, cadastral surveys, various natural resource inventory programmes and the use of the remote sensing images. Further, with the availability of precision, high-resolution satellite images, use of geographic information system, combined with the Global Positioning System, the accuracy and information content of these spatial datasets are extremely high. The Department of Science and Technology is encapsulating these maps and images into a national spatial data infrastructure (NSDI) which would be shared appropriately with the concerned agencies in the government and outside, citizens, society and private enterprise for their works on various applications including disaster management.

The Indian Council of Agricultural Research (ICAR) and a large number of units under its jurisdiction are engaged in research on cereals, vegetables, horticulture and cash crops, particularly in the contexts of changing climate and vagaries of nature. ICAR played a prominent role in ushering in green revolution, and it is now faced with formidable challenges of what has been described as second green revolution, when there are definite indications that tropical agriculture and horticulture would be affected by global warming. Although the share of agriculture in GDP of the country has been declining, agriculture remains the main source of livelihood and subsistence of an overwhelming majority of population. Natural disasters like flood, droughts and cyclone take a heavy toll on agriculture, and therefore research on the development of agricultural crops and practices that are resilient to the hazards continues to remain the focus of some of the activities of the ICAR and its laboratories.

Similarly, the Indian Council of Medical Research (ICMR) and the laboratories under its jurisdiction and networking are working on various aspects of disease surveillance, mass casualty management, management of post disaster trauma, etc. Based on the research many useful guidelines and modules have been developed for application on the ground by the health and disaster management institutions.

4.5 Incident Response System (IRS) as Emergency Response Tool

4.5.1 Genesis and Background of Incident Command System (ICS) in India

The management of response during disasters requires the existing administrative setup, civil society and its various institutions to carry out a large number of tasks. The activities involved in response management would depend on the nature and type of disaster. It has been observed that in times of disaster, apart from lack of resources, lack of coordination amongst various agencies and an absence of role clarity amongst various stakeholders pose serious challenges. If the response is planned and the stakeholders are trained, there will be no scope for ad hoc measures, and the response will be smooth and effective. The objective of these guidelines is to predesignate officers to perform various duties as well as train them in their respective roles. Realisation of certain shortcomings in response and a desire to address the critical gaps led the Government of India (GoI) to look at the world's best practices and adopt the Incident Command System (ICS) (Ministry of Home Affairs 2005).

The ICS incorporates all the duties that may be performed in case of any disaster or event. It envisages a complete team with various sections to attend to all possible requirements. If the ICS is put in place and stakeholders trained in their respective duties and roles, it will help reduce chaos and confusion during actual incident management, and everyone involved will know what all needs to be done, who will do it, where are the resources, who is in command, etc. (Introduction to Incident Command System 2005).

The ICS is a flexible system, and all its sections need not be activated in every situation at the same time. Only required sections may be made operational as and when required. This system envisages that the roles and duties shall be laid down in advance, the personnel earmarked and trained in their respective roles and duties (Wally and Baum 1994).

This system consists of a number of useful features like:

1. Management by objectives
2. Unity and chain of command
3. Transfer of command
4. Organisational flexibility
5. Manageable span of control

6. Area command
7. Unified command
8. Common terminology
9. Personnel accountability
10. Integrated communications
11. Planning and comprehensive resource mobilisation, deployment and demobilisation
12. Incident action plan
13. Information management
14. Proper documentation of the entire response activities through forms and formats
15. Responder's safety
16. Media management
17. Agency coordination

Keeping in mind the Disaster Management (DM) Act, 2005, and the existing administrative structure of the country, the ICS required some modifications and adaptation to the Indian context. In India the main stakeholders in any incident response are the administrators of the various government departments at the national, state, district, union territory and metropolitan city level. The roles of NGOs, CBOs, PRIs, volunteers of civil defence (CD), NDRF, SDRF personnel and communities, etc. also need to be carefully integrated in the response structure. NDMA, therefore, decided to adapt the ICS duly indigenised so that it is in consonance with the administrative structure of the country and in order to strengthen and standardise the response system in India.

4.5.2 Need for IRS

The DM Act 2005 has heralded a paradigm shift in DM from a post-event response to one of pre-event prevention, mitigation and preparedness. Though India has a long history of battling disasters and providing adequate response, it was clearly realised that there were a number of shortcomings like:

(a) Lack of accountability because of ad hoc and emergent nature of arrangements and no prior training for effective performance
(b) Lack of an orderly and systematic planning process
(c) Unclear chain of command and supervision of response activity
(d) Lack of proper communication, inefficient use of available resources, use of conflicting codes and terminology and no prior communication plan
(e) Lack of predetermined method/system to effectively integrate inter-agency requirements into the disaster management structures and planning process
(f) Lack of coordination between the first responders and individuals, professionals and NGOs with specialised skills during the response phase
(g) Lack of use of common terminology for different resources resulting in improper requisitioning and inappropriate resource mobilisation, etc.

4.5.3 Definition and Context

The Incident Response System (IRS) is an effective mechanism for reducing the scope for ad hoc measures in response. It incorporates all the tasks that may be performed during DM irrespective of their level of complexity. It envisages a composite team with various sections to attend to all the possible response requirements. The IRS identifies and designates officers to perform various duties and get them trained in their respective roles. If IRS is put in place and stakeholders trained and made aware of their roles, it will greatly help in reducing chaos and confusion during the response phase. Everyone will know what needs to be done, who will do it, who is in command, etc. IRS is a flexible system, and all the sections, branches and units need not be activated at the same time. Various sections, branches and units need to be activated only as and when they are required.

4.5.4 IRS Organisation

The IRS organisation functions through incident response teams (IRTs) in the field. In line with our administrative structure and DM Act 2005, responsible officers (ROs) have been designated at the state and district level as overall in charge of the incident response management (Fig. 4.2). The RO may however delegate responsibilities to the incident commander (IC), who in turn will manage the incident through IRTs. The IRTs will be predesignated at all levels: state, district, subdivision and tehsil/block. On receipt of early warning, the RO will activate them. In case a disaster occurs without any warning, the local IRT will respond and contact RO for further support, if required. A nodal officer (NO) has to be designated for proper coordination between the district, state and national level in activating air support for response. Apart from the RO and nodal officer (NO), the IRS has two main components: (a) command staff and (b) general staff.

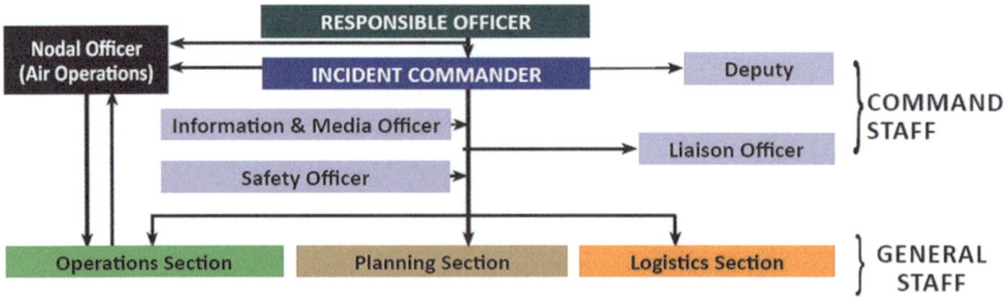

Fig. 4.2 IRS organisation

4.5.4.1 Command Staff

The command staff consists of an incident commander (IC), information and media officer (IMO), safety officer (SO) and liaison officer (LO). They report directly to the IC and may have assistants. The command staff may or may not have supporting organisations under them. The main function of the command staff is to assist the IC in the discharge of his functions.

4.5.4.2 General Staff

The general staff has three components which are as follows:

4.5.4.3 Operations Section (OS)

The OS is responsible for directing the required tactical actions to meet incident objectives. Management of disaster may not immediately require activation of branch, division and group. Expansion of the OS depends on the enormity of the situation and number of different types and kinds of functional groups required in the response management.

Planning Section (PS)

The PS is responsible for collecting, evaluating and displaying of incident information, maintaining and tracking resources and preparing the incident action plan (IAP) and other necessary incident-related documentation. They will assess the requirement of additional resources, propose from where it can be mobilised and keep IC informed. This section also prepares the demobilisation plan.

Logistics Section (LS)

The LS is responsible for providing facilities, services, materials, equipment and other resources in support of the incident response. The section chief participates in the development and implementation of the IAP and activates and supervises branches and units of his section. In order to ensure prompt and smooth procurement and supply of resources as per financial rules, the finance branch has been included in the LS.

4.5.5 *Features of IRS*

4.5.5.1 Management by Objectives

Management by objectives (MBO) covers four essential steps in IRS. These steps should be taken for the management of every incident regardless of its size or complexity:

(a) Understand government policy and directions including relief code, evacuation procedures, etc.
(b) Establishment of incident objectives
(c) Selection of appropriate strategies
(d) Performance of tactical moves (assigning the right resources, monitoring performance, etc.)

4.5.5.2 Unity of Command and Chain of Command

In IRS, unity of command means that every individual has a designated supervisor. Chain of command means that there is an orderly line of authority within the ranks of the organisation with a clear-cut reporting pattern right from the lowest level to the highest. In the IRS, the chain of command is established through a prescribed organisational structure which consists of various layers such as sections, branches, divisions, etc. This feature eliminates the possibility of receiving conflicting orders from various supervisors. Thus it increases accountability, prevents freelancing, improves the flow of information and helps in smooth coordination in operational efforts.

4.5.5.3 Transfer of Command

The command of an incident initially is vested in the highest ranking authority in the area where the disaster occurs. The transfer of command in any incident may take place for the following reasons:

(a) An incident becomes overwhelming for the IC and IRT.
(b) More qualified and experienced senior officers arrive at the scene.
(c) The incident situation changes over time, where a jurisdictional or agency change in command is operationally required.
(d) Normal turnover of personnel in the case of long or extended incidents.

The various processes in IRS of briefing, debriefing and documentation through forms and formats prove very useful during transfer of command. The IAP, assignment list, details of actions already taken, resources deployed, available and ordered, etc. give an immediate and comprehensive view of the incident status to the new comer.

4.5.5.4 Organisational Flexibility

The IRS organisation is a need-based, flexible organisation. All the components need not be activated simultaneously. It would depend upon the nature and requirements of the incident. Each activated section, branch or unit must have a person in charge to perform its role. In some cases, because of lack of personnel, a single supervisor may be made in charge of more than one group, unit or section. It should be clearly understood that in such cases, the groups, units and sections do not get merged/amalgamated. Their functioning would continue to be independent. Only the supervisory work/responsibility will be performed and discharged by the same individual. The organisational elements that are no longer required should be deactivated to reduce the size of the organisation and to ensure appropriate use of resources.

4.5.5.5 Span of Control

Span of control refers to the number of elements (section, branch or unit) that one supervisor can directly manage effectively. Ideally a supervisor should have five organisational elements under his control. However, if the elements increase to more than five or are reduced to less than three, necessary changes in the IRS organisational structure should be carried out.

4.5.5.6 Area Command

Area command is an expansion of the incident response function, primarily designed to manage a very large number of incidents that has multiple IRTs assigned or area being isolated because of geographical reasons. It is established for overseeing response and to ensure that conflicts, jurisdictional or otherwise, do not arise amongst deployed responding teams.

4.5.5.7 Unified Command (UC)

UC is a team effort that allows all agencies with jurisdictional responsibility for the incident, either geographical or functional, to manage an incident by establishing a common set of incident objectives and strategies under one commander. This is accomplished through the UC framework headed by a governor /lt. governor (LG) / administrator/chief minister (CM) and assisted by chief secretary (CS) without losing or abdicating agency authority, responsibility or accountability.

4.5.5.8 Common Terminology

In IRS, the common terminologies that are applied to organisational elements, position titles, resources and facilities are as follows:

(a) *Organisational Elements:* There is a consistent pattern for designating each level of the organisation (e.g. sections, branches, divisions and units, etc.).
(b) *Position Titles:* Those charged with management or leadership responsibility in IRS are referred to by specific position titles such as commander, officer, chief, director, supervisor, leader, incharge, etc. It provides a standardised nomenclature for requisitioning personnel to fill various levels of positions.
(c) *Branch:* The organisational level having functional or geographic responsibility for major segments of incident operations. The branch is found in operations and logistics sections. It is based on various functional requirement of the section.
(d) *Division:* Divisions are used to divide an incident into geographical area of operations. It is positioned in the IRS organisation between the branch and groups. Divisions are established when the number of resources deployed exceeds the span of control of the operation section chief. It is also activated for closer supervision when an area is very distant or isolated.
(e) *Group:* Group refers to only functional responsibilities for major segments of incident operations. Group consists of different functional teams (single resource, strike team and task force).
(f) *Resources:* Resources are grouped into two categories: (i) primary and (ii) support. The primary resources are meant for the responder, and support resources are meant for the affected people. All resources are however designated according to the 'kind' and 'type'. 'Kind' would mean the overall description of the resource like bus, truck, bulldozer and medical team. 'Type' would mean the performance capability of the resource which may be large, medium or small. This helps in ordering the exact and correct resource by the ordering unit. It also helps the deploying agencies to send the correct requirement.
(g) *Facilities:* Different kinds of facilities have to be established to meet the specific needs of the incident. IRS tries to standardise them by using common terminology like incident command post, staging area, incident base, camp, relief camp, helibase, helipad, etc.

4.5.5.9 Symbols for Different IRS Facilities

In the IRS, different symbols are used for the identification of different facilities established for response management (Fig. 4.3). They are as follows.

4.5.5.10 Accountability

In IRS, through a clear-cut chain of command, it is ensured that one individual or group is not assigned to more than one supervisor. Through other procedures and use of various forms, accountability of personnel and resources is ensured. It makes

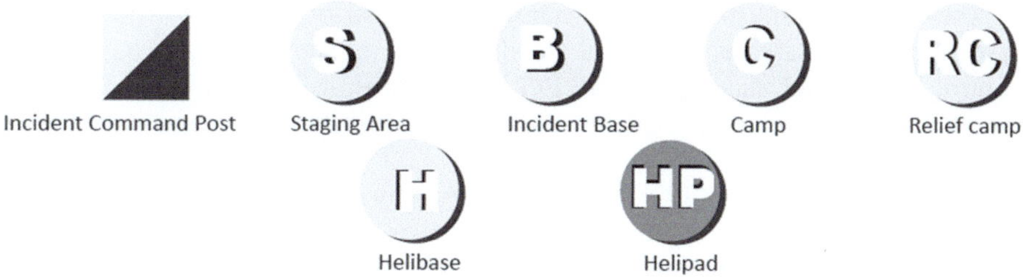

Fig. 4.3 Symbols used for Identification of different facilities

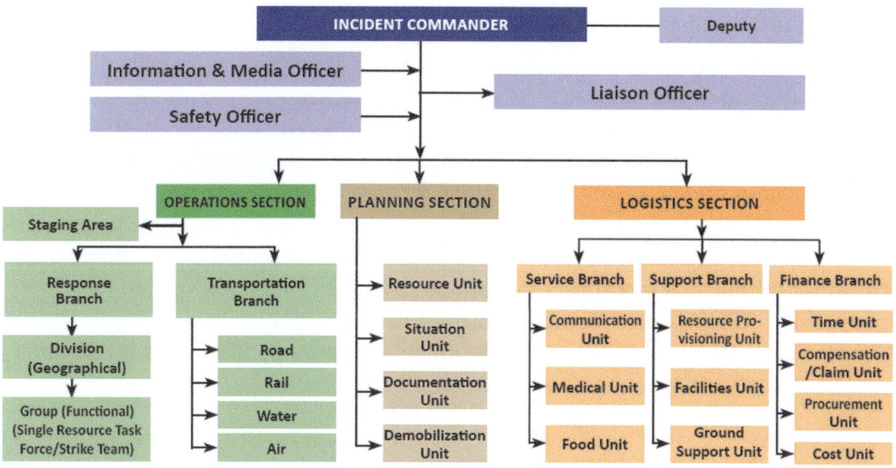

Fig. 4.4 Incident response system (IRS) organisation structure

the response effort absolutely focused and leaves no room for unsupervised activity. It helps maintain a complete record of all activities performed and resources deployed.

4.5.6 The Incident Response Teams (IRTs) at State and District Levels

The IRT is a team comprising of all positions of IRS organisation as shown in Fig. 4.4 headed by IC. The OS helps to prepare different tactical operations as required. The PS helps in obtaining different information and preparing plans as required. The LS assesses the availability and requirement of resources and takes action for obtaining them. IRTs will function at state, district, subdivision and the tehsil/block levels. These teams will respond to all natural and man-made disasters.

The lowest administrative unit (subdivision, tehsil or block) will be the first responder as the case may be. If the incident becomes complex and is beyond the control of local IRT, the higher-level IRT will be informed, and they will take over the response management. In such cases the lower-level IRT will merge with higher-level IRT.

When a lower level of IRT (e.g. block/tehsil) merges with a higher level (e.g. subdivision, district or state), the role of IC of lower level of IRT will change. When the block level IRT merges with subdivision level IRT, IC of the block level may play the role of deputy IC or OSC or any other duty that the IC of higher authority assigns. This process will be applicable at all levels.

To sum up, IRS is a system of management by objectives through IAP. It takes care of any expanding incident through an organisational structure of command staff, sections, branches, divisions, groups, units, resources and span of control. Through unified command (UC) it allows all agencies having jurisdictional or functional responsibilities to jointly develop incident objectives and strategies. It has a clear-cut process for personnel accountability, resource management, integrated communications and transfer of command.

In line with the federal structure of the country, it should be clearly understood that response to any disaster will be carried out by the concerned states and districts. The GoI will play a supporting role by way of assistance in the form of resources, manpower (NDRF, armed and paramilitary forces), equipment and funds. At the GoI level, the NCMC or NEC will coordinate and provide the required resources. NDMA will also help in monitoring the coordination of response.

4.5.7 Illustrative List of Activities Identified as of an Immediate Nature

4.5.7.1 Drinking Water Supply

(a) Repair of damaged platforms of handpumps/ring wells/spring-tapped chambers/public stand posts, cisterns
(b) Restoration of damaged stand posts including replacement of damaged pipe lengths with new pipe lengths and cleaning of clear water reservoir (to make it leak proof)
(c) Repair of damaged pumping machines, leaking overhead reservoirs and water pumps including damaged intake – structures and approach gantries/jetties

4.5.7.2 Roads

(a) Filling up of breaches and potholes, use of pipe for creating waterways and repair and stone pitching of embankments
(b) Repairing of breached culverts

(c) Providing diversions to the damaged/washed-out portions of bridges to restore immediate connectivity
(d) Temporary repairing of approaches to bridges/embankments of bridges, repairing of damaged railing bridges, repairing of causeways to restore immediate connectivity and repairing of damaged stretch of roads to restore traffic

4.5.7.3 Irrigation

(a) Immediate repair of damaged canal structures and earthen/masonry works of tanks and small reservoirs with the use of cement, sand bags and stones
(b) Repair of weak areas such as piping or rat holes in dam walls/embankments
(c) Removal of vegetative material/building material/debris from canal and drainage system

4.5.7.4 Health

Repair of damaged approach roads, buildings and electrical lines of PHCs/community health centres

4.5.7.5 Community Assets of Panchayat

(a) Repair of village internal roads
(b) Removal of debris from drainage/sewerage lines
(c) Repair of internal water supply lines
(d) Repair of street lights
(e) Temporary repair of primary schools, panchayat ghars, community halls, anganwadi, etc.

4.6 Conclusion

India has put in place a credible system of disaster management. A legal and institutional setup has been established, and policies and guidelines for holistic management of various types of natural and man-made disasters have been formulated. This has set in motion demands for allocation of funds for disaster risk reduction across sectors. The National Cyclone Risk Mitigation Project has been launched recently, and similar projects for earthquake, flood, landslide, etc. are on the anvil. Strong institutionalised arrangements for allocation of funds for disaster management are already in place. Sizeable allocations are being made for disaster response, relief, rehabilitation and early recovery, which have helped to reduce the loss of lives and property during disasters. India is also investing huge amounts for social and economic developments across many sectors. Many of these schemes contain elements that have potential for reducing the risks of disasters. The National Policy on Disaster Management and the Five-Year Development Plans have laid considerable emphasis on mainstreaming as the key strategy for risk reduction. Not much effort in mainstreaming disaster risk in various sectors of development has been made. There are tremendous scopes and huge challenges in mainstreaming, and the future success of reducing the risks of disasters would depend on a large extent how the critical needs of prevention, mitigation and preparedness are built into the process of development across all sectors.

References

High Powered Committee (HPC) Report (2001) Government of India Ministry of Agriculture, New Delhi

Government of India (2008) Ministry of Environment and Forests, India's Expenditure on Adaptation Related Activities

Government of India (2009a) Finance Commission, Report of Thirteenth Finance Commission

Government of India (2009b) National Disaster Management Authority, National Policy on Disaster Management

Government of India (2009c) National Institute of Disaster Management, Financing Disaster Management in India

Government of India (2009d) PMO, National Action Plan on Climate Change, 2009

Introduction to Incident Command System (2005) Centre for Disaster Management (CDM), Lal Bahadur Shastri National Academy of Administration (LBSNAA), Mussoorie, India

Ministry of Home Affairs (2005) National Disaster Management Division, Government of India: policy document on the Incident Command System, New Delhi

NDRF (2016) National Disaster Response Force, Coffee Table Book, Ministry of Home Affairs, India

UNISDR (2009) Global assessment report on disaster risk reduction: risk and poverty in a changing climate- invest today for a better tomorrow

UNISDR (2011) Global assessment report on disaster risk reduction: revealing risk, redefining development

United Nations Development Group (2003) Indicators for monitoring the millennium development goals: definitions, rationale, concepts and sources

Vulnerability Atlas of India (1999) Ministry of urban development, building materials and technology promotion council. Government of India, New Delhi

Wally S, Baum JR (1994) Personal and structural determinants of the pace of strategic decision making. Acad Manag J 37:932–956

World Bank (2003) Financing rapid onset natural disaster losses in India: a risk management approach, 2003

World Bank (2010) Natural hazards, unnatural disasters: the economics of effective prevention, 2010

Chapter 5
Disaster Risk Reduction: Importance of Responsibilities of States

Narender Kumar

Abstract Disasters have become serial killers of humanity, and with the growth of population, mankind is becoming increasingly vulnerable. Disasters are not limited to territorial geography of a nation; it has the potential to cause sufferings to millions of people across the region, and the casualties during disaster can be due to injuries, starvation, epidemic, deaths, psychological de-arrangement of the people and destruction of life support system. It is possible to restrict the collateral damages with improved technology for forecasting, warning and swift actions by the government and civil society. Population can be trained, psychologically prepared and relocated prior to the disaster or whenever danger to lives is imminent. The past experience suggests that disasters kill lesser people, but poor infrastructure and ignorance to deal with the calamity kill more number of humans. There is a need to respect sense of geography and sense of history when it comes to the disasters. The empirical studies can outline the nature of disasters in a geographical limit which have caused havoc in the past; it can suggest periodicity of disasters, triggers and scales of destruction. Nature, people and the government invariably are either responsible or victims of the disasters. It is desirable that community awareness programmes relating to disasters and impact of unsustainable development on environment ideally should be debated, discussed and disseminated to the affected section of population. Institutional mechanisms, respect for environmental laws, traditional wisdom, use of technology and drills and procedures to deal with wide-ranging natural and man-made disasters need to be practised and put in place. In the race for development people often forget about the preparation for disasters. There are large number of built-up areas in the mountains as well as in the plains where there are no road communication to gain access into the heavily congested areas. Such unplanned infrastructure development precludes access of fire tender and heavy machinery in case of disasters. In the light of the above facts, infrastructure must be developed in such a manner that is disaster resistant and facilitate access of equipment and heavy machinery during crisis. Prevention and disaster relief are collective responsibility of the government and society.

N. Kumar (✉)
Centre for Land Warfare Studies (CLAWS), New Delhi, India
e-mail: narry13@gmail.com

I. Pal, R. Shaw (eds.), *Disaster Risk Governance in India and Cross Cutting Issues*, Disaster Risk Reduction, DOI 10.1007/978-981-10-3310-0_5

Keywords Disaster • Environment • Traditional wisdom • Climate change • Disaster risk profile and sustainable development

5.1 Introduction

Earth has been going through the geological process, and these can be described as natural events. These events become disasters because they impact humans in a negative manner (Nelson 2014). South Asia has become home to natural disasters, and this has happened primarily due to heavy density of population and lack of understanding of geological process that earth is going through. It is important to understand each phenomenon that is categorised in natural disasters such as earthquake, floods, hurricane, avalanche, heat waves, extreme winters, draught and wind storms. These processes are part of evolution of earth, and such activity becomes disasters when the traditional knowledge and natural phenomenon are overlooked. Some of these processes become serious challenge to the humans and nature per se when human intervention becomes unsustainable. The impact of these disasters are collateral damage, loss of human lives, destruction of flora and fauna, destruction of life support chain and destruction of heritage and culture that could lead to mass relocation of survivors. Such events often put additional strain on natural resources for creation of new population centres and struggle for resource acquisition to support lives. Natural disasters and extreme environmental events have the potential to cause economic, social and political stress that could provoke both civil and international conflicts (Tipson 2013).

On the other side, there are disasters that are initiated by unsustainable human intervention that destabilises the fragile balance of nature. The underdeveloped nations are facing major crisis on this account since resources are meagre, pressure of population is high and exploitation of natural resources is unscientific and unsustainable. This is further aggravated by poor implementation of law and order and laws related to environmental protection. Another factor that leads to the increased number of man-made disasters is corruption and neglect of traditional wisdom.

UN office for disaster risk reduction (UNISDR) revealed that "India was among the top three most disaster-hit countries in 2015, with whopping economic damages worth $3.30 billion" (Jha 2013). In the light of the above facts, there is bigger impact of natural and man-made disasters on human lives; it would require increased efforts to mitigate and reduce the impact of disasters by spreading awareness, education, building resilience, local capacities to deal with disasters and infrastructure development that could sustain humans at the time of disasters. To reduce the direct harm of such disasters will require initiatives in three areas: increasing local resilience, improving relief capabilities and, where unavoidable, facilitating relocation from the most vulnerable areas (Tipson 2013).

5.2 Environmental Education and Disaster Risk Reduction

India has not developed the culture of understanding nature and disasters. The main reason could be environmental education was left to traditional wisdom and not included in modern education system; thus neither the subject is taught in educational institute nor any formal training is imparted to train the trainers, whereas all government officers from village revenue officer to deputy commissioner should be trained in disaster relief and disaster mitigation. Moreover the traditional wisdom has been forgotten because of relocation of population and dependence on modern education system. This has resulted in masses not aware of the responsibility of citizens towards nature and environment. There is a need to create this awareness through education, and children are the best agent of change. The case of disaster awareness is no different. Survival techniques, disaster awareness and building local capacity to fight disaster are still to be made mandatory as per education curriculum.

(a) Risk reduction through education and awareness campaign is a must in a nation that is vulnerable to disasters.
(b) Environmental education is mandatory to bring awareness through children.
(c) It is important to create environmental leaders to bring change in outlook towards nature.
(d) Environmental protection is not only the responsibility of government but also of every citizen.
(e) Every geographical region of India has different disaster risk; thus it is imperative to develop region specific awareness.
(f) Capacity building through collective efforts of public and government participation to mitigate the disaster risks.
(g) Awareness campaigns to make citizen responsible during disasters based on their skills.
(h) Training of volunteers to act as leaders and support staff to deal with disasters during the golden period.

It is imperative to understand the fragile nature of earth and its environment. Imbalance initiates turbulence that manifests into disasters. Unsustainable intervention has consequences to trigger disasters. It is not important for the government agencies alone to have the data bank about the nature of disasters that can strike in a particular zone, but even the citizens living in the fragile region should be aware of the dangers and their responsibilities prior to the disaster and during disasters. A case in point is Latur; it is on the edge due to draught and earthquake. Thus the citizens should be aware of the threats that can manifest from the twin disasters. Earthquake has rapid onset, whereas draught has slow onset. Therefore, the awareness of the citizens of this area must have creation of dwelling unit and infrastructure that can withstand earthquake and at the same time prevent a situation of extreme draught. That means water harvesting measures, afforestation, water conservation and recharging of groundwater are a must. Earthquake cannot be prevented; thus citizens and government can mitigate impact to limit the loss of lives and breakdown of life support system. The significance of understanding nature has the following benefits:

(a) Awareness of fragility of environment and impact of imbalance on nature
(b) Nature of disasters that can strike in a geographical region and triggers that can intensify the impact of disasters
(c) Measures that should be taken to reduce the impact
(d) Adoption of good practices of environment protection

5.3 Traditional Wisdom and Risk Management

Ancient India treated earth as sacred, and humans made no endeavour to change the behaviour of nature including drainage system. Blessings of the earth were sought for prosperity in all endeavours and fulfilment of all righteous aspirations (Environmental Wisdom 2016). Cutting of trees without purpose was considered highly inappropriate. Vegetation remained untouched, and it was considered improper and a sinful act to defecate or urinate in water; similarly trees were planted around ponds and reservoir, and the scientific reason is that it prevents evaporation of water from ponds. Mangroves were not disturbed because they were considered a wall to prevent the wind and hurricanes to cause damage inland. Seashore and riverbeds were not disturbed because floodplains were considered property of "sacred river or river goddess" or ocean. But when greed got better of the old customs and traditions, rivers are encroached, disaster fire walls have been breached and seashores have become new destination for rich and famous.

Every community had developed wisdom to deal with disasters in their geographical regions. But that wisdom has been buried forever. During Uttarakhand flood disaster in 2013, the route to Kedarnath was constructed along the river Mandakini, whereas the old traditional route that was taken by the Saints was along the ridge line, and it was noticed that along the old route, not even a single landslide had damaged the route. Because this was based on traditional wisdom that mountain rivers are unpredictable, therefore, tracks in the mountains should not be along the banks of the river as far as possible. In addition the river basins are in perpetual downward motion due to soil being wet. Similarly in the mountains, old wisdom suggests that don't lose height while climbing up. In Uttarakhand traditional wisdom was violated, and everything that was not supposed to be done was done. The route to Kedarnath was along the river, the rest areas were constructed almost on the riverbed and within 150 m of the river and water channels at Kedar Temple were diverted to facilitate the construction of hotels and shops. Though glacial lake outburst floods (GLOF) are relatively unknown or, perhaps, not adequately recorded in the upper Ganga basin, the last 100 years have witnessed at least 50 glacial lake outburst floods at several places across the Hindu Kush Himalayas, from Afghanistan to Myanmar (Jolly 2013). Historical wisdom suggests that villages/human dwellings are not established below the reservoirs/lakes. But in the instant case, the entire complex of Kedarnath was established below the Chorabari Tal (Chorabari Glacial

Lake) (Fig. 5.1). Satellite-based geographical information system (GIS) and high-resolution remote sensing technology have located nearly 20,000 glacial lakes across the Himalayas, straddling Afghanistan, Pakistan, Tibet, India, Nepal and Bhutan. The worry is that "More than 200 of these have been classified as potentially dangerous (Jolly 2013)". What nature proved was that humans can be bribed but not nature; tempering with the nature is suicidal. There is a need to identify the fragile lakes, and if there are townships or infrastructure along the run of the river or stream, the government must consider relocation of such infrastructure to avoid collateral damage. Lessons from Kedarnath should be taken seriously, and collateral damage can be prevented.

Traditional wisdom was based on the experience of centuries, and accordingly the people developed the architect of the habitat that suited the type of hazards they could face. The houses in hills of North and Northeast India were mainly constructed with woods so that if the earthquake does strike, the loss of life could be reduced; similarly in the mountains, the tracks were along the grain of the country and rarely involved disturbing the slopes. But that has reversed now, and the traditional wisdom has been discarded and resulted in the construction of roads along the shortest possible route along the valley floor.

It is very rare in the mountains that villages are constructed in avalanche-prone areas, and similarly no houses were constructed in floodplains. Water harvesting, environmental protection and preservation of soil were given importance. Even as late as the 1970s in Himachal and Garhwal, people would repair and de-silt ponds

Fig. 5.1 Construction of infrastructure just below the fragile glacial lake Chorabari Tal, Kedarnath (Indian Army Aviation July 2013 post-disaster photograph)

on mountaintop every year prior to rainy season. These ponds prevented water erosion and acted as drip irrigation for the forests around the slopes of the hills and also a source of water to animals and humans alike. These ponds developed an entire system of microorganism, and some of the rare species of medicinal plants and insects are found around these Khaals. Plants such as bhringraj (famous for hair oil) and acorus (known insecticidal that has properties even to repel snakes) are found around these ponds. Now the concept has vanished, and such ponds are rare site. This has resulted in drying up of forests and vegetation and is one of the main reasons for soil erosion.

More than half a million people in Jammu and Kashmir (J&K) were affected by the earthquake that greatly affected 90,000 households in the Kashmir division and 8000 households in the Jammu division. Regardless of this destruction and devastation, indigenous construction techniques helped to save the lives of many individuals (ISDR 2008). Unfortunately, the trend of the construction of houses in hills of Himachal, J&K, Sikkim and Uttarakhand has shifted from traditional wooden structure to concrete multi-storey building. Such dwelling units are at greater risk than the traditional houses constructed with the help of woods.

Traditional wisdom also helps in the prediction of weather. In the entire Himalayas, the Bakarwal and Gujars (the communities of those who rear goats and sheep move up in high mountains during summer and move down during winters to valleys) for thousands of years moved up and down on traditional routes through hazardous regions and snowbound areas but always manage to avoid snow and rain hazards primarily due to traditional wisdom. Weather prediction by these nomadic tribes is done by reading the wind pattern and behaviour of the animals. Before the onset of the winters and closure of the passes, Bakarwals and Gujars always manage to move down to the valleys. Rarely they ever get trapped in avalanche and other weather hazards. Such wisdoms are not restricted only to Indian subcontinents, but these wisdoms are practised over the world by the indigenous people.

The traditional wisdom has been passed from generation to generations by "oral literature". There is a need to preserve the traditional wisdom because it is pure knowledge and has served the mankind for thousands of years. This knowledge was (and is) nature friendly and suggestive of the fact that humanity can achieve sustainable development without jeopardising the balance of nature. There is a need to preserve this knowledge and compile the traditional wisdom prevailing in India and the region for the larger good of the humanity. It will save the nature and the civilisations as well.

5.4 Responsibility of the State

The recent natural disasters in the subcontinent have questioned the ability of the states and robustness of the system to mitigate the impact of events of extreme natural calamities. Collateral damages and human casualties due to wide-ranging disasters are rising, yet there is limited focus on ownership of the nature as the collective wealth of the humanity. The two stakeholders, citizens and the state, ought to have contractual obligation to create the capacity to reduce vulnerabilities and conflict with the nature. Treating disasters as the will of the God will be a costly mistake; it is indeed a nontraditional threat and must be treated as an adversary, because in the last one decade, more people have lost their lives due to disasters than the war. Disasters can spiral instability, mass relocation of demography which may become unsustainable in a given geographical confines, and has the potential to disrupt the life support system. Post-disaster recovery also has a critical need for integrated planning. If ill-planned, recovery can "recreate vulnerabilities, generate new risks and undermine sustainability and security" (ISDR 2008).

The magnitude of each disaster in India, be in terms of deaths, property damage or costs for reconstruction, increases, because the population density is very high; as a result the impact increases manifolds, the best land in rural and urban areas is taken up and those seeking land for building their lives all over again including farming or housing are forced to accept inadequacy (Sena and Michael 2006). India's case is different than any other nation in the subcontinent as well as other nations except China, because it is the second most populated nation with limited resources and capacity. In the event of the disasters, more lives are lost, and more people are pushed to extreme hardship including threat to life. Unmanageable population that would need resources is huge, and states are neither geared to deal with such calamity nor do resources permit. Disasters cause disruption in life support system and break in development, as a result more population is pushed below the poverty line, and some may be further pushed down to the verge of starvation. Past studies have shown that economic growth and poverty alleviation are closely tied, with the latter highly dependent on the former (World Bank Group 2016). This is a chain that triggers mass relocation due to insecurity of life, strain on limited resources and failure of law and order and can disrupt the entire chain of life support system.

In the backdrop of the three major disasters which India and the subcontinent have witnessed in the recent past (Indian Ocean tsunami, Uttarakhand floods, Nepal earthquake and draught across India), there is a need to take a relook at the very idea of disaster response strategy at the national and the regional level. Monitoring and observing environmental factors that signal the onset of a hazard are fundamental to early warning systems. Such systems generally include a mix of space-based or remotely sensed observations, as well as on-site, ground-based monitoring (UNEP 2008). The role of citizens, systemic response and ability of a state to identify and prevent unsustainable human intervention to the environment assume significance. It is important to restore balance at a time when pressure on usage of land and natural resources is rapidly increasing. Casualty ratio in developing or underdeveloped nations is far higher than the developed nation because of the poverty, defiance of the rule of law and wilful disobedience to accept the limit of exploitation of the nature. It may not be possible to determine the timing of the disaster, but the place of event of the natural calamity can be identified, and the mechanism to mitigate the loss can be endeavoured.

Somewhere collectively state and subjects have overlooked the fact that poor infrastructure, violation of environmental laws and ignoring the necessity of technical audit of the infrastructure to withstand the disasters are also some of the reasons of avoidable loss. In the backdrop of the above, the scale of the collateral damage will be circumscribed by the ability of the state to decongest the ground zero, active participation of the citizen to assume the role and responsibility of disaster relief operations and quick restoration of public order. This can be achieved if there is robust disaster-proof infrastructure, uninterrupted communication lines, public awareness and ability of the state to regain control of the situation fastest. The significance of traditional knowledge, technology, active participation of the public especially the youth and strict compliance of environmental laws cannot be ignored for risk reduction.

South Asia has seen the explosion of population, urbanisation, economic opportunity, construction of megacities and exploitation of natural resources. Climate change, proximity to fault lines, political inaction, unsustainable development and absence of law-enforcing agencies are the main reasons for the region being called the home to disasters. Disasters do not respect borders or distinguish between income levels; however, the effect of disasters on human lives tends to be the lowest in high-income countries. In Asia-Pacific high-income countries, about 1 person in every 1000 people was affected by disasters, and 1 in 1 million died during the 10 years from 2001 to 2010; in low-income countries, nearly 30 in 1000 people were affected and 52 in 1 million people killed (Statistical Yearbook 2011). States have two important responsibilities: first, to ensure the protection of environment and second, to implement laws especially related to environment and disaster relief.

5.5 Implementation of Environment Laws

A study was carried out on Chile versus Haiti earthquake response of the governments. Chileans have homes and offices built to ride out quakes; their steel skeletons are designed to sway with seismic waves rather than resist them (Frank 2015), whereas in Haiti, by contrast, there is no building code. “Earthquakes don’t kill – they don’t create damage – if there’s nothing to damage,” (Frank 2015). “Chile has a responsible government”, he said, waving his hand in disgust. “Our government is incompetent” (Frank 2015). Both Chile and Haiti sit atop large, volatile fault lines. In recent decades, Chile has mandated earthquake proofing for new structures, requiring that materials like rubber and features like counterweights be built into the architectural designs to allow buildings to bend and sway rather than break during temblors. Haiti, by contrast, lets its buildings rise with little if any input from engineers and plenty of bribes to so-called government inspectors. Structures have scant reinforcement and are often set on weak foundations (Padgett 2010). In the recent past, collateral damage during the disasters in Uttarakhand flash floods (2013); floods in Srinagar (2014), Chennai (2015) and Mumbai (2005); and earthquake in Latur was primarily due to the violation of environment laws, unscientific infrastructure development, unplanned land use and unsustainable development. The floods in cities were the outcome of choking of drains and rivers that use to drain the rainwater. Casualties in Latur were due to the construction of houses unsuitable for disaster-prone regions. The law enforcement agencies failed to prevent violations in all the above cases.

5.6 Disaster Relief and Mitigation at Local Level

Breakdown in communication links and bad weather often delay or interrupt the deployment of disaster relief forces at ground zero. It happened in Uttarakhand, J&K and Chennai floods (Fig. 5.2). Even the Nepal earthquake also posed similar difficulties. In such a situation, it is always the locals who can take the lead by assuming leadership and responsibility to organise relief and rescue. Towards this end, the following important aspects assume significance in developing collective capabilities of communities at local level.

(a) Community disaster relief plan ideally should not be made in isolation; it should conform to the state and national disaster relief plan. Local/community disaster relief plan should act as a launch pad and should assist state and national efforts for speedy relief and rescue.
(b) Building local capacity for prevention and mitigation of disaster. Disaster strikes at community level and the most vulnerable among the communities are the poor.
(c) Most successful disaster relief operations can be conducted by community at local level since they are at ground zero at the time of disaster. Community can assess and utilise golden period to provide relief immediately.
(d) Direct participation of the community at local level for disaster risk reduction is most significant. Locals would be in a better position to identify the disaster risks and utilise traditional wisdom and cultural heritage.

Fig. 5.2 Floods in Chennai 2016 caused by choking of River Adyar and flood drains (Picture by Indian Army)

(e) Community-based warning system has proved very useful and has minimised the loses.
(f) The involvement of local people promotes self-reliance and ensures that emergency management plans meet local needs and circumstances (World Bank Study 2009).
(g) Local bodies and social organisations are in a better position to demand accountability from the government agencies and are in a position to fill the information and operational vacuum.
(h) Community can play an important role in the prevention of degradation of environment. Chipko Andolan in Uttarakhand led by Mr. Sunderlal Bahuguna is one of the best examples of community effort to prevent degradation of environment.
(i) First responders during disaster relief should be locals since they are at ground zero and have knowledge of the extent of the damages.
(j) Imposition of Essential Services Maintenance Act (ESMA) is vital during disaster because there is a lack of administrative control and invariably unscrupulous businessmen and antisocial elements take advantage of chaos to fleece victims. This was seen in Uttarakhand, Chennai, Srinagar and almost every ground zero.
(k) During disasters, antisocial elements take advantage of the fluid situation and resort to looting of houses and even the victims. Cases have come to light that even dead were not spared. Therefore, along with the disaster relief operations, police and volunteers should assist the administration to restore law and order earliest.

5.7 Adaptation to Climate Change

Climate change is a reality, and there is a need to adapt to this inevitable change. India being an agrarian society, farmers have to be flexible to change the cropping pattern, and farmers need to develop draught-resistant species of food grain and vegetables. India may also have to adopt de-urbanisation of the civilization rather than risking an abrupt collapse (Pandey 2016) of the system. The adaptation to change has to be systematic and scientific; it would need change in farming sector, infrastructure development, slow but scientific method of water harvesting and storage of water as well as food grains. Gradual shift from the hazardous material, gases, chemical and nondegradable material needs to be replaced with more environmentally friendly material. Reducing the direct harm of such disasters will require initiatives in three areas: increasing local resilience, improving relief capabilities and, where unavoidable, facilitating relocation from the most vulnerable areas (Tipson 2013).

5.8 Sub-conventional War Emerging as Major Disaster

Though conventional wars are rare, sub-conventional wars have replaced the natural disasters as the biggest source of collateral damage and loss of life. Conflict in West Asia has displaced close to ten million people, and the impact is felt across three continents, i.e. Europe, Asia and Africa. Eelam War in Sri Lanka brought lakhs of Tamils to Indian shore, and similarly close to 6% population of Kashmir was forced to migrate, and large percentage still lives in camps. Such man-made disasters impact culture, religion, economic avenues and social harmony. Man-made disasters not only displace people, but it also disrupts the harmonious growth of society. It causes breakdown in life support system, stress on infrastructure, environmental degradation, unstructured urbanisation and stress on resources that causes rift between migrant and local population. To deal with such disasters, there is multi-agency response that is needed. First is the restoration of law and order by CAPF and Armed Forces, a response of the state to accommodate displaced population spread over large areas without breaking the families and to avoid confrontation camps be located in such a manner that can support large population without impinging on the meagre resources that supports the locals and rehabilitation of internally displaced population socially and economically. Internal instability also occurred in large number of Muslims who have migrated from Myanmar to Mizoram; similarly Tibetan refugees have settled in various parts of India with a capital at Dharamsala in HP. Economic reasons have brought Bangladeshi and Nepali origin people to India that certainly has put stress on social and economic spheres. Natural and man-made disasters in neighbouring countries such as Nepal, Bangladesh, Myanmar and even POK will find India a home to refugees and IDPs. The government should have plan prior to bigger disasters to avoid internal instability triggered by displaced population.

5.9 Disaster Resilient Cultural Development

Ann Miller and Mike Douglass say that disaster response tends towards infrastructure solution to disasters, whereas the management of disasters and relocated population is far more complex and needs whole of government approach. This also needs the development of culture to prepare and cope up with disasters. Mana and Malari are two villages located at 12,000 ft above sea level in Uttarakhand. These villages are the last inhabited villages closest to the border. Slopes around Malari and Mana are avalanche prone, but these villages have never been struck by an avalanche in spite of very heavy snowfall (Fig. 5.3). The architecture of the houses is such that it can withstand the adverse weather, heavy snowfall and even the earthquake. Till a couple of decades back, the villages were not even connected by road, yet life was usual for the inhabitant. Heavy snowfall, avalanche and even earthquake

Fig. 5.3 Malari village in Uttarakhand at an altitude of 12,000 ft has never been struck by avalanche (Picture by Author)

had no impact on day-to-day life of the people who are basically agrarian/herdsmen. These people mastered the art of reading the slopes and wind pattern (to avoid house getting directly impacted by icy winds) and architecture of the houses to cope up with snow, rains and even earthquakes. The use of wood for interlocking is used extensively to construct the dwelling units. These people have very fine sense of weather prediction, and it is unheard of that any of the village inhabitants is ever caught up in snow or weather storm. One can put it as traditional wisdom, but it has more to do with the development of culture to cope up with the disasters small or big.

Disasters should not be always considered as a natural phenomenon. Disasters are also creation of avoidable human interventions. Disasters are simply the collapse of cultural protection of environment; thus, they are principally man-made (Dombrowsky 1998). In the last 50 years, the temperatures in Himalayas have gone up by almost 2 °C. It is primarily due to the collapse of the culture of protection of environment, deforestation and unsustainable development. A large number of mountain lakes and ponds are either drying or have dried up. It is causing major destruction to rare variety of flora and fauna that is balancing the cycle of nature. More importantly these natural assets were a fireball against the disaster.

5.10 Disaster Risk Governance

Disaster governance is prevention, mitigation of disaster risk, implementation of environmental laws, building capacity to deal with disaster, filling knowledge gap and training of vulnerable population to deal with the disaster. In addition, disaster governance is to assume responsibility at the time of disaster to maintain public order and restoration of law and order that is often neglected. In India focus is more to deal with the disaster relief than to incorporate disaster prevention and disaster mitigation as part of governance. Some of the important issues related to disaster governance that needs to be taken into consideration are as follows:

(a) *Infrastructure norms*: Disasters cause massive collateral damages if the infrastructure norms are flouted. The construction of concrete multi-storey houses in earthquake-prone areas such as the Himalayan fault line is inviting a catastrophe. But infrastructure norms in Sikkim, Himachal and even in Uttarakhand have been completely disobeyed. Unscientific infrastructure causes bigger catastrophe because it causes casualties, it denies access to disaster relief forces and it leads to huge losses (Fig. 5.4). National Disaster Management Authority (NDMA) (2016) must enforce infrastructure norms for various natural and man-made disasters.
(b) There should be preconditions for the development of infrastructure in such a manner that facilitates disaster relief operations so that fire tender, heavy machinery and dozers can gain access to almost every locality and every dwelling unit. Invariably most of the agencies private and public ignore the abovementioned conditions to facilitate disaster relief operations.

Fig. 5.4 Construction of buildings on riverbed in Joshimath, Uttarakhand, violating infrastructure norms (Picture taken by author post Uttarakhand Disaster in 2013)

(c) *Alarm system*: The alarm system as part of civic body responsibility should be put in place. Presently, there is no such system that exists on ground. It should be based on mobile app, FM radio, TV channels and even hooters in localities. This system should be a norm, and local civic bodies should be made responsible for such system.
(d) *Local capacity*: The idea of building local capacity is to utilise the golden period to save maximum victims that can only be done if there is local capacity to deal with disaster. Often the deployment of the disaster response forces will get delayed due to adverse weather and interruption or breakdown of surface communication or even electronic communications.
(e) *Training for disasters*: Disaster survival techniques and disaster relief are a process that needs to be part of education curriculum. Drills and procedures need to be practised and rehearsed.
(f) *Water harvesting*: The maintenance of water table and underground water can be ensured only by water harvesting. There is a need to ensure that water harvesting is made mandatory for all urban and rural infrastructures so that water recharge takes place on regular bases.
(g) *Protection of environment*: Environment and forests are one of the major sources of disaster mitigation. It should ideally be part of disaster governance, and violators should be penalised to avoid the degradation of environment.
(h) *Maintenance of essential services*: The biggest beneficiary of war and disasters are businessmen and unscrupulous vultures in the form of humans. Be it Srinagar, Chennai's flood or Uttarakhand flash floods, the first to collapse is government machinery. Law and order is thrown to winds; there is a sense of anarchy. More often there is looting, arson and fleecing of victims for basic essential items. The disaster is in two forms, first is the natural or man-made disaster that strikes at the heart of public order, and second disaster is an anarchic situation where law and order is a major casualty.
(i) *Sustainable development*: India was hit by super and very severe cyclones four times between 1977 and 1999 that caused massive destruction of infrastructure and human lives. It can be attributed to encroachment of coast and destruction of mangroves and vegetation along the coastline. Because vegetation and forests are the biggest barriers and storm breakers, similarly, occupation of river floodplains is inviting disasters. State governments have often overlooked such encroachments due to political and domestic pressure, but NDMA must declare these areas as no-go areas.
(j) *Professional and academia*: Intelligence agencies, think tanks and professionals and academia should increase their focus on the potential for major disasters in various parts of the country. Disasters cause economic, social and political "ripple effects" that lead to deadly conflicts (Tipson 2013).
(k) *Establishment of traditional wisdom preservation centre*: The knowledge and experience of centuries and generations must be preserved and formally documented for practice. It will not only help in disaster preparation but will also help in saving invaluable lives. In addition, it should become a subject of research in academic institutions.

(l) *Communication*: The biggest problem during disaster relief operations is communication blackout and lack of update on ground situation. Electricity goes off and so are most of the landline and even mobile towers. High-threat zone should have radio connectivity and ability of the state to establish man-portable communication relay system. Even the use of coloured smoke should be resorted to indicate the nature of distress.

(m) *Disaster detection and warning*: Except television and radio, there is no method to forewarn the disasters. Some of the coastal states have the warning system to make the public at large aware of the disasters that have slow to medium onset. There is a need to link up detection and warning of disasters so that people could be evacuated to safer places.

5.11 Conclusion

Natural disasters not only cause collateral damage but also destroy the civility, control over human reactions in a chaotic environment and sense of responsibility as civilised citizens. It obliterates the cultures, societies and at times sense of being humans. Disasters have emerged as the common threat to humanity. Leon Panetta said, "Rising sea levels, severe draughts, the melting of the polar caps, the more frequent and devastating natural disasters all raise demand for humanitarian assistance and disaster relief". Petra Nemcova in a famous quote said, "Humans cannot stop natural disasters but can arm themselves with knowledge: so many lives wouldn't have to be lost if there is enough disaster preparedness".

Nature is uncompromising and merciless. If humans do not respect environment and refuse to understand the impact of tempering with the laws of the nature, disasters will continue to play havoc with the humanity. To develop the deep understanding of environment and process of geological evolution of earth, environment education is mandatory. Max Mayfield, Director of the National Hurricane Centre, said, "preparation through education is less costly than learning through tragedy".

The Simeulueans, living off the coast of Sumatra, Indonesia, and the Moken, living in the Surin Islands off the coast of Thailand and Myanmar, both used knowledge passed on orally from their ancestors to survive the devastating tsunami of 2004 (ISDR 2008). The traditional wisdom passed from one generation to another is an experience of an entire civilisation that needs to be preserved. The earthquake-proof structures constructed in Himachal Pradesh and J&K with wooden horizontal runners embedded in walls have been found far more resilient than the concrete structures. It is high time traditional wisdom is revived with improved technology.

The failure of governance is one of the major reasons of the high rate of casualties during disasters. Faulty structures, unscientific infrastructure development, violation of environment laws, corruption and lack of awareness are the main reasons of the high rate of casualties. It is a criminal negligence, and NDMA should also act as a watchdog to forewarn the government about the wilful omission by citizens and state administration.

There is an old saying, "hope for the best but plan for the worst". The development of disaster mitigation culture is a social and individual responsibility. One need not go to school or education institute, but it could be done in a village or even in an urban colony. Sunderlal Bahuguna could spread the culture of protection of environment by launching Chipko Movement. Whatever forest cover is left in Uttarakhand is courtesy of Mr. Bahuguna and his awareness campaign. There is a need to document the traditional wisdom that could help in preservation of environment, prevention of casualties during disasters and actions that assist population during disaster including survival techniques.

The implementation of environmental laws is unequivocally the most significant aspect of disaster mitigation and disaster prevention. The statutory bodies have failed to strictly implement the environmental laws. Land use planning is also part of environmental protection, but there is hardly any focus on these issues. The establishment of tanneries and chemical factories in densely populated areas or river beds is an invitation for disaster. Bhopal gas tragedy is not an accident but wilful negligence and criminal act. There is a need to implement environment laws ruthlessly to reduce disaster risk.

There is no alternative to build local capacities to deal with disasters. Based on the disaster mapping, local population should be trained and educated on how to reduce effects of disasters and disaster relief at local level. It is not possible for disaster relief forces to reach each and every part of affected region on the occurrence of disaster; thus the best relief during the golden period (initial few hours) can only come from the citizens living at ground zero. This must be implemented, and resident welfare organisations and Village Panchayats should be entrusted with the responsibilities to develop local capacities. We need to nurture nature to make this earth safe for living. It is an obligation and responsibility of the state and its people to reduce and prevent collateral damages during disasters.

References

Dombrowsky WR (1998) Again and again: is a disaster what we call a disaster? In: Quarantelli EL (ed) What is a disaster. Routledge, 1998, p no 19

Environmental Wisdom in Ancient India., http://www.ecomall.com/greenshopping/eastgreen.htm. Accessed on 08 May 2016

Frank B (2015) Chile-Haiti earthquake comparison: Chile was more prepared. Huffington Post. 26 Sep 2015

International Strategy for Disaster Reduction (ISDR) (2008) Indigenous knowledge for disaster risk reduction: good practices and lessons learned from experiences in the Asia-Pacific region. International Strategy for Disaster Reduction. July 2008 P 5

Jha SM (2013) Natural disasters cost India $3.30bn in 2015: here's why we should be very worried. First Post. 12 Feb 2016

Jolly A (2013) How the hills can kill again. India Today. 5 July 2013

National Disaster Management Plan (NDMP) (2016) A publication of the national disaster management authority. Government of India. May 2016, New Delhi

Nelson SA (2014) Natural disasters & assessing hazards and risk. Research Paper for Tulane University. 20 Aug 2014

Padgett T (2010) Chile and Haiti: a tale of two earthquakes. Time. 01 Mar 2010

Pandey, J. M. (2016) Indus era 8000 years old, not 5500; ended because of weaker monsoon. Times of India. 29 May 2016

Sena L, Michael KW (2006) Disaster prevention and preparedness, Ethiopia Public Health Training Initiatives, The Carter Center, the Ethiopia Ministry of Health, and the Ethiopia Ministry of Education, LECTURE NOTES for Health Science Students, 2006

Statistical Yearbook for Asia and the Pacific (2011)

Tipson FS (2013) Natural disasters as threats to peace, United States Institute of Peace. Special Report 324, February 2013. United States Institute of Peace, Washington DC

United Nations Environment Programme (UNEP) (2008) Opportunities in environmental management for disaster risk reduction: recent progress a practice area review: global assessment report on disaster risk reduction prepared by the United Nations Environment Programme in collaboration with the UNISDR Partnership for Environment and Disaster Risk Reduction

World Bank Group (2016) High and dry, climate change, water and economy. International Bank for Reconstruction and Development

World Bank Study (2009) Building resilient, communities risk management and response to natural disasters through social funds and community-driven development operations

Part II
Sub-national and Local Level-Perspectives

Chapter 6
Disaster Governance in West Bengal, India

Himadri Maitra

Abstract Crisis management was carried out throughout India before the enactment of Disaster Management Act, 2005. Relief and rehabilitation works were done following relief codes/manuals. Civil defence had its warden structure for SAR.

In pre-independence period, a high-powered commission appointed by Lord Lytton submitted its report in 1880 for formulating general principles and suggesting particular measures of a preventive or protective character to resolve the defects in relief work. The report suggested for a famine relief code. This was the first attempt towards governance over relief work in India.

Disaster Management Act, 2005, enacted in India, has a provision to construct a disaster management authority at national and state levels in an attempt to create a standardised disaster governance. Apart from establishing the State Disaster Management Authority (SDMA), the state of West Bengal reconstructed its Department of Relief into Department of Disaster Management (DDM) in 2007 with an added responsibility of risk management. The Government of West Bengal continues its disaster risk reduction governance through this department which was actually set up in 1961 following the provisions laid in the *Manual for Relief of Distress* (1959), Food, Relief and Supplies Department (Relief Branch) of the Government of West Bengal. This paper will evaluate the role of the Disaster Management Department of the Government of West Bengal before and after the enactment of D.M. Act, 2005, as an example of governance in DRR and its drawbacks.

Keywords Disaster • Governance • Disaster management • Disaster risk reduction

---

H. Maitra (✉)
District Disaster Management Officer, Govt. of West Bengal, Bengal, India
e-mail: maitra59@gmail.com

I. Pal, R. Shaw (eds.), *Disaster Risk Governance in India and Cross Cutting Issues*, Disaster Risk Reduction, DOI 10.1007/978-981-10-3310-0_6

6.1 Introduction

> There is now international acknowledgement that efforts to reduce disaster risks must be systematically integrated into policies, plans and programmes for sustainable development and poverty reduction...Sustainable development, poverty reduction, good governance and disaster risk reduction are mutually supportive objectives and in order to meet the challenges ahead, accelerated efforts must be made. (World Conference on Disaster Reduction 2005)

A hazard becomes a disaster when communities are not able to cope with it. It is the vulnerability of people that has a direct bearing on the intensity of the disaster. 'Disasters occur because of interaction between natural events that cause them and social, political, and economic vulnerabilities that structure the lives and the livelihoods of different groups of people' (Blaikei et al. 1995).

India is one of the most disaster-prone countries of the world. It has had some of the world's most severe droughts, famines, cyclones, earthquakes and chemical disasters. India was, until recently, reactive and only responded to disasters and provided relief from calamity. It was a relief-driven system.

India also has the world's oldest famine relief codes. A high-powered commission appointed by Lord Lytton submitted its report in 1880 for formulating general principles and suggesting particular measures of a preventive or protective character to resolve the defects in relief work. The report suggested for a famine relief code. This was the first attempt of governance over relief work in India. But they were based mainly on the ideas to maintain law and order and preservation of life. Once social life was secured, the responsibility to the afflicted ceased.

Traditionally the approach towards disaster management was on post-disaster relief and rehabilitation, and very little effort was concentrated on disaster mitigation, preparedness or risk reduction measures. This approach has witnessed a sea change now, recognising disaster management as a development issue. Under modern conditions, not only preservation of life but the maintenance of economic health of the people has devolved upon the state. In recent times, there has been a paradigm shift, and India has become or is becoming more proactive with emphasis on disaster prevention, mitigation, preparedness and risk reduction.

The state of West Bengal, an eastern regional state of India, is vulnerable to natural calamities like flood, cyclone, hailstorm, thunder squall, drought, landslide, erosion and earthquakes because of its varied geomorphological, climatic and seismic conditions. Floods and cyclonic storms occur almost every year in different parts of the state and inflict huge loss of life and property causing myriad hardships and suffering in the lives of the people. These natural disasters strike at the very root of the economic growth of the state.

Apart from establishing the State Disaster Management Authority following the provisions laid in the D.M. Act, 2005, the state government reconstructed its Department of Relief into Department of Disaster Management in 2007 with an added responsibility of risk management. The Government of West Bengal continues its disaster risk reduction governance through this department which was actually set up in 1961–1962, following provisions laid in the *Manual for Relief of Distress* (1959).

This paper will evaluate the role of the Disaster Management Department of the Government of West Bengal before and after the enactment of DM Act, 2005, as an example of governance in DRR and also its drawbacks in managing disasters of the state of West Bengal in the perspective of India's institutional environment.

6.2 Disaster Risk and Governance

6.2.1 Disaster Risk

'Disaster risk' is defined as the probability value of losses (deaths, injuries, property, etc.) that would be caused by a hazard. Disaster risk can be seen as a function of the hazard, exposure and vulnerability. Disaster risk is a vague idea to the people whose livelihood is already at risk. Benefits of development do not reach them. They have no savings, no assets, no entitlements, no endowments, no access to power and no ownership to social structure. But poverty only is not the vulnerability. Poverty can be wiped out with funding the poor people. But vulnerability cannot be reduced in that way. Poverty is one of the crucial factors, but there are other determinants of vulnerability.

Reduction of risk is the focus area to make a society resilient, but disaster resilience is not itself a straightforward concept. Resilience of a part of the community may be achieved by exploiting the other part of the community. The process of building a resilient community without considering the equity creates instability in the society which in turn generates another type of vulnerability in the society. In the broader domain of good governance, this critical issue can be taken care of.

The impact of disasters on economic growth has propelled up the issue of disaster risk management in the international political agenda. Over the next few years, disaster risk and disaster losses are likely to increase as more people and assets are being located in vulnerable areas. These trends are prone to pose significant challenge for achieving the development goals.

Disaster risk management is a strategic management process. Environmental threat is uncertain and cannot be sensed directly. Risks come into being with an unintended consequence. There should be more than one option of opportunities to tackle it. Current governance system of budgeting and 5 years of planning fail to match these uncertainties of hazards. Risks become more complex than previously thought, and modern society demands proper services. The established model seems not serving the purpose, and a better framework is needed. Risk identification, risk reduction and risk transfer are the mainframe of disaster risk management. These should be supported by effective governance (e.g. legislation, policies, planning, legal frameworks, etc.), as well as institutional capacities at national to local levels, supplemented by effective information and knowledge sharing mechanisms among different stakeholders.

6.2.2 Governance

Political assurance manifested through good governance decides the success of disaster risk management efforts. Good governance includes the adoption and promotion of forceful and sound policies, legislation, coordination mechanisms and regulatory frameworks. It is also the creation of an enabling environment that is typified by appropriate decision-making processes to have the same opinion and useful participation of stakeholders, matched by the proper allotment of resources. Governance is generally the mechanism through which people are living in a state, believing in common principles and governing themselves by the means of laws, rules and regulations imposed by the state machinery. Society manages its economic, political and social affairs through a system of values, policies and institutions by interacting among the state, civil society and private sector through the system of governance. It also denotes those processes and institutions through which citizens and groups communicate their interests, exercise their legal rights and meet their obligations and mediate their differences. It encompasses all relevant groups, including the private sector and civil society organisations.

Governance is not government. Governance, as a concept, recognizes that power is not confined to the formal authority and institutions of government. Therefore, government is not the only actor influencing decisions and how they are implemented. It recognizes that decisions are made based on complex relationships between many actors with different priorities. Governance is conceived as encompassing a number of formal and informal arrangements and procedures. 'Governance is the sum of the many ways individuals and institutions, public and private, manage their common affairs. It is a continuing process through which conflicting or diverse interests may be accommodated and co-operative action may be taken. It includes formal institutions and regimes empowered to enforce compliance, as well as informal arrangements that people and institutions either have agreed to or perceive to be in their interest' (Commission of Global Governance 1995).

Good governance creates a responsive environment for effective disaster risk reduction through mobilising the political will and facilitating the broad participation and partnerships to make certain that political, social and economic priorities are based on broad agreement in society. Another critical role of the governance is that the voices of the poorest and most vulnerable are heard in the decision-making processes. 'Principles of good governance include broad-participation, transparency, accountability, efficiency and responsiveness. A central criterion of good governance - namely, the principle of ensuring that the voices of the poorest and the most vulnerable are heard indecisions about the allocation of resources affecting them - is essential for effective DRR and sustainable disaster recovery' (UNDP 2010).

'Supportive governance is necessary to ensure coping-capacities in communities. It is recognised that effective disaster risk reduction cannot be achieved without integrating disaster risk reduction into development planning and development processes. Governance influences the way in which national and regional stakeholders (including governments, parliamentarians, public servants, the media, the private sector, and civil society organizations) are willing and able to coordinate their actions to manage and reduce disaster-related risk. Mainstreaming of DRR is a governance process enabling the systematic integration of DRR concerns into all relevant DRR spheres' (UNDP 2010). Governance systems and processes affect the outcome of disaster risk reduction efforts. Disaster risk governance is, as Lassa pointed out, of recent development. He provides the following working definition of disaster risk governance: 'the way society as a whole manages the full array of its disaster risks. It promotes the notion that there are many overlapping arenas or centres of authority for decision-making and responsibility for disaster risk reduction the arenas may emerge as networks. Risk governance encompasses a broader spectrum of politics policies and polity at different scales and levels from global to local. It recognizes the polycentric nature of disaster risk reduction. Disaster risk governance provides the framework within which disaster risk management is to be implemented' (Lassa 2010).

In many countries, the institutional and legislative arrangements for disaster risk reduction are weakly connected to development sectors, and integration of disaster risk consideration is not facilitated into development. One of the key findings is that poorer countries with weaker institutions and governance are the places where global disaster risk is highly concentrated. Developing countries are still exposed to a great deal of high risk (UNISDR 2009). Another finding is that an institutional 'failure to address the underlying risk drivers will result in dramatic increases in disaster risk and associated poverty outcomes. In contrast, if addressing these drivers is given priority, risk can be reduced, human development protected, and adaptation to climate change facilitated' (UNISDR 2009).

6.2.2.1 Hyogo Framework for Action and Governance

Hyogo Framework for Action outlined five priorities of action of which four priorities are related to the initiatives of the State Government. The first priority is to ensure that DRR is a national and a local priority with a strong institutional basis for implementation. It is assumed that the country that develops and institutional frameworks for disaster risk reduction have greater capacity to manage risks. The second action point stated that the starting point for reducing disaster risk and for promoting a culture of disaster resilience lies in the knowledge of the hazards and the physical, social, economic and environmental vulnerabilities to disasters, and of the ways in which hazards and vulnerabilities are changing in the short and long term, followed by action taken on the basis of that knowledge. The third priority action is that disasters can be substantially reduced if people are well informed and motivated towards a culture of disaster prevention and resilience, which in turn requires the collection, compilation and dissemination of relevant knowledge and information on hazards, vulnerabilities and capacities, and the fifth is that at times of disaster, impacts and losses can be substantially reduced if authorities, individuals and communities in hazard-prone areas are well prepared and ready to act and are equipped with the knowledge and capacities for effective disaster management.

6.2.3 Institutional Weakness: Problem Area in Disaster Governance

The primary responsibility for disaster risk reduction lies with the governments. In all disasters, it is the government which steps in for relief and recovery. The government can make budgetary provisions for ex ante community preparedness and mitigation measures or relocate resources for ex post recovery and reconstructions. These opportunities could be used for disseminating knowledge and understanding of hazard mitigation among people and providing technical solutions.

Vulnerability reduction is a key investment not only to reduce the human and material costs of natural disasters but also to achieve sustainable development. Disasters push back developments. Vulnerability reduction and governability are now justifiably linked in the vision of development strategy. In the past, a great deal of attention has been paid to the physical, social, economic and environmental aspects of vulnerability, and less attention has been paid to political and institutional vulnerability. This vulnerability is understood as institutional weakness of the system. This has often been seen to be one of the prime causes of vulnerability to natural phenomenon and, in turn, even a cause underlying other forms of vulnerability. These vulnerabilities essentially measure the outcome of institutions. 'The governance arrangements for disaster risk governance in many countries do not facilitate the integration of risk consideration into development. The institutional and legislative arrangements for DRR are weakly connected to development issues' (UNDP 2010). The countries with a higher level of institutional quality usually have better governance of disaster risk. 'Good governance is not only an important "pre-requisite" of sustainable disaster risk management but disaster risk management constitutes a core responsibility and capacity of "good" government practices in disaster prone countries. Disaster risk management needs to link up with processes that aim to deepen the equity, accountability, efficiency and responsiveness of governance institutions in relation to the needs of vulnerable communities' (UNDP 2007).

Based on the lessons of institutional vulnerabilities, four overriding objectives have been identified as guiding principles of the International Strategy for Disaster Reduction (ISDR 2001):

- Obtaining political commitment from public authorities
- Growing public awareness and public sharing to reduce vulnerability
- Nurturing better understanding and knowledge of the causes of disasters through the sharing of experiences and providing greater access to relevant data and information

- Motivating interdisciplinary and inter-sectoral partnerships and the expansion of risk reduction networking among governments at national and local levels and greater involvement of private sectors, academic institutions, NGOs and community-based organisations

6.3 The Case of West Bengal: Historical Background of Disaster Governance

The colonial government had no policy to deal with emergencies, except to prevent hoarding and crime, which was followed by ad hoc relief measures. 'A despatch to the Govt. of India in 1867, the British Secretary of State Stafford Northcote began by admitting that although the "melancholy loss of life" was due to "natural and inevitable causes" – "there has been a most unfortunate want of foresight and the energy on the part of those who were charged with the administration of the province where it occurred". "The rest of the despatch is a stinging assessment of the various failures of the Board of Revenue and its faulty recommendations, but it is clear that the disaster of famine, for Northcote, is here an administrative or governance issue"' (Mukherjee 2013).

It was the Famine Commission appointed in 1878 which resulted in the first famine code and to be suitably adapted in different regions of British rule. These codes evolved under the influence of two subsequent Famine Commissions in 1898 and 1901, to provide comprehensive institutionalised guidelines to colonial administrators. These included instructions to anticipate famines and to save life but explicitly at the lowest possible cost to the exchequer, by providing employment at subsistence wage and 'gratuitous' relief to the 'unemployable'.

The Bengal Famine Code, 1913, and the *Bengal Famine Manual*, 1941, were excellent codifications. But they were based mainly on the ideas to maintain law and order and, once social life was secured, the responsibility to the afflicted ceased. British codes were explicit in casting a duty on public officials to spend the minimum that was necessary, only to prevent the loss of lives.

6.3.1 Post-Independent Era: The Case of West Bengal After Partition

As already stated, India has probably the world's oldest famine relief code from 1880s. That time famine was a recurring occurrence, and the British government prepared a relief code giving guidelines to how much calamity relief to be provided to each family after a famine. Over the passage of time, calamity relief system had a lot of experience, and it was well developed and documented in the government records. After India got its independence from the British in 1947, India continued with the relief commissioner system. However, the Government of India has suggested the state governments to redesignate them as secretary of disaster management (erstwhile relief).

Under modern conditions, however, not only preservation of life but the maintenance of economic health of the people has devolved upon the state. After the independence, it has no longer become the policy of the government 'to wait upon events' but to approach events so that distress can be avoided. The *Manual for Relief of Distress* stated in section 101 that 'the Unit of relief organisation should be the Gram-Panchayats [panchayat means assembly (ayat) of five (panch) and gram means village] where Panchayats have been set-up and Union in the rest of the areas; when distress is anticipated, Gram-Panchayats or Anchal Panchayats or Unions in an area (as the case may be) should be formed into groups or charges each under a Relief Officer. Relief Officers should usually be of the rank of Sub-Deputy Collectors to be assisted by Supervisors of Panchayats in areas where the Panchayats have been set-up and by other suitable Officers in the rest of the areas as the District Officer may decide' (Manual 1959). Based on this principle, a separate departmental set-up named as 'Department of Relief and Social Welfare' was created in 1961–1962 for relief management in West Bengal.

6.3.2 Institutional Set-Up: Disaster Management Act, 2005

India has a federal system with the Government of India at the federal level. For the administrative purpose, India has been divided into many jurisdictions known as states and union territories.

Disaster management is the responsibility of the state government, facilitated by the Government of India. The states and union territories are divided into districts. Each district is administrated by a collector and district magistrate (same person performs both the duties). A collector of the district is the administrative head for all matters within the district.

On December 23, 2005, the Disaster Management Act, 2005, was enacted by the Government of India. The Disaster Management Act, 2005, provides for constitution of a number of institutions at national, state and district levels. At the national level, the act establishes the National Disaster Management Authority headed by the prime minister. The model is replicated at the state level with State Disaster Management Authorities. The National Disaster Management Authority is responsible for laying down policies and guidelines on disaster management. State Disaster Management Authorities headed by the respective chief ministers also lay down policies and plans for disaster management in their respective states. Each state has a disaster management cell, located generally in the State Administrative Training Institutes. Major funding for the faculties of the disaster management cell comes from the central government. Each cell is supposed to carry training in disaster management and prepare plans and documents. Only in West Bengal a dedicated relief set-up was restructured into a disaster management set-up.

The act also provides for a National Institute of Disaster Management for human resource development in the field of disaster management. Besides these, a National Disaster Response Force has been constituted.

District Disaster Management Authorities are the planning, coordinating and implementing bodies for disaster management in the district. Under the Disaster Management Act, 2005, each district is supposed to have a district disaster management committee, district EOC and responsibilities for training and drills, disaster prevention, preparedness and mitigation activities.

6.3.3 Role of Panchayati Raj Institution in Disaster Governance

The 73rd Amendment to the Constitution of India in 1992 was a landmark legislation to accelerate democratic decentralisation in the country. The legislation ushered in a new era to strengthen the panchayats by devolving powers and responsibilities to the elected bodies at the district (Zilla Parishad), intermediate (Panchayat Samiti at block level) and the village level (Gram Panchayat). The powers and functions of panchayats emanate from Article 243(G) of the Indian Constitution which empowers the state legislature to endow such powers and functions necessary to enable them to function as institution of self-government. In West Bengal Panchayat Rules, 2004, the Upa-Samiti (subcommittee) of Gram Panchayat on finance and planning has been assigned to deal with the subject of disaster control management. This is a practical approach and integration of disaster with development and planning.

DRR is a governance issue, but it has not been adequately addressed from the institutional point of view, particularly identifying institutional mechanism to manage disasters at local and community level. Panchayats as a constitutional grassroot unit of governance are playing a prominent role in managing disasters at the local level in West Bengal. Moreover, they are an effective institutional arrangement to deal with the emergency response within its jurisdiction.

6.3.4 Department of Disaster Management, Government of West Bengal (Maitra 2015)

Only after the enactment of Disaster Management Act, 2005, the Department of Relief under the Government of West Bengal was restructured into the Department of Disaster Management in 2006. So, only the duties and responsibilities were enhanced, but the institutional mechanism had already been there to accept the changed structure. There had been no requirement for staffing or funding; only it required to build capacity and modernisation. The state government promptly turned attention to capacity building of relief officers (now disaster management officers) at all levels of administration.

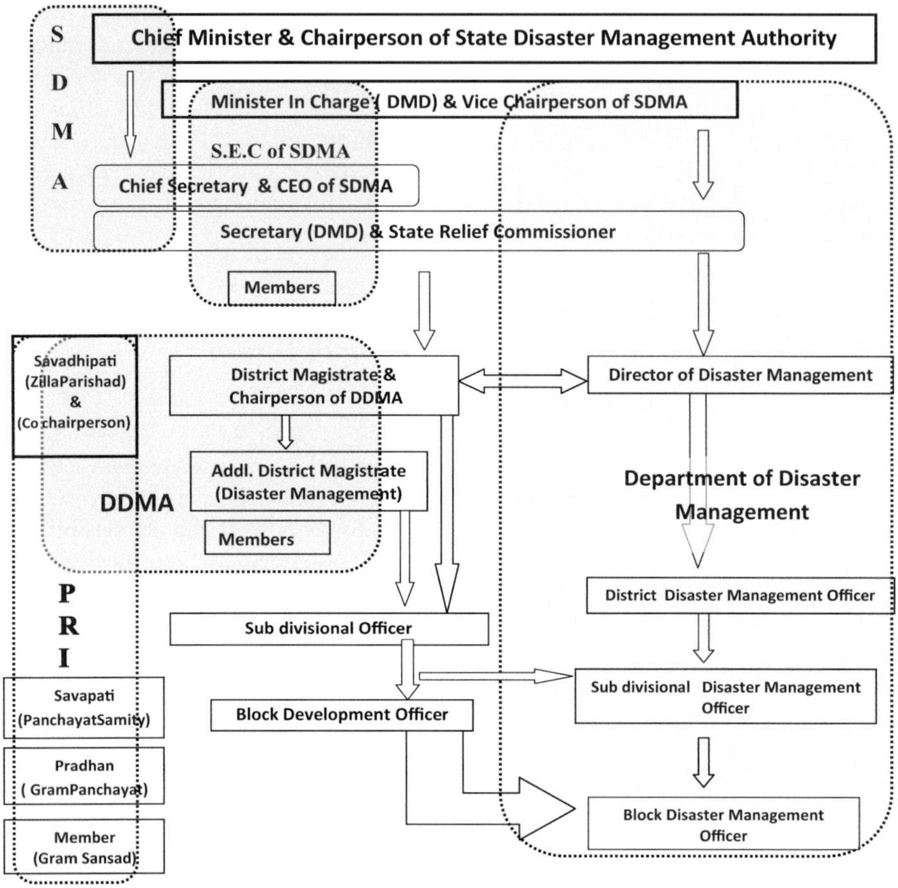

Fig. 6.1 Present set-up of disaster governance in West Bengal

The Disaster Management structure of West Bengal is represented in the framework in Fig. 6.1. As the actors of both State Disaster Management Authority (SDMA) and the Disaster Management Department (DMD) are common, no gap is generated in decision-making and operational management. The peculiarity of SDMA structure in West Bengal is that it provides a space for Panchayati Raj Institution (PRI) in its set-up. With more than 50 years of organisational experience in the field of disaster response, the Government of West Bengal strengthens its mechanism with the reconstruction of the Disaster Management Department (DMD) from the Department of Relief to address the institutional weakness with all its organisational memory of earlier relief set-up remaining intact. It had so far been responsible for crisis management in the aftermath of any natural calamity and providing various kinds of succour to the life of the citizens of this state. Long before the disaster management got its impetus throughout India, the then Relief Department would follow different operational mechanisms which afterwards came under preview of disaster management. In fact, the *Manual for Relief of Distress* (1959), the coded guideline for the Relief Department, was no less than a manual for disaster management, and following different codes of the said manual, the Relief Administration of this state had been functioning years along.

6.3.5 Hazard Profile of West Bengal, A Regional State of India

The northern mountainous terrain of the state covers Darjeeling and part of Jalpaiguri districts. The southwestern peninsular tract is, in fact, the eastern extension of the Chota Nagpur Plateau. The general slope of West Bengal is towards south, though in the southwestern part, an easterly slope is conspicuous. The river Ganga flowing from west to east (as Padma, through Bangladesh) and to south (as Bhagirathi) constitutes the major drainage of the central southern part of the state. The drainage is controlled by a number of rivers and streams.

Bengal is associated with delta. The South Bengal Delta, one of the biggest of its kind, is formed by the combined effect of two major rivers, the Ganga and the Brahmaputra along with their tributaries. An area of approximately 65,000 km^2 of the Bengal Delta is situated in the state of West Bengal. The surface feature of the area is gentle to moderately flat with certain micro-relict of at places. Structurally, the Ganga-Brahmaputra Delta can be divided into two parts. The slope in the north-western part is only 2–3°, but it increased to 6–12° in the southeastern part. Geologists gave the name of this line as 'eocene hinge zone'.

6.3.6 Historical Data of Disasters in West Bengal

6.3.6.1 Flood

Every year the state faced major or minor flood, be it a flash flood at the northern part or normal flood at the southern part of it. Large floods in West Bengal are presented in Table 6.1.

6.3.6.2 Cyclone

There is a huge coastal area in West Bengal on Bay of Bengal. Cyclones play havoc in this area. Table 6.2 provides the history of damaging cyclones in West Bengal.

Table 6.1 Statistics of flood in West Bengal (State Disaster Management Plan 2014)

Period	Description
1978	Major flood affected 13 districts, 1370 human lives were lost, about 14 lakh houses were damaged
1986	Flooding due to heavy rains in some areas of Kolkata, Hooghly, Howrah, 24-Parganas and Midnapore
1988	Monsoonal rains caused flooding in areas of Balurghat and Dinajpur lying under the purview of the Ganges and Churani rivers
1991	Flash floods caused damage 35,000 houses
1995	Flooding triggered by heavy rains caused erosion, severe agriculture damage and outbreak of diseases
1998	Monsoon rains caused flooding of the Ganges River
1999	Tropical cyclones caused destruction of an estimated number of 1500 villages. Floods due to brief torrential rains affected areas of Kolkata, Burdan and Birbhum
2000	Besides flash floods triggered by incessant torrential rains, disaster is also accredited to the opening of sluice gates of dams. The fatalities counted to the tune of 1262, besides affecting millions of people
2002	Flooding in Jalpaiguri, Cooch Behar and Jalpaiguri in North Bengal due to monsoonal rains. Flash floods swamped 10 villages, causing four death and 11,000 displacements
2003	Monsoonal rains caused floods affecting the regions of Darjeeling, Jalpaiguri, Malda and Murshidabad
2005	Heavy rains caused floods in many areas. About 3000 coastal villages were inundated and 60,000 huts and many roads washed away
2005	Heavy monsoon rains triggered flash floods and landslides
2006	The regions of Birbhum, Burdwan and Murshidabad were affected mainly from continuous monsoonal downpour
2006	Monsoonal rains and tropical cyclone-driven storms in the Bay of Bengal hit India and Bangladesh. West Bengal recorded 50 deaths, 300 were injured and 30,000 mud houses destroyed. Heavy rains left large parts of Kolkata City under water, subsequently 2000 people were evacuated from the city
2007	Severe cyclonic storm affected Kolkata and several other districts. Eighty-three deaths were reported and millions of people were marooned in 3000 villages in coastal areas of the state
2007	Heavy rain from tropical depression in the Bay of Bengal caused flooding leading to 51 deaths, affecting 3.2 million people
2013	Moderate flood occurred in 4 districts, affected 21 lakh population, 84,720 houses were damaged

6.3.6.3 Earthquake

All the seismic zones exist in West Bengal. The largest instrumented earthquake in West Bengal was on *15 April 1964* at Sagar Island, West Bengal, Mb 5.2 (4). 21.600 N, 88.700 E, D = 036.0 kms, OT = 08:35:27, felt in southern West Bengal and eastern Orissa. The maximum intensity in Kolkata was V. Historical records of earthquakes are given in Table 6.3.

Table 6.2 Statistics of cyclone in West Bengal (SDMP 2014)

Period	Description
1916	Extensive damage reported, however, no estimation of deaths
1942	About 5 m big surge reported at Midnapore (64 km upstream in Hooghly River). Overall 15,000 deaths reported
1956	Caused flooding in Midnapore District and also damage to agriculture due to saline water intrusion
1971	Sixty people died and thousands of houses collapsed
1974	Cyclone storm over land with maximum wind speed of 139 km/h caused floods in several districts. Seven deaths reported
1976	About 2.5 m high surge along with 1.4 m tide caused 40 deaths
1981	Caused loss of five launches in the bay and damage to many houses in Midnapore District
1984	Caused damage in Midnapore District
1989	Sixty-one persons died and thousands of cattle perished
2002	Caused 78 deaths along with the destruction of agriculture crops and property
2009	Cyclone AILA caused huge damage to the North 24 Parganas, South 24 Parganas, Pubbo Midnapore and Kolkata. Death tolled to 146
2013	Cyclone caused by Phailin. Death of 17 persons in the districts of Purba Medinipur, Paschim Midnapore and Burdwan

Table 6.3 Statistical data of earthquakes in West Bengal (http://nidm.gov.in/PDF/DP/WESTBENGAL.pdf)

Period	Description
23 June 1976	Mb 5.0 – This earthquake was located in the Bay of Bengal off the Ganga Delta
19 Nov 1980	Ms 6.1–8 – People were injured and Gangtok was damaged. Felt throughout eastern India, Bangladesh, Bhutan and Nepal
26 March 1981	Mb 4.9 – This earthquake was located along the India-Bangladesh boarder to the east of Canning, West Bengal
12 June 1989	Mw 5.7–1 – Person was killed and 100 injured in the Banaripara area of Bangladesh. Felt in much of eastern Bangladesh including at Chittagong and Rangpur. It was also felt in Meghalaya
28 Nov 2005	Mb 4.7 – A light earthquake occurred in the Ganga Canyon in the northern Bay of Bengal, off the Sundarbans in India

6.3.6.4 Landslide

Hilly region of the northern part of West Bengal is prone to landslide. Historical records of landslide in West Bengal is given in Table 6.4.

The western part of West Bengal has been affected by drought at regular intervals, mainly due to deficient rainfall and adverse soil conditions.

Table 6.4 Statistics of landslide in West Bengal (http://nidm.gov.in/PDF/DP/WESTBENGAL.pdf)

Year of occurrence	Location	Description
1998	Lodhoma, Labdah, Kolbong, Simbong	Population affected, 40,000; death, 21; house damaged, 3498
1999	Samalbong, Sittong, Tindharia	Population affected, 321; death, 13; house damaged, 1697
2000	Halbalay	House damaged, 50
2002	Goethals, Sepoy Dhura, Tista Valley, Poomung, Lamahatta, Samdong, Daragaon, Gayabari	Population affected, 72; death, 07; house damaged, 05
2003	Dabaipani, Jinglam	
2004	Sakhibhir, Maneydara, Labadh	Population affected, 103,417; death, 25; house damaged, 5499
2006	Bloonfield	Population affected, 3029; death, 13; house damaged, 2375
2007	Gairigaun, Sindepong, Mangalpuri, Nalban, Margaret's Hop, Beachgaon, Singringtam, Rangli, Upper khani, Mangal Basti, Lamini, Bindu, Paten Godak, Panjabi Golai, Shermali, Bagdogra, Kharny, Bhote Basti, Mirik, Kurseong municipal area	
2008	5 blocks and 1 Municipalities are affected	Population affected,16,674 Death, 03; house damaged, 3173
2009	555 villages in 8 blocks and 4 municipalities	Population affected, 1,45,758; death, 41; house damaged, 26,595

6.3.7 The Mechanism of Disaster Governance in West Bengal

The mechanism that was followed is summarised below:

- At block level, in February/March, a surveillance team comprising of a block relief officer (now block disaster management officer) and other concerned officers surveys the banks of rivers and canals and reports it to the block development officer.
- BRO (now, BDMO) would prepare the action plan and contingency plan. The contingency plan of other departments is also consolidated by the BRO into that block action plan.
- Block level pre-monsoon preparedness meeting is held with BDO, Savapati (Panchayat Samiti), BDMO and all concerned departmental officers. The

preparedness measures as taken by the different concerned departments are assessed. A committee is formed to monitor and functionalise the control room.
- The plan is submitted to the district magistrate and the district also conducts pre-monsoon preparedness meeting under the supervision of the district magistrate. A district level consolidated action plan is prepared by the Disaster Management Section of the district.
- These district plans are submitted to the Relief Department (now, DMD).
- Normally in June, at the state level, a meeting is convened by the secretary, DMD. The chief secretary heads the meeting and all concerned departmental secretaries used to be present there.
- In any other disaster, the crisis management committee headed by the chief secretary would sit for a meeting to discuss the emergency response mechanism.
- In post-disaster situation, block disaster management officers (BDMO) prepare FIR/SITRAP under the guidance of BDO after assessing immediate needs with the help of PRI members. It is the duty of BDMO under the supervision of BDO to monitor the whole process of distribution of relief, preparation of reports and returns, etc.
- Post-disaster rehabilitation needs assessment is also done by the BDMOs.
- The top-down chain of command and control from the department to block unit through the district disaster management officer and subdivisional disaster management officer is properly maintained.
- The bottom-up communication chain is also maintained in the same line of command.
- In urban areas, subdivisional disaster management officers play a nodal role in case of any disaster.

This was the system that has been the driving force behind the success story of Government of India-UNDP-sponsored Disaster Risk Management Programme (DRMP, 2002–2009) of the state in the ten project districts. The BDMOs easily adapted the mechanism that DRMP initiated. They were imparted training through the Administrative Training Institute and other resources to become a trainer and facilitator in community-based disaster management.

In three districts, MHA-UNDP-sponsored DRR programme was implemented under the supervision of DMD. In this programme also, the departmental officers have taken the nodal role. Capacity Building Programme initiated by the 13th finance commission has also been implemented. A resource pool has been generated with the departmental officers.

The Disaster Management Policy of the state and the *Manual for Disaster Management* have been published by DMD in 2011. It is a political commitment to the public which is addressed through increased inter-sector coordination, the adoption of risk management strategies and allocation of appropriate resources (State Disaster Response Fund). Disaster reduction is dealt with as a primary policy issue and is pursued as a cross-cutting issue aimed at policy integration among various sectors such as agriculture, food security, health and education. These are not only pursued at the top level but up to the block level which has only been possible for the existence of a set-up at that level. Through this institutional arrangement, programmes related to awareness generation are initiated and addressed through public information, education and multidisciplinary training.

DMD is involved in assessment and analysis of issue-specific (gender, disability) socio-economic impact of disasters, the construction of database on disasters, the formation of suitable coping strategies for different social groups, the introduction of early warning systems and the promotion of relevant scientific research, which takes into account of both indigenous or traditional knowledge and the development and transfer of new knowledge and technologies.

The department is directly involved with the State Inter Agency Group (IAG – a common platform of DMD and NGOs) and promotes its activities for effective coordination mechanism with INGOs/NGOs/CBOs for DRR as well as the incorporation of DRR into planning process. After super cyclone Aila, DMD and State IAG undertook Unified Response Strategy to coordinate the whole response process.

6.3.7.1 Pre-positioning and Distribution of Relief Materials

The department is vastly experienced in pre-positioning and distribution of relief materials in pre- and post-disaster period as a measure of preparedness. It is noted in various studies that relief agencies are confronted with a problem of identification of what relief material has been received and where it is and to determine exactly how useful the supplies are. Situations such as this bring out the following problems:

1. The means of transportation and time are limited.
2. Technical information of supplies is missing,
3. The donor and mass media receive a negative impression.

The directorate of disaster management is designed to operate under the control of DMD. The principal facilities of this system are:

- Defining parameters to be used such as reception site and delivery directions and defining the main users in consultation with and the needs assessment of the concerned districts
- The presence of an established and experienced departmental chain of command
- The integration of information sent by the block level unit
- Supporting consultations and making up reports which serve to lend the support to the decision-making process and promote internal coordination of the structure
- Presence of organisational memory that supports future decision-making process in case of pre-positioning of materials

The block level unit (BDMO) is designed to access needs and to work at points of entry, local warehouses, distribution points, etc.

The main work accomplished at the government relief store under the control of director is:

- The separation, identification and classification of supplies through labels
- Random sample checking of items
- Balance of inventory
- Inventory follow-up of other subordinate warehouses
- Making up various reports regarding stock, distributions, etc.
- Consolidating data to send to the DMD

The block, subdivision and district units maintain stock registers for audit purpose, and also the stock position of these units are shared with the central body on regular basis.

6.3.8 *Institutional Arrangement of Disaster Risk Management Programme (UNDP)*

Development of holistic policies and programmes, strategies, legislation, as well as institutional mechanisms and administrative capacities to manage risks are areas where UNDP can make a difference, drawing on its expertise in key development areas such as democratic governance while making use of its considerable experience in risk management. Furthermore UNDP can play a key role in assisting governments to steer and harmonise the various multilateral and bilateral agencies interested in strengthening DRM capacities.

The Disaster Management Department is based upon hierarchic 'command and control' structures. The challenge was to assist these often inward-looking and closed systems to open up and make more use of existing resources in the government, civil society and private sector. The Government of West Bengal has designated the Disaster Management Department as the State Nodal Agency for implementing this Government of India-UNDP-sponsored DRM programme (2002–2009) in the ten selected districts of the state. Subsequently, a State Steering Committee has been constituted under the chairmanship of the chief secretary of the state in May 2003. Appropriate policies, institutional systems, legislation and regulatory frameworks for DRM are not created overnight. This project has been successful in generating and sustaining a range of outputs across the geographical region on different aspects of disaster risk management at various levels. More than 26,000 villages are targeted by the programme in ten project districts.

Following a multidimensional approach in pursuit of the programme objectives, achievements have been made in institutionalising disaster management at various levels. While the programme pervades all sections of the society in rural and urban habitats, it is sharply focussed on more vulnerable disadvantaged groups including women. It found a way to engage with a wide cross section of stakeholders and build vertical linkages between local, regional and national levels while simultaneously strengthening communication and cooperation between them.

Some notable highlights of programme impact includes significant formulation of disaster management teams (DMTs) and disaster management committees (DMCs) at various levels, training of its members, satisfactory involvement of women in programme activities and witnessing community preparedness by holding mock drills at village level.

Human resources from the state officials up to the block level have been designated under the leadership of the joint secretary of the Nodal Department in implementing the programme at various levels. Up to the block level, there are departmental officers. One hundred ninety-four blocks have redesignated block disaster management officers, subdivisions have subdivisional disaster management officers, and ten districts have district disaster management officers; all are fully discharging their tasks for successful implementation of the programme. UNDP played 'a key role in supporting communication between local level initiatives and thereby increase the relevance of policy-making, planning and resource allocation of governments' (UNDP 2007). In replication of this programme, W.B. Disaster Risk Management Programme (2009–2011) has been launched in six districts by the Government of West Bengal itself (separate budget allocation) after understanding the impact of the DRMP. This was only possible with the institutional setup of DMD up to the block level and horizontal linkage with PRI bodies and CBOs/NGOs at grass root level.

6.4 Challenges

- DRR is a concept- and a knowledge-based work. Accordingly, the progress of DRR depends upon the conceptual framework of the workforce involved in this department. But the recruitment of workforce is done through generalised recruitment system which is meant for rendering day-to-day administrative service.
- DRR is not only an operation level job; it is a strategic management process. The understanding of the whole process requires specialised skill and awareness.
- The visibility of the DRR is less than any other social development work. So, there is always a chance that these works are viewed as departmental works, not mainstream government works, and as such it is neglected by other sector departments.
- Conducting the relief work successfully, the responsibility of the administrative authority ceased after a certain point when chance of breaking law and order is minimised.
- The control mechanism of state based upon diplomacy and management rather than traditional process of coercion and enforcement is practiced at the grassroot level by this block level setup using strategies like shared norms and habits, informal agreements, negotiations, etc. with the help of local elected bodies and voluntary sectors.

6.4.1 Bureaucracy

The Global Governance Survey (GGS) project defines *bureaucracy* as an arena that refers to 'all state organizations engaged in formulating and implementing policy as well as in regulating and delivering services' (Hyden et al. 2003). Bureaucracy can either be disabling or enabling conditions for overall disaster risk management performance with its action or inaction, and embedded within it is a certain institutional quality. Takeda and Helms (2006) argue that 'the main feature of bureaucracy is based on clearly defined objectives where in the case of a disaster; it is also designed to facilitate "rational" response in a highly irrational and chaotic set of circumstances' (Lassa 2010). Rationalised disaster management cycle ceases to play in times of disaster. In reality, 'bureaucrats involved in disaster management follow different rules or simply not follow the rules'. One attribute to disaster governance relates to strategic planning, which requires creative thinking in planning and decision-making. 'By chance, there may be few creative bureaucrats exist out there in a particular context of time and place' (Lassa 2010). Therefore, it can be said that creative bureaucracy is often not a by-design product but rather an accidental fact within a specific context. 'In the context of disaster risk, bureaucracy is defined as a form of mental model and rationality that shapes the structures that comprise a set of regulations put in place to control disaster risk and disaster risk management' (Lassa 2010). Preconceived mental model and rationality tend to inaction. Generally, bureaucrats are dedicated to hierarchal orders of authority, not to the fulfilment of institutional objectives, and therefore have less ownership in DRR. The government is the main stakeholder of DRR, but bureaucrats are not; normally they have no incentives to work towards the common good of the community.

In this background, the success of the department depends upon the creative planning of the policy makers and the political will to make decision; it does not depend on the system. At the operational level, if codes and standard procedures are maintained, the immediate relief work is not hampered. But, in the long run, sustainability of the whole DRR process will be in question if it is confined to regular departmental work and try to address it with an administrative outlook brought forward from colonial system of administration.

6.4.2 Institutional Plurality

'Risk Management strategies to avoid or to limit adverse effects of hazards is in the form of engineering approaches and non-engineering measures such as land-use planning, building codes, and risk transfers and sharing. For every DRR option, there is a legal context in which such risk reduction activities take place and have legal propositions' (Lassa 2010). The context of legality is different from that of the developed world on account of the legacy of the colonial legal system which still coexists with our traditional, religious or new legal reform systems. The significance of understanding institutional plurality for risk reduction strategy is very much related to the issues of governance and institutions, termed by the HFA as a 'strong basis for implementation'. For instance, the issue of climate change by all accounts could not be managed through one single form of institution such as formal law but may coexist with traditional laws. In addition, risk management exercises such as risk assessment and mapping are powerful tools but are ignored by technical persons. If we ignore the fact of institutional plurality, it will lead to institutional clashes, which would hold back the implementation of risk reduction.

In a democratic system, PRI bodies are elected, and not necessarily one political party with same ideological background is elected at all the layers of the PRI body. This may create ideological and policy differences among different layers in implementing of DRR processes. After all, DRR is mainly a subject related to political economy.

The success of community-based planning and implementation depends upon the lowest level of PRI body. If this is not integrated into mainstream planning process, and only a top-down approach is taken, defying the plurality of the democratic system, the community for whom the activity is planned, will not get ownership and the sustainability will be in question.

6.4.3 *Risk Governance*

The DRMP was a one-dimensional non-structural project, and considering the context of that time, it had a great impact on the community. It was possible for the DMD to make it a success at that time. Those activities best supported disaster management, in particular preparedness and response. The Relief Department is progressively replaced by DMD. But the drawback is that other line departments use to assume that disaster management is the concern and responsibility of only this particular department. They use to involve themselves at the time of crisis. It becomes a challenge to sensitise all the development sectors that DRR is for all. DRR is multidimensional – not only preparedness and response or sensitisation and awareness. DMD is multilayered, but it is not a multi-sector organisation. DRR has various cross-cutting issues, the responsibility for which lies within different sector-wise departments. Unless the department creates a good networking with different stakeholders and pursue other national and local departments to integrate the DRR issues into their plans, impact and influence on national or local sector developments may be minimal.

The organisational structure of DMD lacks in technical capacity, if not supported by the provisions laid in SDMA. The ultimate responsibility is vested at the highest possible political level through SDMA structure.

6.5 Lessons Learnt

Governance defined by political commitment and strong institutions is identified as a key area for the success of effective and sustained disaster risk reduction. Good governance will:

- Elevate disaster risk reduction as a policy priority – disaster reduction is dealt with as primary policy issue.
- Allocate the necessary resources for disaster risk reduction – the department from its own budget provision allocates resources, e.g. funding for WBDRMP-II; there is also the State Disaster Response Fund (SDRF).
- Enforce implementation of disaster risk reduction measures and assigns accountability for failures – involved PRI at grass root level implementation of DRR.
- Facilitate participation from civil society – DMD is directly involved in State Inter Agency Group. The major components of governance for disaster risk reduction which is maintained by DMD are:
  - Policy and planning
  - Legal and regulatory frameworks
  - Resources
  - Organisation and structures

Disasters are not viewed as isolated events which break 'normalcy' for a while. The policy makers of this state did not adhere to the dominant perspective belief, even from the early 1960s, that disaster would supposedly be a provisional break that could efficiently be reversed with emergency interventions largely based on rescue and relief. Contrary to the prevailing practice in other parts of India to manage emergency situation by deputing officials from other departments, this state administration created this dedicated set-up up to the field level. This extended set-up addresses the operational management even in post-disaster build back period and ensures the governance to the community. The field level platform smooths the interfaces between the government and the communities and helps the communities to have direct to access the government and feel a sense of ownership over government.

6.6 Conclusion

The disaster governance of West Bengal evolved through a culture of decentralised practices. From its inception, the DMD extended its branch down to block level, which is the lowest level administrative setup. After the 73rd and 74th constitutional amendments, the department collaborates with local bodies with a view to enhance the last mile connectivity to shoulder the responsibility to the afflicted properly. The annual expenditure on this set-up during the past several years is due to this responsibility which governments have taken upon themselves. This expenditure on relief is more in the nature of social service payments. The state, though more about making policy decisions, is no less about service delivery with this set-up. The question of (a) whether this set-up continues to handle the changes in wider society and has a positive impact on it with a normative or positive orientation, as the demand is for better ways of handling the risk challenges, and (b) whether in a developing country like India (and more specifically, a regional state like West Bengal) a mix-and-match system of linear chain of command and a more complex structure based on networks (Walker et al. 2010) is more useful or not is a matter of an in-depth study which is beyond the scope of this paper.

References

Blaikei P. et al. (1995) At risk: natural hazards, people's vulnerability and disasters Routledge, 1st, 4, Commission of Global Governance Our Global neighbourhood, Oxford University Press.

Commission on Global Governance (1995) Our global neighbourhood. Oxford University Press

Hyden G et al (2003) "The bureaucracy and governance in 16 developing countries", world governance survey discussion paper 7. Overseas Development Institute, London

Hyogo Framework of Action 2005–2015 (2005) Extract from the final report of the World Conference on Disaster Reduction, 18–22 Jan, (A/CONF, 206/6), p.1.

ISDR, (2001), Inter-agency task force, framework for action for the implementation of international strategy for disaster redution

Lassa, J.A., (2010) Institutional vulnerability and governance of disaster risk reduction: Macro, Meso and Micro Scale Assessment, Nov 2010 [hss.ulb.uni-bonn.de > 2451-engl (Accessed on 4th Apr 2016)

Maitra H (2015) Good practice in disaster governance. Disaster Response Manag LBSNAA, Mussoorie. 2015 3(1):73

Manual for Relief of Distress, (1959), Food, relief & supplies department (relief branch) of Government of West Bengal

Takeda MB, Helms MM (2006) "Bureaucracy, meet catastrophe": analysis of the tsunami disaster relief efforts and their implications for global emergency governance. Int J Publ Sect Manag 19(2):204–217

Upamanyu Pablo Mukherjee (2013) Natural disaster and victorian empire, Palgrave Macmillan, p.30

State Disaster Management Plan (SDMP) (2014) Disaster management Department, Govt. of West Bengal

UNDP (2007), Governance for disaster risk management, 'How to Guide', BCPR, Conference Draft

UNDP (2010) Disaster risk reduction, governance and mainstreaming, Bureau of Crisis Prevention and Recovery

UNISDR (2009) Global assessment report on disaster risk reduction. United Nations, Geneva

Walker, Gordon et al., (2010), CapHaz-net WP 2 report risk governance (Version 3, Apr 2010)

World Conference on Disaster Risk Reduction (2005, 18–22 January) Kobe, Japan

Chapter 7
Multi-stakeholder Support in Disaster Risk Governance in India

Priya Namrata Topno

Abstract Governance is for effective decision-making process which is usually done with the collaborative governance. The multi-stakeholder domain in disaster management includes representatives from different ministries and departments. A separate institutional structure is established that looks upon the disaster management framework having legislative measures for managing disaster and government to execute decisions. Disaster governance is an emerging concept which is rudimentary at this stage. It is influenced by the social, economic, and political forces. Forming a governance system is a protracted process. Disaster governance is related to environmental governance and risk governance. In disaster context, governance includes the legislations, regulations, norms, and governmental planning. The governance functions are mostly carried out by the public entities but this approach has shifted toward the public-private partnership domain. Governance is the key for disaster risk reduction and sustainable development while there is a need to strengthen the disaster risk governance. The preventive measures and shift from disaster response to disaster risk reduction help in the sustainable development. This ensures that the decision-making is by mainstreaming disaster risk reduction among all the sectors for the planning processes, national policies, and plans.

Keywords Governance • Disaster risk reduction • Risk governance • Stakeholder

7.1 Introduction

> "The more government, UN agencies, organizations, businesses and civil society understands risk and vulnerability, the better equipped they will be to mitigate disasters when they strike and save more lives"- Ban Ki-Moon, Secretary General, United Nations

P.N. Topno (✉)
Jamsetji Tata School of Disaster Studies, Tata Institute of Social Sciences, Mumbai, Maharashtra, India
e-mail: priyatopno18@gmail.com

I. Pal, R. Shaw (eds.), *Disaster Risk Governance in India and Cross Cutting Issues*, Disaster Risk Reduction, DOI 10.1007/978-981-10-3310-0_7

Governance or good governance is the management of resource and policy making by means of exercising power and authority. The policy's stakeholders exercise their legal rights to achieve political, cultural, economic, and social objectives. Therefore, governance is used to denote a complex set of structures and processes of both public and private level which are associated with national administration. The Report of the Commission on Global Governance defines governance as the sum of many ways individuals and institutions, public and private, manage their common affairs. It is a continuous process through which conflicting or diverse interests may be accommodated and cooperative action may be taken. It includes formal institutions and regimes empowered to enforce compliance as well as informal arrangements that people and institutions either have agreed to or perceive to be in their interest (GDRC 1995).

7.1.1 Components of Governance

Governance includes participation, rule of law, transparency of decision-making, accountability, responsiveness, consensus seeking, inclusiveness, predictability or coherence, efficiency, and effectiveness. Participation refers to equal participation by all the members of the society as a key element of good governance, with everyone having a role in a process of decision-making. Rule of law is maintained through the impartiality and effectiveness of the legal system. It also means the protection of human rights (particularly of marginalized group), independent judiciary, having a good legal framework, verified dispute mechanism, and equal access to justice. Good governance is also based on transparency in decision-making process, which ensures availability and accessibility of free information to those involved and affected by the decision taken. The key requirement of governance also includes accountability and responsibility (EDRS and KBF 2007).

Natural hazards pose a greater threat to the entire world, in the future having devastating impact upon life and property. Over the last decade, more than 780,000 fatalities and a minimum of $960 billion property and infrastructure were destroyed due to natural disasters (UNISDR 2010). According to (UNESCO 2010), the cost of natural disaster will exceed $300 billion per year by 2050, while the projection of lives lost each year is around 100,000. In developing countries, the poor are the most affected as they possess the least resource to cope during disasters. The vulnerability of all the communities varies from place to place depending on the availability of intrinsic resources. The increase in the frequency, intensity, and scale of natural hazards leading to disasters has intense consequence for the governance of disaster risk. Many countries have disaster risk reduction (DRR) measures and established disaster-specific agencies to coordinate and manage the resources during disaster (Bang 2013). Disaster's severity is the impact of hazard upon the society and environment. The scale varies based on the choices, decisions, and the action taken by the government, professional, and community.

7.1.2 Disaster Risk Reduction Framework

After the devastating Indian Ocean Tsunami, in January 2005, the World Conference on Disaster Reduction focused on taking proactive measures to address systematically the main institutional, legal, technical, and human aspects to ensure the disaster preparedness, recovery, and disaster proof development. The Hyogo Framework for Action (HFA): Building the Resilience of Nations and Communities (2005–2015) was adopted as the outcome of the conference to deal with disaster in holistic manner. The aim is substantial reduction in disaster risk and actions taken to reduce human and socioeconomic losses due to disaster. The strategic goal of the framework is to develop and strengthen the institutions, mechanisms, and capacities to build resilience to disaster. Its first priority is to ensure disaster risk reduction as a national and local priority having strong institutional basis for implementation. Prior to HFA, the United Nations had established the International Strategy for Disaster Reduction (ISDR) to coordinate, promote, and facilitate disaster risk reduction (Darwanto 2012). The Sendai Framework for Disaster Risk Reduction (2015–2030) is a successor agreement to the Hyogo Framework for Action. Table 7.1 shows the priorities for action for both the Hyogo and Sendai Framework.

In recent years, disaster governance has become an important approach for risk reduction. The existing risk governance capacities and arrangements are unable to achieve their objectives, which increases the losses due to the existing vulnerabilities. It points out the failure of existing governance structures which includes policy, administration, and regulatory mechanism for risk management (IRDR 2014). According to Tierney (2012), disaster governance has an interrelated set of norms, organizational and institutional actors, and practices to reduce disaster impact and losses. The area of disaster governance is not only limited to the governmental setup but also includes all the stakeholders from local to global level.

Table 7.1 Priority for action for the Hyogo and Sendai Framework (Source: UNISDR 2010, n.d.)

Priorities for action	
Hyogo Framework for Action (2005–2015)	Sendai Framework for Disaster Risk Reduction (2015–2030)
Ensure disaster risk reduction as national and local priority with a strong institutional basis for implementation	Understanding disaster risk
Identify, assess, and monitor disaster risk and enhance early warning	Strengthening disaster risk governance to manage disaster risk
Use knowledge, innovation, and education to build a culture of safety and resilience at all levels	Investing in disaster risk reduction for resilience
Reduce the underlying risk factors	Enhancing disaster preparedness for effective response and to "build better back" in recovery, rehabilitation, and reconstruction
Strengthen disaster preparedness for effective response at all levels	–

7.2 Risk Governance

Traditionally, disaster risk governance has been divided among local, state, and national entities, various sectors, and bureaucratic structures. Risk governance is mostly considered for the department of disaster or emergency management, agencies, or organizations. These departments or agencies have limited interaction with other governmental department or corporate entities and civil societies. The risk governance is rarely visible during normal situation (IRDR 2014).

Risk governance includes both institutional structure and policy process which guide and control the activities of groups or communities in the societies. The trend to handle risk has been decentralized to various governmental agencies. Civil societies and nongovernmental organizations (NGOs) also play a vital role in risk governance. This ensures the multilayered and diversified sociopolitical actors to be included in the process of decision-making, risk analysis, and risk management. The diversified and multilayered actors can offer significant advantage while dealing with risk. Adequate risk governance can address:

1. The management of risk at different level
2. Overlapping between nonhierarchical adaptive and integrative risk governance system to reduce the vulnerability
3. Increase networking and enhance the knowledge of different actors

The functionality of risk governance can be understood by the classical model of risk analysis which includes risk assessment, management, and communication. Risk governance includes various experts, stakeholders, and people participation (Renn et al. 2011).

UNDP has promoted the governance for disaster risk management in a framework of "institutional and legislative systems" to support risk reduction. The role of governance is to elevate disaster risk management as a policy to mainstream risk reduction into developmental process to have effective decisions that minimize risk and vulnerabilities, generate political commitment for decision-makers to have participatory approach, and promote disaster risk management as a multi-sector responsibility where inter-sectoral cooperation from various agencies, civil societies, and private sectors along with state are involved to ensure community-based governance in decision-making processes and foster participation (UNDP 2007).

7.3 Governance and Disaster Risk Reduction (DRR)

According to the (UNDP, ISDR & IRP 2010), governance is a concept and a process that holds the power of formal authority and government institution. The government is the major actor that influences the decision-making and its implementation. Other than the government, there are various other actors that help in governance which include religious organizations, unions, cooperatives, private enterprises, financial institutions, and political parties. Governance is required during disaster situation because at this phase the complexities make the condition more chaotic. The resource allocation needs to be coordinated and tracked along with the limited information for decision-making. The dynamic and unpredictable situation impacts the social system, economies, and recovery processes, and at that period of time, governance is everything (UNDP et al. 2010).

The decisions of private actors from individuals to households and firms will have an effect on the risk exposure of the society. In the present scenario, the government's response to public and private sector roles during disaster has focused more on reducing vulnerability and increasing resilience. To manage the risk of uncertain losses, an approach of risk transfer is very useful. During any catastrophic situation, the mechanism of insurance plays a vital role to mitigate the risk of properties having huge monetary value or financial risk associated with occurrence of natural hazard. Adaptation measures are useful in disaster preparedness, while governance plays an important role by providing political leadership and protecting and supporting public infrastructures. Communities and household individuals along with local government and other actors create enabling structures of local adaptation choices to deal with the emerging risk. Governance has diffused as well as softened the mode of governing which includes policies, programs, best practices, and information disseminations. The shift from disaster to disaster risk reduction has brought together different actors like government organizations, private firms, NGOs, civil societies, communities, and individuals to have an effective response to deal with disasters. By understanding the risk involved, the public-private partnership has specified roles for different agencies to adapt a risk reduction strategy. The shift in governance responsibility from the state to multiple actors has brought the multilevel governance in the forefront (Fisher and Surminski 2012).

7.3.1 *Disaster Risk Management*

UNDP definition says that the disaster risk reduction is the conceptual framework of elements considered with the possibilities to minimize vulnerabilities and disaster risks throughout a society, to avoid (prevention) or to limit (mitigation and preparedness) the adverse impacts of hazards, within the broad context of sustainable development. According to ISDR, disaster risk management is the systematic management of using administrative decisions, organizations, operational skills, and capacities to implement policies and strategies and coping capacities of the society and communities to lessen the impact of natural hazards and related environmental and technological disasters. This comprises all forms of activities, including structural and nonstructural measures to avoid (prevent) or limit (mitigation and preparedness) adverse effects of hazards (UNDP 2007).

The government's approach toward disaster has shifted from relief-centric approach to holistic approach which focuses on prevention, mitigation and, preparedness to build effective response, relief, and rehabilitation mechanism. The

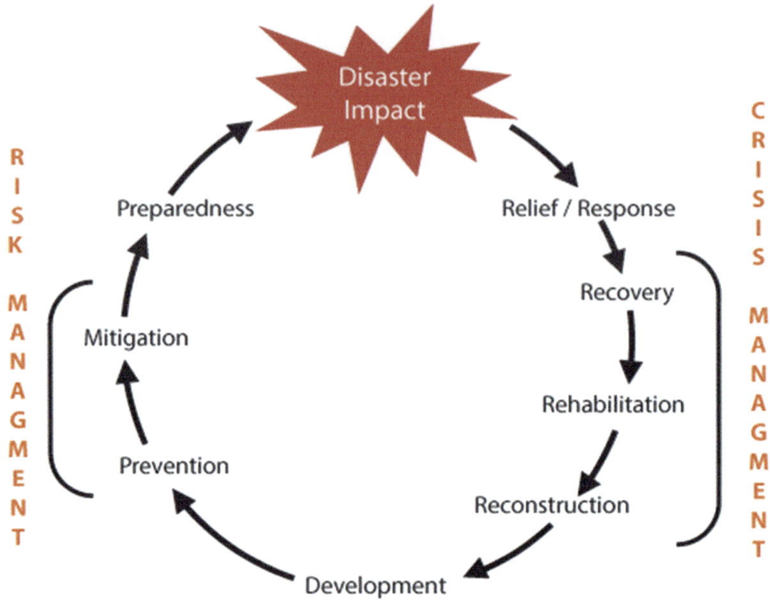

Fig. 7.1 Disaster management cycle (Source: Municipal Corporation of Greater Mumbai (GMDMA 2016))

institutional mechanism provides a coordinated platform to reduce and manage risk. Figure 7.1 shows the disaster management cycle which is a process that guides the government, civil society, and other actors what steps should be taken during and after the disaster and also in the recovery phase. It helps in structuring the plans and public policies to avoid, modify, prevent, or mitigate the impact of disaster. The action taken at every phase of the disaster management cycle leads to reduce vulnerability and enhance preparedness. The steps of risk management generally focus upon pre-disaster phase (before a disaster), while the steps of crisis management focus upon post-disaster phase (after a disaster).

The risk management is generally known as disaster risk reduction which broadly has three components or measures – preparedness, mitigation, and prevention (GMDMA 2016).

7.3.2 Preparedness and Mitigation Measures

It includes the awareness generation among public through education, incorporation of disaster management in school curriculum, sensitization and training of government officials, community preparedness through trainings, strengthening early warning system, preparation of contingency plan, construction of shelters, and evaluation of building codes to ensure safety (Kumar 2015).

7.3.3 Prevention Measures

It focuses on the activities that help to prevent the impact of disaster through vulnerability analysis, hazard mapping, hazard and vulnerability assessment, use of building codes, and zoning.

Governance is the determining factor for the success of disaster risk management to achieve sustainable human development. Governance is defined as the system of values, practices, and institutions by which a society manages its economic, political, and social affairs through interactions within and among the state, civil society, and private sectors. This is the way in which a society organizes itself to make and implement decisions, achieving mutual understanding, agreement, and actions. It comprises the mechanisms and processes for citizens and groups, articulates their interest, mediates their differences, and exercises their legal rights and obligations. It is the rules, institutions, and practices that set limits and provide incentives for individuals, organizations, and firms. Governance includes social, political, and economic dimensions and operates at every level of human enterprise be it household, village, municipality, region, and nation (UNDP 2007).

According to UNISDR, disaster risk reduction initiatives are to mitigate and prepare for the hazards through systematic development and application of policies, strategies, and practices (Bang 2013), and it also includes disaster mitigation and disaster preparedness along with sustainable development to reduce the risk. Disaster risk reduction constitutes every part of government, professional and private sectors, and society. According to Gaston et al. (2012), approaches were taken up in Cameron to improve the local governance which includes adequate distribution of roles, allocation of human and financial resources, building capacity of local community, and having a participatory approach to disaster risk reduction.

According to UNISDR (2004), governance in disaster risk reduction is:

1. To elevate disaster risk reduction as a policy priority
2. To allocate the necessary resources for disaster risk reduction
3. To enforce implementation of disaster risk reduction measures and assign accountability for failure
4. To facilitate participation from civil society

The four major components of governance for disaster risk reduction include policy and planning, legal and regulatory frameworks, resources, and organization and structures.

(a) *Policy and planning*: It has a multi-sectoral framework with coordination among institutions and government, civil society participation, and collaboration of private sectors including all stakeholders. For a comprehensive disaster risk reduction policy, risk identification and assessment are important. The disaster risk management plan has a legislation support in the activities and programs which need to be regularly updated. The plan should have details of resources required, allocated responsibilities, preparedness measures (awareness, knowledge management, early warning), mitigation measures, vulnerability, and risk reduction activities.

(b) *Legal and regulation framework*: For coordination among disaster risk management activities and to have good policy and planning, there should be political commitment. Legislation has defined roles, institutional arrangements to ensure implementation of policy, and framework to enforce building codes and resource allocation.

(c) *Resource*: Government's commitment toward disaster risk reduction is seen in good resource allocation and mobilization. The policy and plans are futile if proper resource implementation is not done. The disaster risk structure should have proper finances and human and material resources from the government. The increased demand and limited supply of resources create problem in disaster situation. It is due to insufficient capacity, prevailing corruption, reduced resources, and weak governance structure. To support the disaster risk reduction, the government should rearrange the developmental priorities to identify resources. The regional and international institutions play an important role in advocacy and facilitation of resource mobilization.

(d) *Local government and decentralization*: Often the local communities are more aware of the disaster risks and the existing resources along with opportunities to recognize and manage the risks. Disaster risk management must be built based on the community networks, municipals, and institutions of the local government, as it is a key element of national risk reduction strategy. The participation and involvement of various sections of society form the decentralized structure. The local community-level participation in decision-making, planning, and implementation process is an effective initiative of national development and disaster risk reduction strategies. Creation of a favorable political environment at the community level promotes participation and even provides opportunities for women involvement in disaster risk reduction activities. District, regional, and national levels should have integration of risk reduction activities and community planning with the government structure and resource capabilities (UNISDR 2004).

In the twenty-first century, capacity building focused on institutions' legal and organizational capabilities (FIG 2006). Capacity building includes the resource, means, and strengths of the community that helps prepare, cope with, withstand, prevent, mitigate, and recover from the devastating effect of disaster (Khan et al. 2008). The capacity building is the long-term, continuous process which creates awareness among people. Capacity building is an effort to reduce the disaster risk through skill development and to improve the societal infrastructure. According to UN/ISDR (2004), capacities include education, training, transfer of technical expertise, technical assistance, infrastructure strengthening, public information, access to technology, and knowledge dissemination (Fig. 7.2). Capacity building processes must be taken up by different sectors at all levels (FIG 2006).

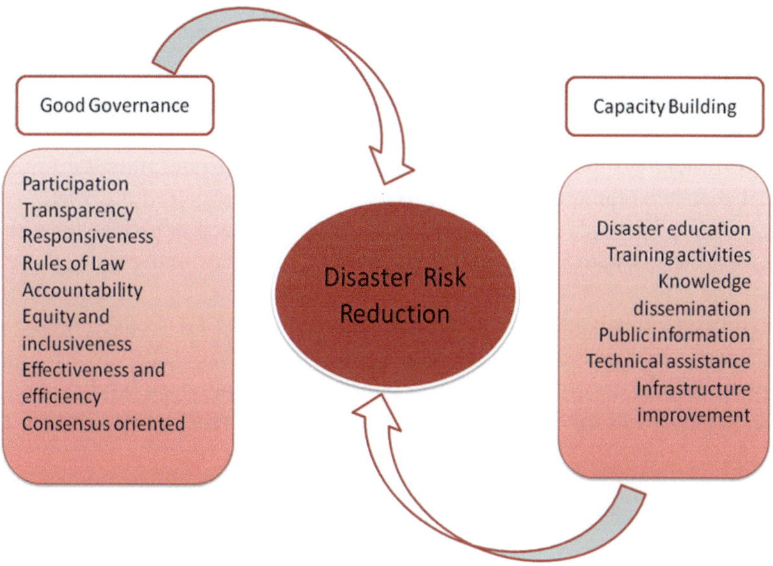

Fig. 7.2 Governance and capacity building for disaster risk reduction (Source: The Contribution of the Surveying Profession to Disaster Risk Management (FIG 2006))

7.3.4 Stakeholders Response

According to Rao, the community is the first responders to initiate rescue and relief operations. The states/union territories need to encourage their participation in decision-making and train them about various aspects of response as first aid, search and rescue, management of emergency shelters, psychosocial counseling, and seeking support from government/agencies. All the stakeholders participate together to reduce the risk with the existing or available resource within the community, society, and organization. It incorporates all the policy and legal frameworks, community participation including women, institutional development, and strengthening of managerial system.

Partnership is another measure to create networks to work together and exchange information within other agencies and partners to reduce disaster risk. It connects all the governmental departments, research organizations/institutes, and statuary agencies to share their knowledge through physical interaction, experiences, workshops, and conferences.

The collaborative or coordinated approach of disaster response will reduce confusion, duplication, and overlapping of activities between all the stakeholders. It leads to the sector-specific focus of different actors. In case of Orissa super cyclone, the state had universalistic approach, while the NGOs took community-oriented approaches (Behera 2002).

7.4 Role of Various Stakeholders

7.4.1 *Role of Government*

The government provides the institutional mechanism to monitor the implementation of disaster management plans to prevent and mitigate the disaster effects. Its response is holistic, coordinated, and prompt toward any disaster situation. The government's role is to protect lives and properties and enhance the well-being of the citizens. The government identifies, defines, and allocates roles and responsibilities to private agencies, civil societies, and NGOs based on the severity and spread of disaster to avoid duplication of response. The government provides critical infrastructure facilities, communication network, and financial and logistic support along with rescue, relief, and rehabilitation during major disasters.

7.4.2 *Role of Nongovernmental Organizations (NGOs)*

NGOs play an essential role in providing vital support services to promote recovery and support response. It provides shelter, emergency food supplies, and counseling, trains the volunteers, provides language assistance, provides transportation and logistics support, and identifies basic needs of elderly and disabled person. NGOs work together with the first responder, the government, and other agencies to provide relief services, promote recovery, and reduce stress of disaster victims.

7.4.3 *Role of Media*

Media plays an important role to provide information regarding disasters. It generates knowledge with the help of print and electronic mode among the people. It leads to community awareness, education regarding disasters, and dissemination of early warning information (Rao). Social media is the most commonly used communication system during crisis period using multitudes of mobile-based and web-based technologies like short message services (SMS) and blogs (START 2012). Various social media platforms like Twitter, Facebook, Viber, WhatsApp, Instagram, and YouTube are extensively used for helping, mapping, and locating loved ones and photo tags to find missing people, express support, notify the authorities, and send status reports along with fund raising, donations, and gathering volunteers for help. Social media has become the integral part of disaster response (Topno 2016).

7.5 Case Study from India

Different actors or stakeholders understand disaster differently. Some private player and NGOs consider disaster mostly in its response phase to distribute the required items like clothes, water, medical assistance, food, cash, and other essential products. The government considers the pre- and post-disaster response as early warning, evacuation, rehabilitation, reconstruction, and compensation. Earlier there were limited steps taken to reduce susceptibility of people, disaster impact, and vulnerability. The disaster risk reduction measures as collaborative action have helped the government, community, private organization, as well as NGOs to adapt the disaster preparedness (Krishna). n.d.

7.5.1 Orissa Super Cyclone (1999)

During Orissa super cyclone (1999), the NGOs played a vital role in relief operations (emergency response) and rehabilitation along with the government. NGOs were active from the immediate disaster phase till the short- and long-term phase. In the immediate phase, the local and international NGOs established an emergency response network named as Orissa Disaster Mitigation Mission (ODMM) which was used to share information about the affected area with the government. ODMM was set up to coordinate the relief and restoration work. The Orissa Development Action Forum was the network formed by the NGO for emergency response. NGOs played a major role in establishing community kitchen and restoration work such as clearing village roads, school, water sources, and carcass disposal in partnership with the local community. NGOs along with the state health department provided medical aid through mobile health camps; they also distributed medicines and spread health awareness. The government and NGOs worked collectively; they provided temporary building and shelter materials to the affected families. Various NGOs provide psychosocial aid to overcome the depression and trauma brought by the cyclone through organizing cultural shows in participatory manner and street theater along with trauma centers. The NGOs also set up a "legal aid center" to provide information regarding government compensation and people's rights. By establishing a "volunteer's hub," NGOs mobilized volunteers for relief activities during cyclone.

In the short- and long-term phase, NGOs focused on rehabilitation and restoration work by encouraging children's participation for trauma relieve activities; reviving schools and social mobilization; creating functional groups including women, youth, and farmers; constructing cyclone shelters and disaster preparedness initiatives; restoring livelihoods (farm and nonfarm); and promoting microcredit schemes for women. NGOs also initiated the Food-for-Work (FFW) programs. The government and other resource providers gave assistance to the NGOs for backyard plantation and construction of community-dwelling units for the poor. These activities strengthened the community-based disaster preparedness measures by providing training and volunteer mobilization and raising awareness, institution building, and contingency planning (Behera 2002).

7.5.2 *Bhuj Earthquake (2001)*

The Bhuj earthquake (2001) forced the government of India to have the institutional and legal framework for disaster situations. At that time, disaster management was moved from the Ministry of Agriculture to the Ministry of Home Affairs. The Bhuj earthquake was the path-breaking event. During Bhuj earthquake challenges, lessons were learnt, and they were taken into consideration for future references. It was also put into practice for preparation and mitigation strategy for future disasters. The knowledge management played a key role in disaster risk reduction which involved all the stakeholders to understand the disaster risk.

7.5.3 *Indian Ocean Tsunami, 2005*

The Central Board of Secondary Education (CBSE), India, has begun integrating disaster risk reduction components into Indian school curriculum in 2003. The disaster risk reduction was incorporated in the social science syllabus in high school education. The textbook was designed and developed by the help of the United Nations Development Programme (UNDP) personal, academician, and teachers. The content of the textbook includes chapters on hazard, development of preparedness and response plan, search and rescue, first aid, and mock drills at school. The teachers are trained to deal with gender-sensitive issues during disaster; this enhanced their knowledge about disaster management.

In Tiruvallur district of Tamil Nadu, India, a puppet show was organized to create awareness about disaster among school children. The role of men and women during disaster scenario was demonstrated to avoid gender discrimination (UNDP 2010).

In Tamil Nadu after the Indian Ocean tsunami, the South Indian Federation of Fishermen Societies and NGO (SNEHA) established the NGO Coordination and Resource Centre (NCRC) to improve coordination of local NGOs and to exchange information among government and communities regarding disaster. In this initiative, the important information and data from the affected communities were shared with the NGOs and the district administration to get attention and adequate support. The information policies were translated into vernacular languages to circulate the information among the local communities. Information which was collected from communities was used for assistance and support such as compensation and allotment of houses at the village level (UNDP et al. 2010).

According to the Oxfam Model of Disaster Preparedness, the coastal district of Andhra Pradesh, India, has taken steps to reduce the community vulnerability by establishing a team of task force and self-help groups (SHGs) to improve the health and sanitation conditions in the villages. The common pool resources like ponds and tanks are controlled efficiently; simultaneously, women participation in public affairs has been encouraged. The initiative has been started to train ex-servicemen and government officers at the local level, and training sessions were conducted at schools to facilitate disaster preparedness activities. The program also includes the disaster insurance. It has resulted in reduction of vulnerability of 250 villages in Andhra Pradesh and has created 3000 trained volunteers from the community for preparedness and response toward disaster (Krishna n.d.).

7.5.4 Chennai Floods, 2015

In 2015, during Chennai floods, the lack of coordination among different stakeholders was experienced. Mitigation mechanism was not much effective during Chennai floods as the instruction from the state government was not on time. The response and rescue team including the National Disaster Response Force (NDRF), the Indian Armed Forces, paramilitary forces, state police, and local people were prepared, but due to the inefficient decision-making process, they were not coordinated by the state and local authorities (Joseph and Krishnan 2015). According to ACAPS (2015), assistance was provided through international nongovernmental organizations as "international response capacity." In the last 40 years of flood history in Chennai, prevention and mitigation measures were not adequately addressed (Bhandari 2015).

7.6 Conclusion

Disaster has created unexpected risk for the society due to its intensity and causalities associated with human and economic losses. India being traditionally vulnerable due to its geo-climatic conditions and higher susceptibility to natural hazards, there is an increased need of emergency preparedness or disaster preparedness in the country. Disaster management plays an important role in reducing the disaster impacts through preparedness measures like continuous and integrated planning, organizing, coordination, and implementation. The paradigm shift from relief-centric approach to proactive approach in the regime of disaster risk reduction has proved to be effective.

The risk reduction includes prevention of threat, mitigation of risk, capacity building, preparedness, prompt response, risk assessment, evacuation, relief and rescue, rehabilitation, and reconstruction. In Indian context, there is a need for adequate preparedness and effective mitigation measure to deal with the impact of disasters. Governance role in these two phases of disaster management cycle is essential. Mitigation includes the regulations and building codes, vulnerability analysis, land use management, safety codes, health care, and public education. It also includes hazard and emergency risk information and its countermeasures along with national and regional developmental planning. Preparedness is the readiness to respond in an emergency situation through technical and managerial capacity of government, organizations, and communities. It includes short-term as well as long-term strategies such as public education, early warning systems, and resource reserves (food, water, and medicines). The role of government, organizations, and individuals during preparedness is to develop a strategy to save lives and minimize damage along with effective response operations. It also includes emergency training, planning, warning systems, media support, mutual aid, public information, education, evacuation plan, resource inventory, emergency communication, and contact inventories. The Disaster Management Act (2005) includes range of responsibilities for the authorities. The National Disaster Management Authority (NDMA) is a body that implements a holistic and integrated approach to disaster management in India. It works to develop effective plans to tackle various kinds of disasters. The Indian disaster management system works in different levels such as central, state, district, block, village, and municipal or local authorities. Local authorities are the first responders during this situation. When the capacity and resources exhaust to tackle the catastrophic situation, the district steps in; if the district is unable to control it, the state steps in. If the state cannot control it, then the central government responds to the disaster. The state, district, and local department coordinate to mitigate the impact of disaster. The resilience relies on the capacity of individual, communities, NGOs, private agencies, local authorities, state government, and central government (Rao 2016). Hence, the collaborative approaches of different stakeholders along with the government structure play an effective role while dealing with disasters.

7.7 Recommendations

To have a sustainable disaster risk reduction, more focus should be on awareness generation through education on disaster risk reduction. Knowledge-sharing networks like solution exchange forums are required. Preparation of specific disaster as well as multi-hazard preparedness plans to mitigate the impact at all levels; capacity building activities to institutionalize DRR in the government system are also needed (Krishna).

The bottom-up approach to have effective decision-making should be encouraged. The frequency of mock drills at the grassroot level has to be increased to spread the awareness among the community who used to be the first responders.

There is a need for proper institutional coordination mechanism and legal framework to manage disaster risks. Consultation among all the stakeholders (government, NGOs, civil society) is necessary for information sharing, preparation of future action plan, and advice on problems to strengthen disaster risk reduction measures. The coordination increases the efficiency to respond to disasters and to have a better policy and plans in place (Behera 2002).

All the above activities would be done with the support of the government as well as other actors including the NGO, private players, civil society, institutions, various organizations, and community. The time gap between the disaster and its response must be effectively utilized.

References

ACAPS (2015) India: floods in Tamil Nadu. Retrieved 15 Apr 2016, from ACAPS briefing note. http://acaps.org/img/documents/b-acaps-start-bn-india-floods-in-tamil-nadu-15-nov-2015.pdf

Bang HN (2013) Governance of disaster risk reduction in Cameroon: the need to empower local government. J Disaster Risk Stud 5(1):1–10

Behera, A. (2002) Government-NGO collaboration for disaster reduction and response: the India (Orissa) Experience. Regional Workshop on Networking and Collaboration among NGOs of Asian Countries in Disaster Reduction and Response, (pp 1–13)

Bhandari, R. (2015) Chennai flood disaster catastrophe of December 2015-are you listening? Retrieved 15 Apr 2016, from Vivekananda International Foundation. http://www.vifindia.org/article/2015/december/17/chennai-flood-disaster-catastrophe-of-december-2015-are-you-listening

Darwanto, H. (2012) Preliminary examination of existing methodologies for allocating and tracking national government budget for disaster risk reduction (DRR) in Indonesia. UNISDR ADB and ADPC

EDRS and KBF (2007) Good governance: the concept. In EDRC, KBF. Good governance in multiethenic communities (pp. 11–17). Ethnocultural Diversity Resource Centre and King Baudouin Foundation

FIG (2006) The contribution of the surveying profession to disaster management. The International Federation of Surveyors, Denmark

Fisher S, Surminski S (2012) The roles of public and private actors in the governance of adaptation: the case of agricultural insurance in India. Centre for Climate Change Economics and Policy, munich Re Programme, Grantham Research Institute on Climate Change and Environment, United Kingdom

Gaston BW, Tongwa AF, Burnley C, Isabella ZT (2012) Local governance in disaster risk reduction in Cameroon. Jamba: J Disaster Risk Stud 4(1):1

GDRC (1995) Our global neighborhood: report of the Commission on Global Governance

GMDMA (2016) Disaster management cycle. Retrieved 16 Mar 2016, from Municipal Corporation of Greater Mumbai. http://dm.mcgm.gov.in:9080/gmdma/?page_id=469

IRDR (2014) Governance in disaster risk management. Integrated Research on Disaster Risk International Centre of Excellence, Beijing

Joseph, J., Krishnan, V. (2015) Disaster response: Time to stop making excuses. Retrieved 15 Apr 2016, from The Hindu. http://www.thehindu.com/news/national/post-chennai-floods-disaster-response-time-to-stop-making-excuses/article7983967.ece

Khan H, Vasilescu GL, Khan A (2008) Disaster management cycle: a theoretical approach. Manag Mark 6:43–50

Krishna, N. H. (n.d.) Approaches in disaster risk reduction: experiences, present status and future requirements. Retrieved 12 Apr 2016, from Dhan. http://dhan.org/acedrr/papers/[01]-Approaches-in-Disaster-Risk-reduction-[Mr.-Harikrishna].pdf

Kumar, V. (2015) Role of government in disaster management at central, state and district level. Retrieved 30 Mar 2016, from Slide share. http://www.slideshare.net/Vijz1/role-of-government-in-disaster-management-at-central-tam-201320?qid=e2243747-62c8-4aad-a329-2871165ac28e&v=&b=&from_search=4

Rao TS (2016) A study on disaster risk governance in India. FRM Consultants, Hyderabad. Retrieved March 12, 2016 from Wcdm. http://www.wcdm.info/Compenium/other/5.FULL%20LENGTH%20PAPER%20FROM%20Mr.T.P.SURYA%20CHANDRA%20RAO--HYDERABAD.pdf

Renn O, Klinke A, Asselt MV (2011) Coping with complexities, uncertainity and ambiguity in risk governance: a synthesis. J Hum Environ (AMBIO) 40(2):231–248

START (2012) Social media use during disasters: a review of the knowledge base and gaps. Department of Homland Security Sciences and Technology Centre of Excellence, Maryland

Tierney K (2012) Disaster governance: social, political and economic dimensions. Ann Rev Environ Res 37:341–363

Topno PN (2016) Social media: an effective tool for disaster response. International Council on Social Welfare. Regional Coorporation Newsletter, South Asia

UNDP (2007) A global report: UNDP support to institutional and legislative systems for disaster risk management. United Nations Development Programme, New York

UNDP (2010) Disaster risk reduction, governance and mainstreaming. United Nation Development Program, New York

UNDP, ISDR & IRP (2010) Guidance note on recovery: governance. International Recovery Platform Secretariat, Japan

UNESCO (2010) Towards improved public awareness for climate related disaster risk reduction and management in the Arab region. United Nations Educational, Scientific and Cultural Organization, Cario

UNISDR (2004) Disaster risk reduction, governance and development, 2, 4, Cape Town: UN/ISDR Africa.

UNISDR (2010) Earthquakes caused the deadliest disasters in the past decade, Retrieved 12 Mar 2016, from The United Nations Office for Disaster Risk Reduction. http://www.unisdr.org/archive/12470

UNISDR (n.d.). https://www.unisdr.org/we/coordinate/hfa, http://www.unisdr.org/we/coordinate/hfa-post2015

UN/ISDR (2004) Living with risk: A global review of disaster reduction initiatives, vol 1. New York and Geneva

Chapter 8
Disaster Risk Governance and Response Management for Flood: A Case Study of Assam, India

Indrajit Pal and Siddharth Singh

Abstract Flood and erosion in the State of Assam, India, is menacing and probably the most acute and unique in the country. Every year due to successive waves of floods, most of the areas in the valley of Assam remain submerged for a considerable numbers of days. Regular flooding added with persistent erosion causing land loss of thousands of hectares resulting to hundreds of people landless virtually destabilize the socio-economic development of the state. It has been observed that every year, the mighty Brahmaputra River is eroding more than 2000 ha of land. Subsequent to the National Policy for Flood in 1954 by the Government of India, flood control activities in the State of Assam started taking place. As envisaged in the National Policy for Flood, the state could take short-term as well as long-term measures for flood mitigation, but to get the immediate relief to the flood-ravaged state, construction of embankments as short-term measures had been widely adopted. In the state as a whole, the total area eroded by Brahmaputra, Barak and their tributaries since 1954 is 3.86 lakh hectares, which constitute 7% of the total area of the state.

The recurrence incidence of extensive floods takes place because of the occasional failure of the existing flood prevention structures, which have outlived their lives. Regular flooding added with unabated erosion causing land loss of thousands of hectares resulting to hundreds of people landless virtually destabilize the socio-economic development of the state. It has been observed that every year, more than 2000 ha of land is being eroded by the Brahmaputra annually.

I. Pal (✉)
Disaster Preparedness, Mitigation and Management (DPMM), Asian Institute of Technology, Khlong Luang, Pathum Thani, Thailand
e-mail: indrajit.pal@gmail.com

S. Singh
Public Health Engineering Department, Indian Administrative Service (IAS), Dispur, Assam, India
e-mail: siddharthsingh17ias@gmail.com

I. Pal, R. Shaw (eds.), *Disaster Risk Governance in India and Cross Cutting Issues*, Disaster Risk Reduction, DOI 10.1007/978-981-10-3310-0_8

The present study will analyse the comprehensive approach towards disaster risk reduction for effective disaster governance that is a combination of actions including mitigation activities for specific hazards. A comprehensive approach towards disaster risk reduction for effective disaster governance is a combination of actions including mitigation activities for specific hazards (Pal et al. 2013, 2017). The present study highlights the flood management and response mechanisms already in existence in the State of Assam, one of the most multi-hazard-prone states in India.

Keywords Disaster risk reduction (DRR) • Disaster governance • Flood • Brahmaputra Valley • Erosion • Siltation • Resilience

8.1 Introduction

Floods are recurrent phenomena in India from the time immemorial. Every year, a number of states in India are affected by floods of varying magnitude. Different regions of the country have different climates and rainfall patterns, and, as such, it is also experienced that while some parts are suffering under devastating floods, another part is suffering under drought (UNISDR 2008). With the increase in population and developmental activity, there has been a tendency to occupy the flood plains, which has resulted in more serious nature of damages over the years. Because of the varying rainfall distribution, many a times, the areas which are not traditionally prone to flood are also experiencing severe inundation. Thus, flood is the single most frequent disaster faced by the country. It has been estimated by the National Commission on Flood that the area prone to the floods in the country is of the order of 40 m.ha, out of which it is considered that 32 m.ha could be provided with reasonable degree of protection. Achievement so far attained is for an area of about 18.22 m.ha.

The State of Assam covers an area of 78,438 km^2 and lies in the middle of the Brahmaputra and Barak basins. The Brahmaputra basin is one of the largest river basins in the northeast region of India. The Brahmaputra River originates from the semiarid region of South Tibet. The Brahmaputra basin covers an area of 5,80,000 km^2 out of which 70,634 km^2 falls within the State of Assam. The Assam basin has a length of about 1540 km in east-west direction and a maximum width of 682 km in north-south direction. The unique geographical location criss-crossed by a vast network of 48 major and 128 small rivers originating from the hills and mountains surrounding the state is largely responsible for the recurring floods and erosion of river banks. When the discharge in the rivers along with their tributaries synchronizes during monsoon, the state faces havoc flood, and the damage caused is colossal. Further, deforestation in upstream and downstream areas of surrounding Hill States and Assam, respectively, has caused excessive siltation, resulting in abnormal rise in the surfaces of major rivers. The siltation problem is acute in the case of rivers of Upper Assam and Central Assam.

This region consists of the rivers Brahmaputra, Barak and their tributaries and covers the states of Assam, Arunachal Pradesh, Meghalaya, Mizoram, northern parts of West Bengal, Manipur, Tripura and Nagaland. The catchments of these rivers receive very heavy rainfall ranging from 110 to 635 cm a year which occurs mostly during the months of May/June to September. As a result, floods in this region are severe and quite frequent. Further, the rocks of the hills, where these rivers originate, are friable and susceptible to erosion and thereby cause exceptionally high silt charge in the rivers. In addition, the region is subject to frequent earthquakes which cause numerous landslides in the hills and upset the regime of the rivers. The predominant problems in this region are the flooding caused by the spilling of rivers over the banks, drainage congestion and tendency of some of the rivers to change their courses. In recent years, the erosion along the banks of Brahmaputra has assumed serious proportions.

The Brahmaputra Valley in Assam is one of the most hazard-prone regions of the country, with more than 40% of its land (3.2 million hectares) susceptible to flood damage. This is 9.4% of the country's total flood-prone area. About 7% of land in the state's 17 riverside districts has been lost because of river erosion over the past 50 years (ADB 2009).

8.2 Major River Systems in Assam

All the rivers in Assam are liable to floods, mainly because they receive heavy rainfall within a short time. These rivers are in their early stage of maturity and are very active agents of erosion. The river waters collect a tremendous amount of silt and other debris and raise the level of the river beds. Therefore, it becomes impossible for the main channel to cope with the vast volume of water received during the rains. The Brahmaputra and the Barak are the two major river systems of the state. The Brahmaputra River originates at an elevation of about 5000 m above mean sea level (AMSL) in Tibet. The Brahmaputra River, known as Tsang Po in Tibet, after a long eastward course of 1600 km abruptly veers towards south around Namcha Barwa peak (7710 m) in Eastern Himalaya. This southward course of the river flowing through Arunachal Pradesh is known as Siang River. It passes through tortuous course across the mountains of Arunachal Pradesh and then emerges onto the plains of Assam, where downstream it is met by the Dihang, which is known as the largest tributary of the Brahmaputra, further fed by tributaries like Dibong, Sessiri, Lohit and Noa-Dihing around Saikhowaghat. The river known as Brahmaputra in Assam initially flows south-westward and thereafter towards the west in the Brahmaputra Valley.

Further downstream, the river swings towards the south and passes onto the plains of Bangladesh. The Brahmaputra River between Namcha Barwa and the confluence with Dihang descends by about 2200 m, and its water power resources have been estimated to be the third biggest in the world coming after Congo and Amazon basins. Along the northern bank, the Brahmaputra River is joined by the tributaries like Subansiri, Ranganadi, Dikrong, Gabharu, North Dhansiri, Pagladiya, Manas, Aie, Beki, Champamati, Gangadhar and Raidak. All these tributaries more or less flow in straight courses up to the junction of the main river. On the south bank, tributaries like Burhi-Dihing, Disang, Dikhau and South Dhansiri originate from Naga-Patkai Hills.

The Kopili River originates from North Cachar Hills, while the Digaru, Bharalu, Kulsi, Singra, Dudnai and Krishnai originate from Meghalaya Plateau. Some of the rivers and tributaries originating from the south flow for quite a distance almost parallel to the Brahmaputra River before joining the main river. The often changing meandering course of the Brahmaputra and its tributaries is not only due to lateral erosion because of the low gradient of the rivers but also due to periodic, local and sudden changes in the basement levels due to the neotectonic activity (Geology and Mineral Resources of Assam, GSI, 2009).

8.3 Administrative Arrangements and Governance

Given the state's geographical location, population size, natural resources, richest biodiversity zones and complexity of hazard risks, there can be no single approach directed towards protecting the people and elements at risk (physical, social and economic resources). A wide range of government institutions operate to govern the state and manage the assets. The state is among nations largest producer of crude oil, significant share in India's agricultural production (tea and rice), and has built up infrastructure system (lifeline buildings/highways/road/rail/airport/heliport/power/communication) that serves as a critical link within the state and as a vital link to the northeastern part of India.

Considering the geographical location, access issues, population exposure, scale and diversity of resources, there exists an urgent need for implementing and expanding statewide comprehensive disaster management strategies encompassing preparedness, prevention and mitigation, emergency response, and rehabilitation. Initiatives on these fronts if taken by all departments of the state will result in minimizing the loss of life, reducing disruption time of basic services vital for society to function, and protecting assets/infrastructure which are vital for the state economy.

The DM Act (2005) forms the legal basis for DM activities within all levels of the state government and it includes:

- Establishing disaster management authorities at the state, district and local level
- Assessing of the vulnerability to different forms of disasters and specific measures to be taken for their prevention or mitigation
- Undertaking preparedness activities to respond to any threatening disaster situations and giving directions where necessary for enhancing such preparedness
- Coordinating emergency response in the event of any threatening disaster situation or disaster

- Promoting general education, awareness and community training including drills in regard to the forms of disasters and the measures that may be taken to prevent, mitigate and respond to such disaster
- Giving direction and ensuring roles and responsibilities are clearly outlined for the departments of the government functioning at all levels in responding to any threatening disaster situation or disaster
- Assisting and protecting the community affected by disaster or providing relief to such community or preventing or combating disruption or dealing with the effects of any threatening disaster situation
- Developing mechanisms in which the mitigation measures shall be integrated with the development plans and projects
- Monitoring the implementation of the plan and annual review and updating the plan at all levels

As per the Disaster Management Act (2005) and the Assam Disaster Management Rules (2010), disaster management arrangements in the state are based upon partnerships between national, state, district and local authority. This partnership recognizes each level of disaster management arrangements. Levels of disasters have already been categorized and disseminated as L0, L1, L2 and L3, based on the ability of various authorities to deal with them:

- **L0**: L0 denotes normal times which are expected to be utilized for close monitoring, documentation, prevention, mitigation and preparatory activities. This is the planning stage where plans at all levels from community to the state shall be put in place. Training on search and rescue, rehearsals, evaluation and inventory updating for response activities will be carried out during this time.
- **L1**: L1 specifies disasters that can be managed at the district level; however, the state and centre will remain in readiness to provide assistance if needed.
- **L2**: L2 specifies disaster situations that may require assistance and active participation of the state and the mobilization of resources at the state level.
- **L3**: L3 disaster situations arise from large-scale disasters where districts and the state may not have the capacity to respond adequately and require assistance from the central government for reinstating the state and district machinery.

The partnership across authorities is to work collaboratively and ensure coordination and planning at all times, information sharing and resource mobilization that are necessary for disaster management. In any response situation, initial efforts would always be taken by the district administration. However, when the district is overwhelmed in any situation, the support necessarily has to come from the state and national level. The responsible officer (RO) within the jurisdiction control will trigger the activation for various levels of disaster.

DM arrangement in Assam is based on a four-tired structure and recognizes all four levels (L0, L1, L2 and L3) (Table 8.1), including the support mechanism from the national level.

Disaster management structure is established across the State of Assam (local/district/state) along with the creation of support structures such as Disaster Response

Table 8.1 Operational structure for disaster response

Overall structure		
Statutory	Authority arrangement	Level of operations
Local	Local authority	L0
	Panchayati Raj institutions	
	Municipalities and urban local bodies	
State	District disaster management authority	L1
	State disaster management authority	L2
	State executive committee	
National	National disaster management authority	L3
	National executive committee	

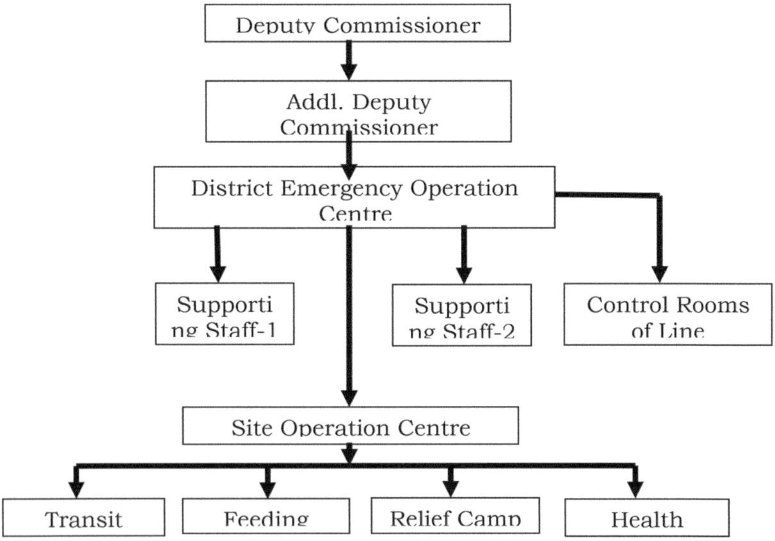

Fig. 8.1 Framework for disaster response management

Information Centre, State Disaster Response Force (SDRF), among others, so as to conduct operations for each level of disaster management arrangements (Fig. 8.1).

8.4 Flood Hazard in Assam (1998–2007)

The floods are caused by the runoff of extremely heavy rainfall during the monsoon and high sediment loads from upper watersheds that are geologically unstable and degraded because of deforestation and changing land use. The flood combined with river erosion has significant impacts each year (Fig. 8.2).

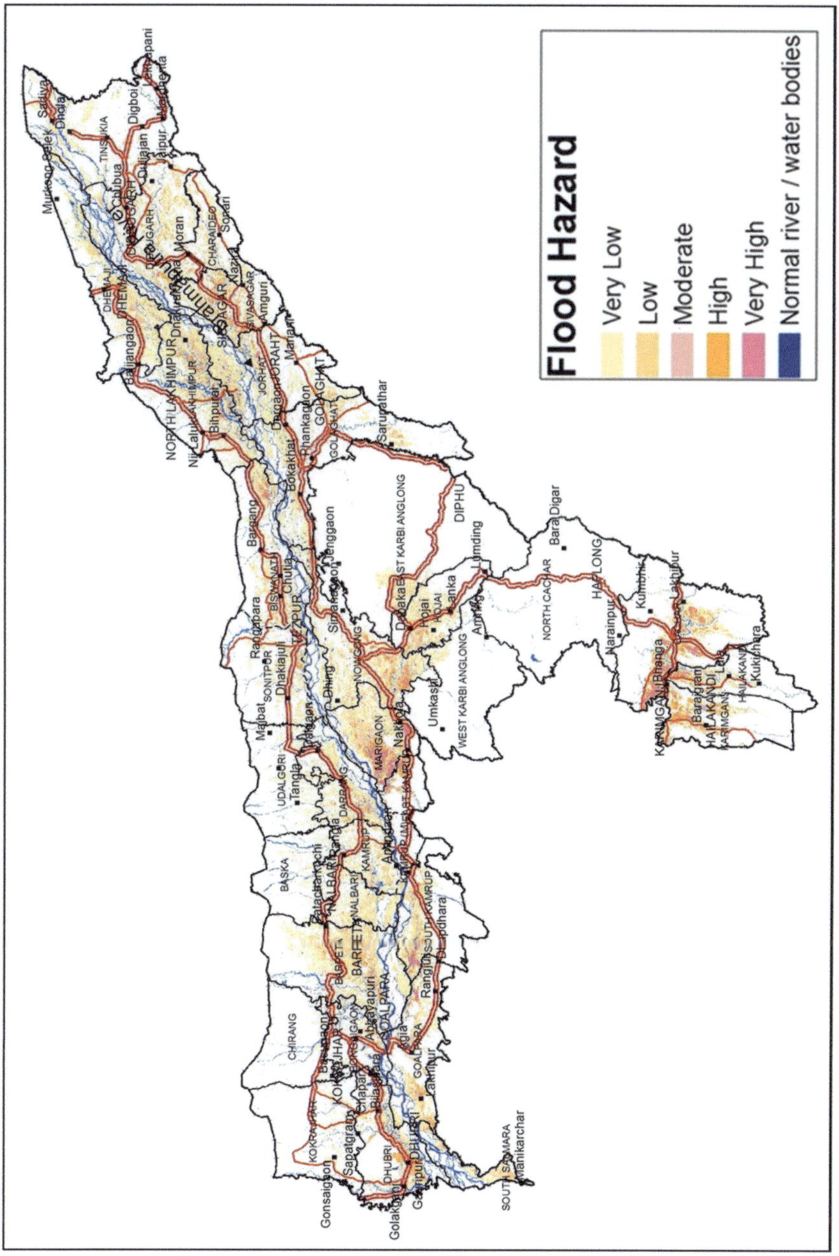

Fig. 8.2 Flood hazard risk of Assam (ISRC 2011)

According to the Flood Hazard Atlas of Assam (ISRO 2011), approximately 28.31% (22.21 lakh hectares) of land in the State of Assam was affected by flood hazard between the period 1998 and 2007 (Tables 8.2 and 8.3). In the above study, the flood-prone areas were divided into categories based on frequency which range from very high to very low. Very high indicates nine to ten times inundation during the last 10 years. This amounts to nearly 1.64% of total geographical areas of the state (5.79% of total flood-affected area). 'High' indicates occurrence of seven to eight floods over the last 10 years. This amounts to approximately 2.86% of total geographical areas of the state (10.11% of total flood-affected area). The rest of the flood-affected zones, i.e. moderate, low and very low, the percentage area of each flood hazard category accounts to 4.48% (15.83% total flood-affected area), 6.27% (22.14% total flood-affected area) and 13.06% (46.13% total flood-affected area), respectively.

Apart from the geo-climatic setting, high rate of population growth in the form of high birth rate and immigration from border countries has led to unplanned settlements. Human activities like deforestation, accelerated rated change in land use, filling up low-lying areas for the construction of buildings, urban development and temporary flood control measures are some changes which do contribute to the

Table 8.2 Flood hazard-affected area in Assam (1998–2007)

Hazard severity	Flood hazard area (ha)	% Flood hazard (w.r.t. state geographic area)	% Flood hazard (w.r.t. total flood hazard area)
Very high	1,28,687	1.64	5.79
High	2,24,629	2.86	10.11
Moderate	3,51,667	4.48	15.83
Low	4,91,761	6.27	22.14
Very low	10,24,584	13.06	46.13
Total	22,21,328	28.31	100

Source: Flood hazard atlas of Assam, ISRO 2011

Table 8.3 Extent of flood hazard in Assam (1998–2007)

Affected area (%)	No. of districts	Name of districts
0–10	4	Baska, Chirang, North Cachar, Karbi Anglong
10–20	2	Kokrajhar, Tinsukia
20–30	6	Cachar, Golaghat, Hailakandi, Kamrup (Metro), Karimganj, Udalguri
30–40	6	Bongaigaon, Dhubri, Dibrugarh, Golpara, Kamrup(Rural), Sonitpur
40–50	5	Dhemaji, Jorhat, Nalbari, Nagaon, Sibsagar
50–60	1	Lakhimpur
60–70	1	Barpeta
70–80	2	Darrang and Morigaon

Analysis is based on 10-year (between 1998 and 2007) data used in the above-mentioned study overall vulnerability of the state to floods (Dhar and Nandaragi 2003). The reliability and effectiveness of the embankments from the Brahmaputra flooding are generally insufficient because of structural deterioration and ongoing riverbank erosion.

8.5 Statewide Strategic Hazard Risk and Vulnerability Assessment

Assam accounts for about 2.4% of the country's geographical area. Its 31.17 million people (2011 Census) are 2.6% of the country's population, and its population density of 397 persons per square kilometre is marginally higher than the average density for the country (382 persons per square kilometre). With the increasing population, the need for critical analysis of vulnerability becomes important. With the distribution of land in Assam already being skewed towards large land owners, it is quite likely that the future population (increase) is to be pushed towards vulnerable areas, i.e. near the rivers or towards the landslide-prone hilly areas. Based on the current population estimates and projecting the same towards the future, it is likely that in spite of limited land availability in some cities (especially cities located near the hills), there will be a substantial increases in the urban population and thereby the vulnerability within those areas.

The percentage of population of the age group 60 and above (5.5% in Assam), who are usually more vulnerable to the natural hazards, are found to be less in Assam in comparison to India (7.4%). But, on the other hand, the proportion of children of age group 6 or below is around 17% in Assam in comparison to India's average of 16%. Assam ranks among the top ten states in India with high proportion of children of age 6 years or below. The proportion of disable persons in Assam is quite less (10 per 1000 people) in comparison to India's average of 20 per 1000 people. While the proportion of children is high in Assam indicating the possibility of demographic dividend, the children of this particular age group are also more vulnerable to the hazard risk of even minimum intensity or magnitude. Further, the requirement of medical facility to handle children around this age group will also be quite different from that of adults.

Apart from the urban population and children in general, the presence of schedule caste, schedule tribes and other backward castes in the state should be analysed for critical circles/districts with respect to specific hazard-risk scenario. According to the provisional census (2011), there are about 445 other tribes per 1000 households (in rural areas) and ranks fourth in India. In some hazard incidences, there may be a possibility of not being able to address these vulnerable groups in time. This may be due to either lack of accessibility to their location or knowledge of their whereabouts and conditions. While this segment of population may have developed some resilience towards much of the natural hazards, with the changing land use and land cover, demographics, and occupation, the vulnerability of this segment of population should be addressed in detail within the district-level disaster management plan, taking into consideration their lifestyle and socio-economic characteristics (Djalante et al. 2011).

The critical area of concern is the vulnerability of the housing stock to the varying hazard. Based on the census statistics, the current housing stock in Assam is in a poor condition in comparison to the rest of the country. This state of the housing stock is especially bad in the rural regions of the state in comparison to its urban counterpart. Only 24% of the total rural housing stock is in a good condition compared to the national average of 45%. Over 11% of the housing stock is in a dilapidated condition (around 4.5 lakh houses). This current condition of housing stock is highly vulnerable and may lead to disasters during hazard events with low magnitude or intensity. A district-wise analysis of different roof-wall type and their respective condition is provided in the subsection below.

In absolute terms, the number of poor people in the state has increased from 7.8 million in 1983 to 9.5 million in 1999–2000. Around 36% of the state's population continues to live below the poverty line, a figure appreciably above the national average of 26%. While the planning commission has identified the general unrest in the state as one of the reasons for the slow growth, much of the lag in growth may also be due to the hazard history and potential hazard-risk profile of the state. Further, with much of the people being involved in this sector and with the current landholding pattern being skewed (less than 1.5 ha per person), the people involved in the agricultural sector are directly affected by the variations in the monsoon and the changes in the river course. The monsoon followed by flooding dominates around one fourth of the productive days adding to the existing (housing, health, etc.) vulnerabilities of people working in this sector (loss of assets and livelihood) (ADB 2009).

8.6 Flood Management Measures by Assam

The government of Assam has executed projects which include construction of 4459 km embankments on Brahmaputra and Barak river (including their tributaries), completion of 689 nos. of anti-erosion/town protection schemes, construction of 851 km of drainage channels, construction of 85 nos. of sluices, and construction of 3 nos. of raised platforms in Brahmaputra and Barak Valley as per the recommendations of Task Force – 2004 (ASDMA 2013; Walia et al. 2013).

8.6.1 Short-Term Structural Measures

The state government of Assam has suggested short-term measures like new embankments, gap closure, raising and strengthening of existing embankments, anti-erosion works, drainage development, and raised platforms.

8.6.2 Long-Term Structural Measures

The state government has suggested for early completion of surveys, investigations, and DPRs and expediting the seven major projects as identified by Brahmaputra Board which are given below:

- Pagladiya Dam
- Tipaimukh Dam
- Subansiri Dam Project
- Dehang Dam Project
- Kameng Dam Project
- Lohit Dam Project
- Debang Dam Project

8.6.3 Brahmaputra Board

Considering the fact that Northeast India with its geographical area of 26.52 million hectare is endowed with enormous water resources and the combined annual water resources potential of Brahmaputra and Barak rivers is 586 BCM, which is the highest among all river basins in the country, the Ministry of Water Resources constituted a Brahmaputra Board in 1980 with the objective of planning and integrating implementation of measures for control of floods and bank erosion in Brahmaputra and for matters connected therewith. Its jurisdiction includes all the states of Northeast and a part of West Bengal falling within the Brahmaputra basin. Its main functions are preparation of master plans for the control of flood and bank erosion and improvement of drainage and development and utilization of water resources of the Brahmaputra Valley for irrigation, hydropower, navigation and other beneficiary purposes.

8.6.4 Brahmaputra River Valley Authority

The mandate of the proposed Brahmaputra River Valley Authority shall be the coordinated development and management of water, land and related sources in order to maximize the resultant economic and social welfare in an equitable manner without compromising the sustainability of the vital ecosystem of the Brahmaputra Valley. The proposed Authority will have two parts, a policy making apex council and an executive wing. The council shall be headed by the Union Minister of Water Resources of the NE States, subject matter experts and other senior representative of the state and central government. One of the chief ministers shall be the vice-chairman of the council on rotational basis in alphabetical order. The executive board will be responsible for implementing the decision of the council. The

Authority shall be a body corporate by the name aforesaid having perpetual succession with powers to acquire, hold and dispose of property, both movable and immovable, and to contract and shall by the said name sue and be sued. The working relationship shall be modelled.

The sphere of the work of the new Authority is briefly stated as follows:

- Integrated Multidisciplinary Basin Planning ensuring their implementation by member states
- Investigation, planning and design, appraisal, clearance, monitoring and implementation of works in consultation with states
- Promotion of sustainable water resources management
- Integrated flood management, flood forecasting
- Hydropower development to the extent provided for national interest

8.7 Flood Mitigation Measures for Disaster Risk Reduction (DRR)

The Brahmaputra Valley in the State of Assam is one of the most hazard-prone regions of the country, with more than 40% of its land (3.2 million hectares) susceptible to flood damage. This is 9.4% of the country's total flood-prone area. About 7% of land in the state's 17 riverside districts have been lost because of river erosion over the past 50 years.

Various strong general flood management/mitigation measures are practised as below.

8.7.1 Structural Measures

The general approach about flood management has been in the form of physical measures to prevent the flood waters from reaching potential damage centres. The main thrust of the flood protection programme undertaken in India so far has been in the nature of structural measures as follows:

8.7.1.1 Embankments

Embankments, including ring bunds and town protection works, confine the flood flows and prevent spilling, thereby reducing the damage. These are generally cheap, quick and the most popular method of flood protection and have been constructed extensively in the past. These are reported to have given considerable protection at comparatively low costs, particularly in the lower reaches of large rivers. In many places, embankments may be the only feasible method of preventing inundation.

Embankments are designed and constructed to afford a degree of protection against floods of a certain frequency and intensity or against the maximum recorded floods till the time of their planning only (in the absence of detailed hydrological data for longer periods) depending upon the location protected and their economic justification. The raising and strengthening of existing embankments have also been taken up in many of the flood-prone states (Fig. 8.3).

8.7.1.2 Dams and Reservoirs

Dams and reservoirs can moderate the intensity and timing of the incoming flood. They store the water during periods of high discharges in the river and release it after the critical high flow condition is over, so as to be ready to receive the next wave. Their effectiveness in moderating floods would depend on the reservoir capacity available at that time for absorbing the flood runoff and their proximity to the likely damage centre. They are operated with a carefully planned regulation schedule which takes into account both the safety of the dam and related structures and the safe carrying capacity of the lower reaches of the river in their present condition.

Fig. 8.3 Strengthening of embankments in flood-prone areas

8.7.1.3 Natural Detention Basins

Natural detention basins are usually formed by utilizing natural depressions/swamps and lakes by improving their capacity by constructing encircling embankments and providing suitable devices for regulating the release of stored waters. Since the land under the marshes or low depression may hardly require much compensation and rehabilitation measures, these methods are relatively inexpensive. Moderation of floods by utilizing natural depression, swamps and lakes to which a portion of flood water could be diverted, can also be an effective method of protection from floods depending on topography of the particular area.

8.7.1.4 Channel Improvement

Adequate channel can be made for carrying its flood discharge at levels lower than its prevailing high flood level by improving the discharge capacity of the channel. This measure for channel improvement by increasing the area of flow or the velocity of flow (or both) can be taken up as flood protection measures so that the banks filling and releasing floods will be controlled. Channel improvement has not been resorted widely in India mainly because of the high cost involved. However, it may be of advantage to take up such work for local reaches.

8.7.1.5 Drainage Channels

Surface water drainage congestion, due to inadequacy of natural or artificial drainage channels, results in floods in many areas. In such cases, drainage improvement by construction of new channels to improve the drainage capacity from the affected area constitutes an effective means of flood control. The increased drainage by such measures reduces the extent of duration for which flood waters remain in the area. The possibility of drainage congestion and flooding in the downstream area as a result of adoption of drainage improvement scheme taken up in upstream area is to be kept in mind while formulating such schemes.

8.7.1.6 Diversion of Flood Waters

Schemes for permanent diversions of flood waters come under river diversions. These can be employed for lowering water levels by diverting all or a part of the discharge into a natural or artificially constructed channel, lying within or in some cases outside the flood plains. The diverted water may be taken away from the river without returning it further downstream, or it may be returned to the river channel some distance downstream or to a lake or to the sea. Diversion of floodwaters takes a part of the flood discharge to another basin or to the same basin downstream of the problem area or to a depression where it could be stored for subsequent release.

8.7.1.7 Watershed Management/Catchment Area Treatment

The watershed management measures include developing and conserving the vegetative and soil covers and undertaking structural works like check dams, detention basins, diversion channels, etc. This helps in serving as an effective measure in reducing the flood peaks and controlling the sudden surge of the runoff except during large flood. The watershed management through above-mentioned measures in the upper catchment helps in arresting the silt from migrating which has been proven to be one of the aggravating factors of floods in certain rivers.

8.7.1.8 Anti-erosion Works

Alluvial rivers are usually meandering in nature and, therefore, raise problems of erosion and silting at various locations. This process is a natural phenomenon and results in loss of land at one location and gain at some other. Generally, there is a tendency of the meander to shift progressively downstream. The process of bank erosion is, therefore, consistently active and measures for protection of banks are a recurring necessity (Fig. 8.4).

Fig. 8.4 Anti-erosion measures

8.7.2 Nonstructural Measures

Nonstructural measures strive to keep the people away from flood waters, bearing in mind the stark reality that the flood plains in fact belong to the river and that the flood perceived only as a curse could be turned into a blessing in disguise in some ways. It contemplates use of flood plains judiciously, simultaneously permitting vacating of the same for use of the river whenever the situation calls for. This technique allows the use of flood plains reducing the disaster dimension, while retaining its beneficial efforts. In view of cost-effectiveness of the nonstructural measures and speedier implementation, and in recognition of the fact that more and more human encroachments and activities are taking place in the flood plains in our country, the main thrust is now on the nonstructural flood management measures (Flood and Flood Control Measures 2011). The nonstructural measures are broadly grouped as follows:

1. Flood plain zoning
2. Flood proofing
3. Flood forecasting and warning
4. Regulation of reservoir
5. Capacity building
6. Research and development

8.7.2.1 Flood Plain Zoning

It is natural for a river to overflow its banks in the event of heavy rainfall in its upper catchment and spill into the flood plains which are basically its domain. Extensive and often unplanned use of flood plains by man disregarding the basic fact that it is a part and parcel of the river leads to damage. Thus, the uncontrolled and indiscriminate development of flood plains due to pressure of population can be considered as one of the main factors responsible for the flood damage reported from different parts of the country in spite of substantial investments in the flood sector during the last four or five decades.

The basic concept of flood plain management is to regulate the land use in the flood plains to reduce the damage due to floods and enhancing the benefit opportunities from the same. This is done by determining the locations and the extent of areas likely to be affected by floods of different magnitudes/frequencies and by developing those areas in such a fashion that the zoning, therefore, aims at disseminating information on a wider basis so as to regulate indiscriminate and unplanned development in flood plains and is relevant both for unprotected and protected area (Flood and Flood Control Measures 2013).

8.7.2.2 Flood Proofing

Flood-proofing measures help greatly in the mitigation of distress and provide immediate relief to the population in flood-prone areas. It is essentially a combination of structural change and emergency action not involving any evacuation. The

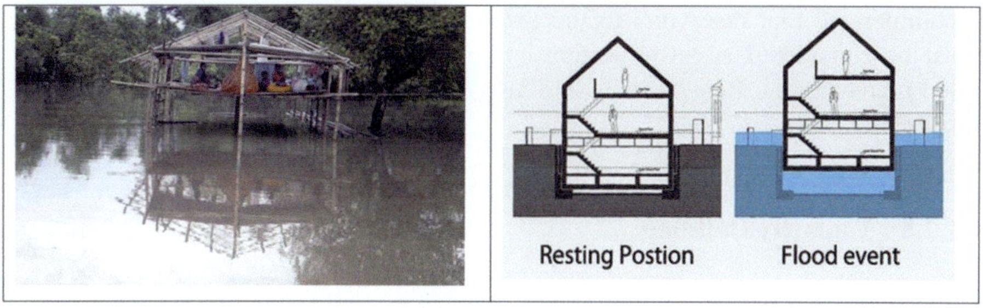

Fig. 8.5 Flood-proofing measures

techniques adopted consist of providing raised platforms for flood shelter for men and cattle and raising the public utility installation above flood levels (Fig. 8.5).

In case of urban areas, certain measures that can be put into action as soon as a flood warning is received involve installation of removable covers such as steel or aluminium bulk heads over doors or windows, permanent closure of low-level windows or other openings, keeping store counters on wheels, closing of sewer well, anchoring machinery, and covering machinery with plastic sheet, seepage control, etc.

8.7.2.3 Flood Forecasting and Warning

Of all the nonstructural measures for flood management which rely on the modification of susceptibility to flood damage, the one which is gaining increased/sustained attention of the planners and acceptance of the public is the flood forecasting and warning. Flood forecasting enables forewarning as to when the river is going to use its flood plain, to what extent and for how long. As per strategy of laying more emphasis on nonstructural measures, a nationwide flood forecasting and warning system has been established by the Central Water Commission.

8.7.2.4 Regulation of Reservoirs

The Working Group on Flood Control Programme for the Tenth Five-Year Plan had observed that the reservoirs provide long-term solution to the problem of floods, and it had recommended that a reservoir is more effective for flood control if a designated space is reserved and a reservoir is more effective for flood control if regulated in accordance with laid down procedures.

The National Water Policy (2002) has recommended provision of adequate flood cushion in water storage projects and flood control to be given overriding consideration in reservoir regulation policy.

However, the reservoirs would involve submergence, and their effectiveness decreases if the flood-prone area is located much downstream of the dam. The RBA recommended that reservoirs to the extent technically and economically feasible must be considered as an important component in any package of measures for flood management. Reservoirs would also ensure the optimum utilization of water resources.

8.7.2.5 Capacity Building

If the flood problem of Assam is to be managed effectively, a larger capacity for understanding the problem and planning mitigation measures would have to be provided for. The strategy would have to consist of:

1. Building a scientific database
2. Building a large R&D capacity
3. Human resource development

Fluvial hydraulics is a branch of engineering where the complex mechanics of water sediment transport and morphological changes are studied. Physical hydraulic models are useful but not sufficient. Mathematical models dealing with both the phenomena are increasingly coming into operational use, but these require access to large computers and computational skills.

A major gap in flood management activities is the lack of availability of data/information, which is required in different situations. For example, information on flooded area showing extent of inundation due to floods, depth and duration of flooding are important parameters which depend on the plan and design of measures required to solve the problem. Similarly, precise and correct information of flood damages is very essential for planning appropriate measures and for deciding priority for their implementation. The Rashtriya Barh Ayog has also emphasized the need for scientific system for flood damage assessment. There is ample scope for collecting the flood-related data/information through remote-sensing technique. As this entire work is voluminous and is required to be seen in totality, exclusive flood data centres are required to be established as state and central levels.

8.7.2.6 Research and Development

Flood inundation mapping serves many purposes. Apart from showing the critical spread of flood-affected area, these identify areas needing post-flood all-relief measures. It also brings out the effectiveness of the flood control works and their effect on river flow. This information can be used for further planning, strengthening, or additional measures/works. The RBA recommended that further studies should be carried out towards developing watershed models suitable for using remotely sensed information as inputs, in order to predict flood flow under ‘inadequate’ or ‘no data’ situations. It also recommended that efforts should be intensified to evolve more and more mathematical models and use them to remove empiricism and introduce better rationality in decision-making process.

It is recommended that for flood management works of permanent nature, e.g. embankment, spurs, revetments, etc. involving huge costs and having significant impact on river behaviour, mathematical model studies for long reaches complemented by physical model studies for problem reaches should invariably be undertaken. The model study facilities in the research stations of the flood-prone states should be suitably upgraded and requisite funds provided. Adequate administrative and financial autonomy needs to be given to the research organizations such as CWPRS for speeding up of these studies. Considering the importance of the studies, the research organizations should be exempted from ban on recruitment of staff.

8.8 Recommendations and Conclusions

Disaster management studies have pointed out the importance of participation and collaboration by different stakeholders at different stages of the planning and management of disasters (Berke et al. 1993; Warner et al. 2002; Sharp 2007; Warner 2008; Tran et al. 2009).

Unless suitable steps are taken, Assam will move relentlessly towards a catastrophe: the persistent problem of flooding and its multitude devastating effects. The unruly floods hit hard the river basins of Brahmaputra and Barak, a constant reminder of Assam's unfortunate geotectonic positioning on the surface of the earth. Yearly monsoon brings devastating floodwaters to the doorsteps of the people, bringing immense loss to life and properties of poor populace.

When the waters recede and the cheerful dry harvesting season begins, people tend to forget about the flood damage until it happens again the following summer. The startling fact is that after about half a century of flood control measures and billions of rupees spent on them, the fury of flooding remains largely undiminished. Nothing so far has effectively worked with the enormous flood control plans and programmes. Clearly, it is time to take a hard look at the whole situation and to move quickly ahead with a new course of action, with all the past learning.

The most popular flood control measure in Assam undertaken so far is the construction of over 4000 km of earth embankments. The track record of these embankments leaves a lot to be desired in terms of effective and beneficial flood control. They have been largely unsuccessful and much of them have outlived their design life. These embankments have been put together without consideration of short- and long-term adverse effects, like waterlogging and loss of soil fertility. It's time to revisit the design and to embark upon a new reconstruction programme providing strength, stability, flow control devices and enhanced elevation.

The second strategy of flood mitigation is dredging the river beds; this is especially significant with respect to the Brahmaputra and its tributaries since situation and other factors have raised their beds over the years and exacerbated flooding around their banks; however, no significant dredging has been attempted thus far. Execution of dredging projects without potential corruption is a prerequisite for dredging to be successful. Additionally, a plan must be designed for the disposal of dredged material. The dredged material can be programmatically used for filling low-lying areas, and they can be reclaimed for building housing on higher grounds.

Any consideration for a dam should be subject to extremely careful technical review from the viewpoint of safety in a credible earthquake shock environment. For Assam, the risk of building big dams in any state in the northeast region far outweighs benefits because of the fact that the whole region is characterized by high seismicity. It is well known that in the entire recorded history of mankind, Assam has experienced two of the most severe earthquakes, each with a magnitude of more than 8.5 on the Richter scale.

The last but perhaps the most significant aspect of flood control is implementation of projects without corruption. Corrupt practices in implementing flood control projects should be avoided at any level. Sacrifice of personal enrichment is a prerequisite to saving Assam and gives it a solid foundation for progress, peace and prosperity.

In reference to yearly flooding, the disaster management authorities can easily identify households/beneficiaries who are affected yearly and give them an identification card, mentioning their prospective place of camping (with certain number), which would help the administration to arrange for relief resources and other logistics, without having much difficulties. Apart from these, the recommendations proposed for effective mitigation measures should be considered seriously to create a sustainable flood-mitigated society. All district disaster management plans should have a uniform format or pattern and should highlight the plans or strategies to respond to various disasters and not report a single piece of incidents (which can be formed part of the individual reports). The plans should be updated at least once in a year, and if possible, once in 6 months is recommended.

References

ADB (2009) Environment assessment report, India: Assam integrated flood and riverbank erosion risk management investment program, ADB, June 2009

ASDMA (2013) Assam state disaster management plan. Assam State Disaster Management Authority, 2013

Berke PR, Kartez J, Wenger D (1993) Recovery after disaster: achieving sustainable development, mitigation and equity. Disasters 17(2):93–109

Dhar ON, Nandargi S (2003) Hydrometeorological Aspects of Floods in India. Nat Hazards 28(1):1–33

Djalante R, Holley C, Thomalla F (2011) Adaptive governance and managing resilience to natural hazards. Int J Disaster Risk Sci. 2011 2(4):1–14

Flood and Flood Control Measures (2011) Chapter – VII, Economic survey, Assam, 2010–2011

Flood and Flood Control Measures (2013) Chapter – VII, Economic survey, Assam, 2012–2013

Flood Situation Meeting Minutes from Barpeta District on 20th–23rd May, 2014 held at DC's Conference Hall, Barpeta, Dhubri, Dhemaji, and Jorhat Assam

ISRO (2011). Flood Hazard Atlas for Assam State, 2011.

Pal I, Singh S, Walia A (2013) Flood Management in Assam, INDIA: a review of Brahmaputra Floods, 2012. Int J Sci Res Publ 3:10. 2013 (ISSN: 2250-3153)

Pal I, Ghosh T, Ghosh C (2017) Institutional framework and administrative systems for effective disaster risk governance – perspectives of 2013 Cyclone Phailin in India. Int J Dis Risk Reduct, Elsevier

Sharp B (2007) Food aid and development in southern Sudan: implications of the comprehensive peace agreement for response planning. Disasters 31(S1):104–123

Tran P, Shaw R, Chantry G, Norton J (2009) GIS and local knowledge in disaster management: a case study of flood risk mapping in Vietnam. Disasters 33(1):152–169

UNISDR, (2008), Briefing note on "Climate Change and Disaster Risk Reduction" by the UN International Strategy for Disaster Reduction, Geneva, September 2008

Walia A, Pal I, Singh S (2013) Assam floods 2012 – a case study of District Barpeta. Dis Resp Manag 1(1). ISSN: 2347–2553

Warner J (2008) Emergency river storage in the Ooij polder—a bridge too far? Forms of participation in flood preparedness policy. Int J Water Resour Dev 24(4):567–582

Warner J, Waalewijn P, Hilhost D (2002). Public participation in disaster-prone watersheds. Time for multi-stakeholder platforms? Wageningen: Disaster Studies Wageningen University. http://www. disasterstudies.wur.nl/NR/rdonlyres/E08690FD-1923-4358-B3B4-FBA2F7734B91/25010/no6mspindisaster.pdf

Chapter 9
Lessons Learned from 16/17 June 2013 Disaster of Uttarakhand, India

Piyoosh Rautela

Abstract Early arrival of monsoon, its collision with westerlies, ensuing abnormally high precipitation resulting in fast melting of fresh snow accompanied by breach of a glacial lake, Chorabari Tal; stage was thus set for a major hydrometeorological disaster in Uttarakhand in June 2013. With no specific warning, a disaster of this magnitude was not anticipated, and everyone was taken by surprise. The incidence also coincided with the peak of the tourist season when pilgrims, tourists, and others from across the country and abroad had gathered in the region in large numbers. The incidence thus resulted in a major tragedy of recent times in which human death toll surpassed 4000. The disaster was a major setback for the economy of the state to which tourism and pilgrimage are major contributors. Burden on public exchequer was further exacerbated by tax waiver and assistance provided to disaster-affected population even on losses that are not generally covered by State Disaster Response Fund (SDRF). Important lessons were however learned in managing this disaster, and these would go a long way in strengthening the disaster management system not only in Uttarakhand but also elsewhere.

Keywords Uttarakhand • Kedarnath • Higher Himalaya • Monsoon • Landslide • Toe erosion • Flash flood • Glacial lake outburst

9.1 Introduction

Even after more than 3 years, what exactly transpired in Uttarakhand, particularly in Kedarnath area, on 16/17 June 2013 remains to be convincingly settled, and hypothesis of various sorts are often put forth and keenly debated even today, not only by scientists and researchers but also by environmentalists, media personnel, and masses (Dobhal et al. 2013; Rana et al. 2013; Rautela 2013; Uttarakhand Flood

P. Rautela (✉)
Department of Disaster Management (Government of Uttarakhand), Disaster Mitigation and Management Centre, Dehradun, Uttarakhand, India
e-mail: rautelapiyoosh@gmail.com

I. Pal, R. Shaw (eds.), *Disaster Risk Governance in India and Cross Cutting Issues*, Disaster Risk Reduction, DOI 10.1007/978-981-10-3310-0_9

Disaster 2013; Bandyopadhyay and Sekhar 2014; Chopra 2014; Dube et al. 2014; Expert Committee Report 2014; Kotal et al. 2014; Allen et al. 2015; Chattoraj and Champatiray 2015; Singh et al. 2015). Such is the power of this suspense that the tragedy still hits the headlines of local and national media, particularly around 16/17 June; what happened, what was done, what went wrong, what could have been done, what lessons were learned, and the like. No dimension of this tragedy has gone uninvestigated, and yet there remain a number of unanswered questions.

Environmentalists often attribute this tragedy to anthropogenic interventions, particularly hydropower projects (Uttarakhand Flood Disaster 2013; Chopra 2014; Expert Committee Report 2014). The script of the disaster was however written in sparsely populated Higher Himalayan region of Uttarakhand that is bereft of anthropogenic pressure of any sort. Moreover, the Forest Conservation Act, 1980, and other related legislations prohibit and restrict human interventions in this region. Though strategically important, this region at the same time has limited road connectivity, which further limits human interference. Furthermore, large portion of the region remains snowbound, particularly during winters when even the inhabitants of the frontier villages retreat to lower elevations.

Most people are aware of devastation in Mandakini valley, particularly around Kedarnath. This is attributed largely to high media coverage it received due to massive loss of human lives. The entire Higher Himalayan region of Uttarakhand from Kali river valley in the east to Yamuna river valley in the west was however devastated by this disaster, and 5 of the 13 districts of the state, namely, Rudraprayag, Chamoli, Uttarkashi, Bageshwar, and Pithoragarh, were worst hit (Fig. 9.1).

This disaster manifested itself in the form of flash flood, landslide, debris flow, and toe erosion that are owed to excessive rainfall in the Higher Himalayan region, which generally receives major portion of precipitation in the form of snow. In 2013, monsoon arrived early in the region, and there was excessively heavy rainfall right in the beginning. As per the Indian Meteorological Department (IMD), the rainfall in the state between 15 and 18 June 2013 was measured to be 385.1 mm against the normal rainfall of 71.3 mm, augmented by 440 %. This is attributed to the confrontation of the SW monsoon front with westerlies (IMD 2013).

Heavy rainfall in the upper reaches resulted in water levels of all major rivers to rise, and fast melting of winter snow due the impact of falling rain drops only worsened the situation. Major devastation was largely caused by toe erosion by fast-flowing debris-laden mountain rivers.

9.2 Disaster-Affected Area and Its Vulnerability

Major portion of the state of Uttarakhand is located in the Himalayan terrain and has an altitudinal range of 200–7784 m above sea level (asl). The state shares its border with Nepal in the east and Tibet (China) in the north. The state has 2 administrative divisions, Garhwal and Kumaun, and 13 districts. Of these, five northern districts,

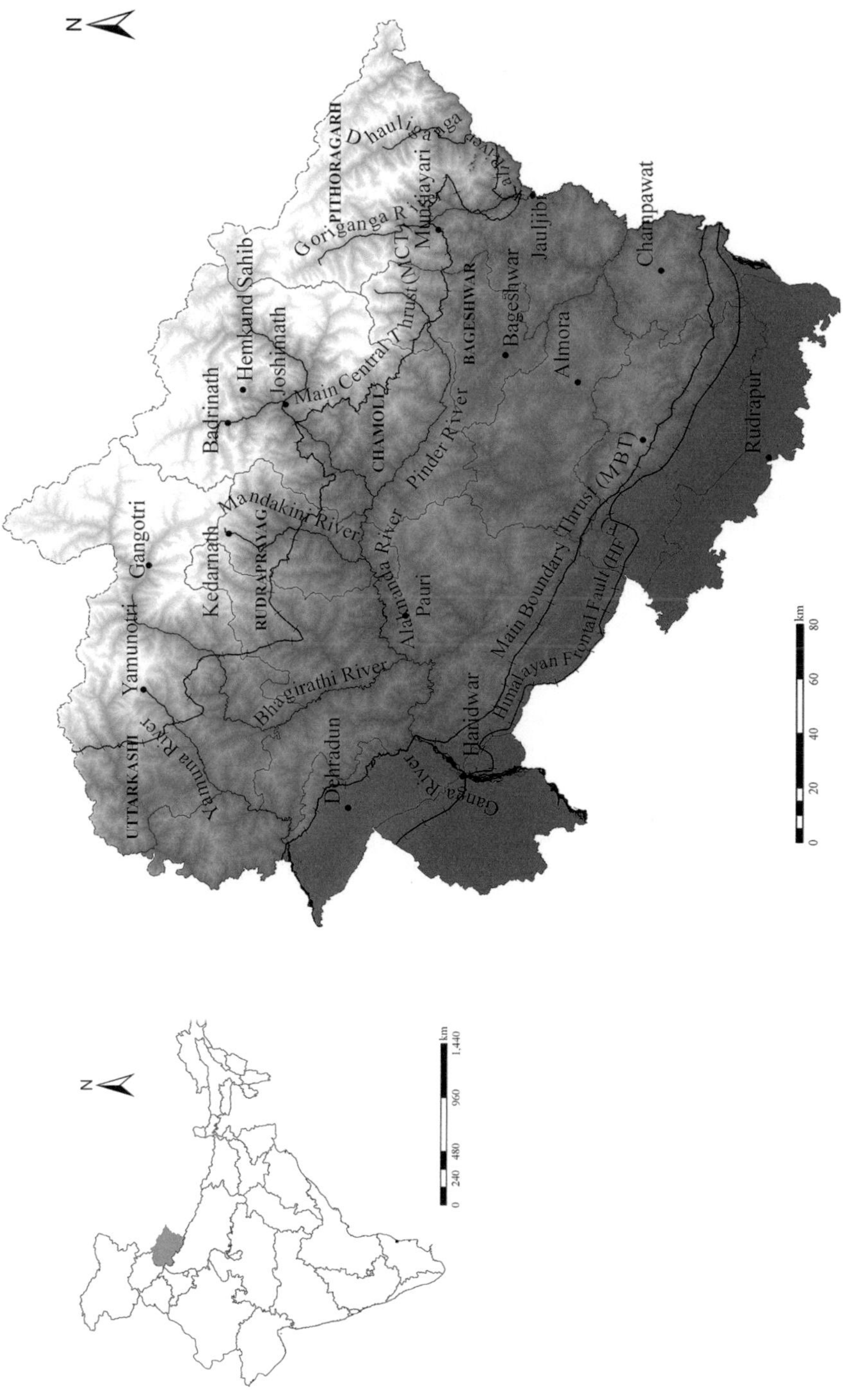

Fig. 9.1 Map depicting the location of the disaster-affected districts of Uttarakhand

namely, Bageshwar, Pithoragarh, Uttarkashi, Chamoli, and Rudraprayag, were worst affected by the disaster of 2013 (Fig. 9.1).

Geologically the disaster-affected area falls in Lesser Himalaya, Central Crystallines, and Higher Himalaya. The Main Central Thrust that is a major tectonic discontinuity of the Himalaya and along which Central Crystallines are juxtaposed against Lesser Himalaya along a NNE dipping thrust traverses through it. The area has particularly high relative relief that promotes mass wastage and erosion. Except for Uttarkashi, some portion of which falls in Zone IV, all the disaster-affected districts fall in Zone V of seismic zonation map of India (IS 1893, 2002). Geological history, ongoing tectonic activities, and high relative relief coupled with peculiar meteorological characteristics make the area vulnerable to a number of hazards of which earthquake, landslide, and flash flood are common.

The disaster-affected area of the state is source to major glacier-fed Himalayan rivers that include Alaknanda, Bhagirathi, Mandakini, Yamuna, Kali, Dhauli, and Pindar. Alaknanda and Bhagirathi confluence at Devprayag, and thereafter the river is known as Ganga.

Mandakini valley of Rudraprayag district that was hit the hardest by the disaster of June 2013 houses the sacred Hindu shrine of Kedarnath that is dedicated to lord Shiva, the god of death and destruction. The temple township (Fig. 9.2) is located on glacial outwash deposits at an altitude of 3581 m asl. The main shrine is located on raised middle portion of the deposit that is 20–25 m above the level of Mandakini (3562 m asl). For reaching Kedarnath, one had to trek upstream along the course of Mandakini from Gaurikund for a distance of 14 km (Fig. 9.3).

Originating from Chorabari glacier, Mandakini forms the western boundary of the temple township, while the abandoned channel of Saraswati that had confluence with Mandakini to the south of the temple forms the eastern boundary. Dudh Ganga meets Mandakini to the south of Kedarnath, and thereafter till Gaurikund, Mandakini maintains a tectonically controlled NNE-SSW course (Fig. 9.3).

A moraine-dammed lake, Chorabari Tal, was present little downstream of the snout of Chorabari glacier. This lake was located in the depression formed in the glacial material to the west of the right lateral moraine and was fed by the seepage of the glacial melt. The lake did not have a well-defined outlet, and its water used to

Fig. 9.2 View of the temple township of Kedarnath, with camera looking east

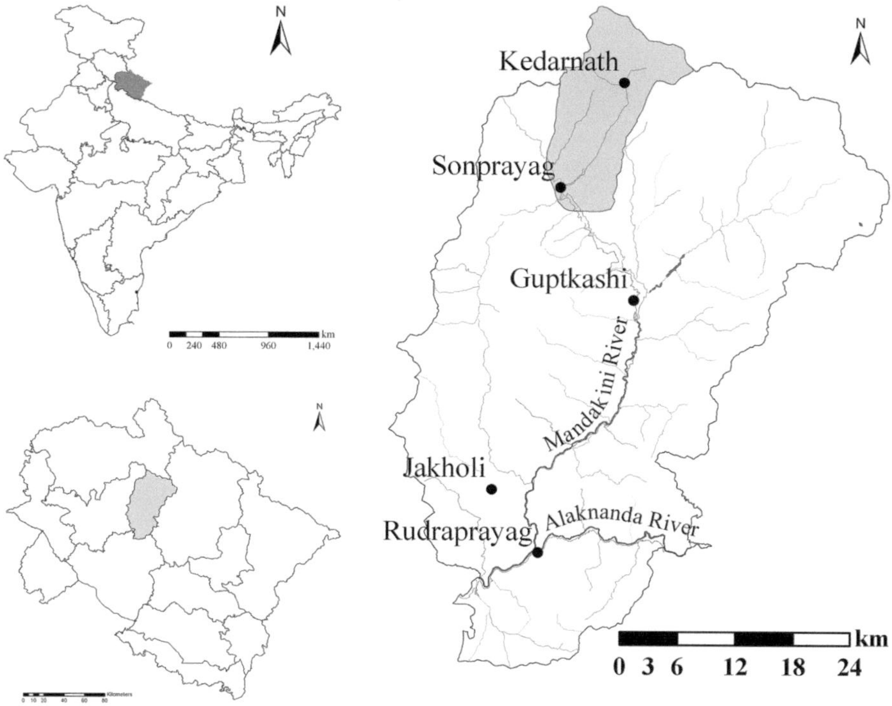

Fig. 9.3 Location map of Kedarnath area. To the left are maps of India, Uttarakhand, and Rudraprayag district

seep out along the moraine slope to the NNW of Kedarnath. Even though the depression was around 200 m long, 100 m wide, and 15–20 m deep, not more than 2–3 m water used to be there in the lake.

The disaster-affected area houses a number of sacred shrines and pilgrimage routes. Besides Chota Kailash-Kailash-Mansarovar and Hemkund Sahib, these include Chardham route leading to Badrinath, Kedarnath, Gangotri, and Yamunotri that is the biggest and most cherished Hindu pilgrimage circuit of the country (Fig. 9.1). Being located in Higher Himalaya, the duration of the pilgrimage is however restricted between June and November; exact dates of opening and closing of the sacred shrines are decided according to Hindu tradition.

The area also has a number of picturesque tourist destinations that include Joshimath, Auli, Chopta, Gopeshwar, Bageshwar, and Munsyari. People from across the country and abroad thus visit the area in large numbers. Tourism and pilgrimage are thus a major source of income for the people of the area as also for the state.

Being associated with the abundance of water, both landslide and flash flood are common in the region during the monsoon period, rainy season over the India subcontinent that extends between mid-June and mid-September. This period coincides with the ingress of both pilgrims and tourists in the area in large numbers. Road

Fig. 9.4 View of road disruption in June 2013 due to bank erosion in the proximity of Tawaghat, Pithoragarh (*left*), and debris slide in Dharali, Uttarkashi (*right*)

connectivity is often disrupted for long periods during this time due to landslides and toe erosion by the rivers (Fig. 9.4; Table 9.1). It is a cause of inconvenience, discomfort, and misery for pilgrims, tourists, and others who are often forced to change their travel plan. At the same time, it is a cause of major concern for the state that has to resort to extraordinary measures for evacuating the stranded people and ensuring supply of essential items.

9.3 The Disaster of 16/17 June 2013

There was heavy rainfall in the entire state with the onset of monsoon that arrived early in 2013. This is attributed to the clash of the SW monsoon front with the westerlies. Prolonged and unprecedented heavy rainfall for consecutive days between 14 and 18 June, 2013, over a large area, resulted in flash flood and landslides at many locations, which eventually turned into a massive disaster.

The rainfall in the state between 15 and 18 June 2013 is measured to be 385.1 mm against the normal rainfall of 71.3 mm, which is in excess by 440 %. In the period of 5 days between 14 and 18 June, the state received approximately 2000 mm of rainfall, which is more than what it receives during the entire monsoon period (Table 9.2).

Percent deviation in rainfall clearly shows that the rainfall during the week ending on 12 June (6–12 June 2013) was more than 100 % in all the districts except Pithoragarh. The rainfall however increased enormously in the subsequent week when it was measured to be 997 % higher than normal over the state. Except for Pithoragarh and Rudraprayag, deviation from normal in other three districts was more than 1000 % in the week ending on 19 June (Table 9.3).

Fast melting of fresh snow due to rainfall impact added to the discharge of the streams and rivers that crossed the danger level. The level of Mandakini at Rudraprayag was 7.5 m above the danger level on 18 June 2013, and around this time, most rivers in the region were flowing well above their normal levels (Fig. 9.5). Gushing debris-laden water through the high-gradient mountain streams thus breached the banks and washed off roads, bridges, habitations, and other infrastructure on their way.

Table 9.1 Disruption of major highways due to landslides and toe erosion during the monsoon period between 2010 and 2015

Sl. no.	Highway	Year	Number of days when traffic was disrupted on the highway					Road disruption (in percent)
			June (30 days)	July (31 days)	August (31 days)	September (30 days)	Total (122 days)	
1.	Rishikesh to Badrinath (NH 58)	2010	1	17	25	9	52	42.6
		2011	2	13	14	10	39	32.0
		2012	2	6	18	7	33	27.1
		2013	15	20	17	13	65	53.3
		2014	1	5	4	4	14	11.5
		2015	5	15	17	0	37	30.3
		Total	26	76	95	43	240	32.8
2.	Karnaprayag to Kedarnath (NH 109)	2010	3	12	9	15	39	32.0
		2011	5	9	17	4	35	28.7
		2012	5	8	6	5	24	19.7
		2013	13	21	18	13	65	53.3
		2014	2	11	2	5	20	16.4
		2015	5	12	5	0	22	18.0
		Total	33	73	57	42	205	28.0
3.	Dharasu to Yamunotri (NH 94)	2010	0	7	40	19	66	54.1
		2011	5	5	14	12	36	29.5
		2012	4	5	16	13	38	31.2
		2013	12	20	22	18	72	59.0
		2014	1	10	4	5	20	16.4
		2015	0	4	2	0	6	4.9
		Total	22	51	98	67	238	32.5
4.	Rishikesh to Gangotri (NH 108)	2010	1	9	38	19	67	54.9
		2011	7	14	16	19	56	45.9
		2012	5	4	22	10	41	33.6
		2013	17	16	18	17	68	55.7
		2014	0	19	7	3	29	23.8
		2015	1	4	6	0	11	9.0
		Total	31	66	107	68	272	37.2

Data source: State Emergency Operations Center, Uttarakhand

Devastation was particularly severe in the Mandakini valley, particularly in Kedarnath-Rambara-Gaurikund area. This is attributed to the breach of Chorabari Tal that had accumulated enough water to force the moraine barrier to give way (Dobhal et al. 2013; Rautela 2013).

Late in the evening of 16 June 2013, debris brought down by Dudh Ganga blocked Mandakini river in the proximity of Kedarnath. The embankment on the left bank of the Mandakini soon gave way, and the abandoned channel of Saraswati to the east of Kedarnath became active. This resulted in washing off of some people in the evening of 16 June 2013 from Kedarnath that subsequently became waterlocked. Sankaracharya Samadhi, Jal Nigam Guest House, and Bharat Seva Sangh Ashram

Table 9.2 Precipitation as recorded by IMD stations in the disaster-affected areas between 14 and 18 June, 2013

Sl. no.	Location	Precipitation (in mm) 14 June	15 June	16 June	17 June	18 June
1.	Bhatwari	20.0	18.0	35.0	70.0	50.0
2.	Barkot	10.0	15.4	112.6	20.0	20.0
3.	Chamoli	1.0	40.0	58.0	80.0	100.0
4.	Jakholi	30.0	70.0	121.0	110.0	70.0
5.	Joshimath	0.0	31.4	41.9	113.8	80.0
6.	Karnaprayag	8.2	7.0	88.0	90.0	82.3
7.	Munsyari	4.0	25.0	44.0	85.0	75.0
8.	Pithoragarh	0.0	0.0	11.2	90.0	120.0
9.	Purola	30.0	40.0	170.0	60.0	104.0
10.	Rudraprayag	4.0	11.8	89.4	92.2	59.2
11.	Tharali	0.0	15.0	58.0	173.0	80.0
12.	Uttarkashi	15.0	50.0	130.0	162.0	19.0

Data source: India Meteorological Department, Government of India

Table 9.3 Weekly percentage deviation of actual rainfall from normal in the disaster-affected districts of Uttarakhand in June–July 2013

Sl. no.	District	Percent deviation in rainfall in the week ending on 12 June	19 June	26 June	3 July	10 July	17 July	24 July
1.	Bageshwar	137	1387	−58	59	127	−45	−22
2.	Chamoli	185	1302	111	37	187	96	96
3.	Pithoragarh	13	238	−41	−50	−4	49	−10
4.	Rudraprayag	213	580	74	−35	−21	−51	16
5.	Uttarkashi	112	1356	−1	−12	−22	−12	24

Data source: India Meteorological Department, Government of India

were also washed off in this event. Rising level of the landslide-dammed lake forced the barrier to give way, and the ensuing floods late in the evening of 16 June 2013 completely washed off Rambara and devastated Gaurikund. All connectivity with the area was thus snapped (Rautela 2013).

Persistent heavy rains caused the level of water in Chorabari Tal to rise continuously. With the recession of the glacier, the lake had a weak moraine barrier that could not withstand mounting hydrostatic pressure. Stage was thus set for a major disaster in Kedarnath, and the barrier ultimately gave way around 7 AM on 17 June 2013. The volume of water was enormous, and it carried with it huge glacial boulders and outwash material that choked the course of Mandakini, and the flow of water and debris got diverted toward the temple township that was thus ravaged (Fig. 9.6).

There was absolutely no warning and most people were taken by surprise and had no time to respond. Besides Kedarnath, this event caused devastation in Gaurikund, Sonprayag, and other places.

Fig. 9.5 View of the motor bridge over Alaknanda at Rudraprayag in April, 2013 (*left*), and on 17 June, 2013 (*right*), with camera looking NNW

Fig. 9.6 View of the Kedarnath township ravaged by the flood of June 2013 with camera looking south

9.4 Human Congregation in Kedarnath

Higher Himalayan shrines in Uttarakhand witness high influx of devotees in the beginning of the pilgrimage season due to pleasant weather conditions, less rains and road disruptions, and summer vacations in schools and colleges. The number of people that had gathered in the Gaurikund-Rambara-Kedarnath area when the disaster took place in June 2013 was unusually high by all standards. Interplay of number of factors was responsible for this.

Due to Uttarakhand High Court order on the issue of sanitation on pedestrian route to Kedarnath, movement of horses and mules was temporarily suspended in 2013. This slowed the pace of movement, and people who could easily travel back to Gaurikund on mule or horseback were forced to stay overnight at Kedarnath or Rambara.

Moreover, helicopter service from Guptakashi to Kedarnath was suspended due to continuous rainfall resulting in poor visibility. The ones intending to travel by air thus had no option but to travel on foot and stay overnight at Kedarnath or Rambara.

Apart from this, continuous heavy rainfall severely slowed down the pace of movement, and many people were forced to change their travel plan and stay overnight at Kedarnath or Rambara.

It was the peak pilgrimage season and Gaurikund-Rambara-Kedarnath area was highly overcrowded. With no provision of registration in place at that time, no one really has an idea as to how many people were actually there in this stretch when the disaster struck.

9.5 Losses

Landslides, flash flood, and toe erosion induced by incessant heavy rainfall in the Higher Himalayan region led to massive loss of human lives, infrastructure, and property. The details of the same are summarized in Table 9.4. It is to be noticed that apart from loss of human lives, other losses in the disaster-affected districts are comparable. It is due to the loss of human lives in Mandakini valley that most people consider 2013 disaster to be restricted to Rudraprayag district.

Besides direct losses and cost of search, rescue, and evacuation, the state incurred heavy loss of revenue due to the disaster. Economy of the state is highly dependent upon pilgrimage and tourism, and there was steep decline in the number of people visiting the state on the aftermath of the disaster (Table 9.5). Resumption of normal pilgrim and tourist inflow is sure to take some time, and this would require dedicated effort to convince the potential visitors that Uttarakhand is safe to visit.

On the aftermath of the disaster, the state had also to waive off tax dues, particularly on tourism and hotel industry. Moreover, in view of severity of the disaster impact, the state also extended relief on items that are not generally covered by State Disaster Response Fund (SDRF) and raised the quantum of the admissible relief by many times. In order to help them sustain during this crisis, relief was also provided to commercial enterprises that are not generally covered by SDRF norms.

Table 9.4 District-wise losses incurred in the state due to the disaster of June 2013

Head	Rudraprayag	Chamoli	Uttarkashi	Pithoragarh	Bageshwar	Total
Persons dead	30	33	14	19	4	100
Persons missing	3998	0	0	21	0	4019
Persons injured	71	31	29	49	3	183
Farm animals lost	2771	1119	474	5263	665	10,292
Fully damaged houses	445	566	249	702	25	1985
Severely damaged houses	374	647	527	340	130	2018
Partially damaged houses	905	2188	1967	397	133	5590
Silted agriculture land (in hectares)	1	379	341	12	8	741
Agricultural land lost (in hectares)	4279	242	339	5575	234	10,669
Crop loss (in hectares)	8	245	103	113	28	497

Data source: State Emergency Operations Center, Uttarakhand

9.6 Aftermath of the Disaster

Though there was general forecast of heavy rains over the entire state, none had expected such a large area to be affected in one go. Communication was hit hard and there was literal information vacuum, and no one had the slightest clue as to what had transpired. In the initial phase, this caused confusion and delayed response and resource mobilization. There was also severe disruption of surface transport network, and almost all the major highways of the state were blocked due to landslides and washing away of roads. More than 150,000 persons were thus stranded at various places across the state.

The magnitude of human loss together with large number of persons stranded at various places in Mandakini and Alaknanda valleys, particularly at Kedarnath, Bhimbali, Gaurikund, Hemkund Sahib, Ghagharia, Govindghat, and Badrinath, resulted in high media attention in these areas. Ease of access due to the concentration of aerial rescue operations in these areas further promoted media to pay attention on this very area. All rescue and relief efforts were thus initially focused on

Table 9.5 Pilgrim footfall over major shrines in the disaster-affected area in the period 2011–2014

Name of place	Month	Number of pilgrims				
		2011	2012	2013	2014	Average fall in percentage in the year 2014
Badrinath	May	225,558	4,138,812	238,116	53,798	96.5
	June	398,260	358,662	251,808	42,676	87.3
	July	43,952	55,383	–	7560	77.2
	August	48,473	43,366	–	3368	89.0
	September	132,794	54,916	–	17,070	72.7
Kedarnath	May	245,821	298,182	149,689	13,823	94.0
	June	249,386	196,830	182,551	14,091	93.3
	July	29,216	27,712	–	3041	84.0
	August	11,759	11,496	–	944	87.8
	September	20,746	12,823	–	3,796	66.1
Gangotri	May	146,870	195,618	105,617	20,193	86.5
	June	233,190	175,272	104,136	15,656	90.8
	July	57,355	40,124	–	6857	78.9
	August	9,229	416	–	675	79.0
	September	21,843	3,970	–	3,337	61.2
Yamunotri	May	170,126	206,545	115,786	15,316	90.7
	June	196,833	142,182	136,997	13,709	91.4
	July	44,864	34,312	–	3725	85.9
	August	16,396	1001	–	612	89.4
	September	32,083	5,231	–	2393	80.8
Hemkund Sahib	May	–	–	12,430	4071	1.7
	June	257,133	178,049	61,867	19,466	88.3
	July	137,253	57,879	–	6907	89.4
	August	92,653	27,658	–	2213	94.5
	September	40,430	29,175	1753	3075	87.1
Total		2,862,223	6,295,614	1,360,750	278,372	92.1

Data source: Uttarakhand Tourism Development Board, Uttarakhand

Mandakini and Alaknanda valleys, and the people in the other disaster-affected areas of the state could not get immediate and adequate attention.

All communication links with Mandakini valley were snapped in the evening of 16 June 2013. Adverse weather and terrain conditions did not provide opportunity of resorting to alternative probes. The outside world and also the district administration therefore remained unaware of the events in Mandakini valley till the afternoon of 17 June 2013.

With improved weather and visibility, aerial rescue operations were however initiated in early hours of 18 June 2013, and 2007 persons were evacuated on the same day from Kedarnath. Realizing the constraints being faced in aerial rescue operations, 69 permanent and temporary helipads were quickly activated across the state, and besides 17 civilian choppers hired by the state government, 54 Indian Air Force (IAF) choppers and fixed-wing crafts, 7 Army Aviation choppers, and 4 choppers of other states were pressed into service. Due to limited stock of aviation turbine fuel (ATF) in the disaster-affected area and inability to immediately move in bowsers due to road disruption, precious time was however lost in refueling. Despite constrains of sorts with more than 39,165 persons evacuated by air alone, this proved out to be the biggest and most successful aerial evacuation ever attempted.

Ground search and rescue operations were slightly delayed due to washing off of motor roads and also pedestrian track leading to Kedarnath and Hemkund Sahib at many places. Despite best efforts and intentions, terrain conditions made it difficult even to air-drop food and water at many locations, and this added to the sufferings and trauma of the affected persons.

Besides civil administration and state police National Disaster Response Force (NDRF), Indo-Tibetan Border Police (ITBP), IAF, Indian Army, and Indian Navy joined rescue operations.

The evacuated persons were first brought to the operational roadhead at Guptakashi, Chinyalisaur, and Joshimath from where they were subsequently evacuated safely by road. Five hundred eighty-six buses and 1,440 taxies were thus requisitioned for evacuation. Seventy-one relief camps were organized across the state that catered to food, shelter, medical, and other needs of 151,629 persons for different durations.

Large number of evacuees had lost all their belongings and did not have resources to manage their journey back home. Arrangements were therefore made for the return journey of the evacuated persons who were also provided cash assistance to cover their en route expenses.

Despite best efforts, evacuation could only be completed on 23 June 2013. Rescue operations were however not risk-free and four choppers crashed during rescue operations. Forty-eight persons engaged in rescue operations including 5 of IAF, 9 of NDRF, 6 of ITBP, and 28 of state police, administration, forest, and civil aviation lost their lives while engaged in rescue operations.

9.7 Lessons Learned

Every disaster exposes the shortcomings of the system and highlights key elements of the vulnerability of the affected community. It thus provides an opportunity to plug the holes and be better prepared for the next disaster. Sharing of these experiences is vital as taking lead from these effective and context-specific measures can be taken for strengthening resilience and avoiding disaster-induced loss, misery, and panic. Issues related to the same are discussed in the sections below.

9.7.1 Telecommunication

All communication links with the disaster-affected area, particularly in the Mandakini valley, were disrupted in the evening of 16 June 2016. The last received message from Gaurikund-Rambara-Kedarnath area was hurriedly communicated through police wireless network at Rambara before its being washed off. It certainly indicated an emergency situation, but what exactly has transpired could not be comprehended from it. Mobile towers in the area also became dysfunctional soon due to electricity disruption, shortage of fuel to run the generators, and other technical snags. Moreover, batteries of most mobile phones were also drained off while still trying to search the network.

With communication having been disrupted, there was no information coming to State Emergency Operations Center (SEOC) from the disaster-affected area. Besides delaying response, this added to the confusion as different versions started being aired by the media due to lack of authentic information.

Steady inflow of updated, authentic, and reliable information is vital to effective management of any disaster. There thus has to be a mechanism for ensuring regular inflow of updated information from the disaster-affected area under all circumstances. The communication system should therefore be robust and reliable with at least triple redundancy so as to ensure functional alternative communication under all circumstances.

Disruption of communication at the same time adds to panic and trauma of disaster-affected people, and therefore alternative power arrangements have to be put in place for running mobile towers. The mobile service providers should therefore be persuaded to maintain adequate stock of fuel to operate the generators and install solar power backup facility, particularly in the remote and disaster prone areas. Solar-powered mobile recharging facilities should at the same time be created, particularly in areas that are visited by people in large numbers. Together with this, people venturing to these areas should be educated on ways of delaying discharge of mobile battery through awareness campaigns.

Though facing information blackout, SEOC was flooded with requests for updated information from officials, media, and next of kin of the persons who had come over to Uttarakhand. To cope up with this situation, a number of new telephone lines were hurriedly activated at the SEOC. Many telephone numbers, including some personal mobile numbers of the officials, were thus circulated through various modes, and this amounted to confusion.

SEOC should therefore have a single telephone number with multiple lines and with capability of being upscaled during major disaster incidences. Dedicated four-digit toll-free number of SEOC (1070) can be used for this purpose. For the convenience of the masses, this number should be publicized through various modes. The use of one number would be convenient for all concerned.

9.7.2 Media Briefing

Arrangements for media briefing and providing information to next of kin of disaster-affected persons were not in place. This resulted in overcrowding at the SEOC that often disrupted routine functioning.

After any disaster, media personnel are under immense pressure to report the latest updates, and lack of information from authentic sources often results in rumors that add to the trauma of affected population. This at the same time demoralizes the ones engaged in post-disaster operations.

Special care therefore needs to be taken for briefing of media persons at regular intervals by duly authorized persons having access to authentic and updated information. The media personnel should at the same time be provided access to video feeds and photographs of the disaster-affected area, particularly those depicting progress of rescue and relief measures. Besides satisfying the next of kin of the affected persons, this would boost the morale of the persons engaged in rescue and relief operations. If possible, arrangements should also be made for taking the media personnel to the disaster-affected area.

The persons involved in operations should not be overburdened with the responsibility of interacting with media. Media should at the same time be discouraged from venturing into the SEOC.

9.7.3 Public Information

Responding to public queries, particularly from the next of kin of the persons perceived to be present in the disaster-affected area, overburdened the SEOC. Large number of callers had come to know of the incidence through media and not being conversant with the geography of the state they were not sure as to which area was actually affected by the disaster and in which area their next of kin were present. Most callers were in an emotionally disturbed state and required counseling and consoling. Talking to them was no less than a trauma for people manning the SEOC. Good number of callers could not communicate either in Hindi or English.

On the aftermath of any disaster, it is normal for next of kin of the affected persons to seek information on the welfare and whereabouts of their loved ones. Separate arrangements have therefore to be made for responding to their queries. If possible, persons with special tele-counseling skills should be engaged for this. Special care should be taken in incidences where there is possibility of linguistic differences between the potential callers and the ones responsible for responding to public queries. In such cases, people conversant with the language of the potential callers should be engaged. Assistance of pre-registered volunteers could be taken for this purpose.

9.7.4 Information Management

After a while, the information being received at the SEOC from different sources became so voluminous that it became difficult to process, segregate, and use it for decision-making, planning, and resource mobilization. Moreover, both officials and non-officials soon started to seek synthesized and specific information on various aspects of the disaster. It often became difficult to respond to their queries.

SEOC should therefore be adequately manned and have the required information handling and processing capabilities to ensure that the information is quickly analyzed and used for decision-making for response and resource mobilization. The information received at SEOC has also to be segregated to promptly address any specific information requirement.

9.7.5 Relief and Rescue

Though highly specialized, professionally trained, and well equipped, the response forces called in for search and rescue on the aftermath of the disaster were not conversant with local terrain- and weather-related peculiarities. They at the same time did not have knowledge of alternative routes, locally available resources, and hardships likely to be faced while undertaking rescue. Moreover, induction of rescue workers in the disaster-affected area was not easy due to transport disruption coupled with bad weather conditions.

Knowledge of local ground realities is often critical to the success of search and rescue operations, particularly in the mountainous terrain. Local people therefore enjoy a distinct advantage, and in almost all disaster incidences, local people and other survivors are the first responders. It is therefore necessary to train local people in search and rescue and provide them required equipment so that they are better prepared to face emergency situations and help their community.

The specialized response forces being raised by the states should at the same time be exposed to local ground realities through regular ground familiarization exercises. This would ensure their effectiveness in the event of a disaster.

9.7.6 Governance

On the aftermath of the disaster taking clue from the Comptroller and Auditor General Report (CAG 2010), the issue related to ineffectiveness of the State Disaster Management Authority (SDMA) was repeatedly highlighted by the media. Besides maligning the image of the state government, it adversely affected the morale of the officials engaged in post-disaster operations.

For effective disaster governance, it is therefore a must to strengthen and empower institutions responsible for disaster risk reduction, particularly SDMA, State Executive Committee (SEC), and District Disaster Management Authority (DDMA). Most states are not paying due attention to the very fact that the Disaster Management Act, 2005, requires all executive actions to be taken either by SEC or DDMA with SDMA being a policy making and supervising institution. Adequate attention is therefore required to strengthen and empower SEC in the spirit of DM Act, 2005.

Disaster management being a multi-departmental affair, coordination and unity of command are critical to the success of post-disaster response. Apart from state administration, police and other state government departments a number of other agencies were involved on the aftermath of the disaster. These included Army, IAF, NDRF, ITBP, and Indian Navy. The uniformed services have their peculiar command structure and reporting procedure due to which problems were faced in information exchange and coordination.

SOPs and protocols pertaining to command structure, reporting formats, procedures, and information exchange have to be therefore laid down, circulated, and rehearsed well in advance to rule out possibility of lapses and confusion on the aftermath of any disaster.

9.7.7 Registration of Tourists/Pilgrims

Estimating the number of persons involved in the disaster of June 2013 was a major challenge, and there were varying claims from various quarters that added to the confusion. It is therefore required that the pilgrimage be regulated and persons be registered. Besides keeping track of exact number of visitors, this would help in communicating with them in case of any exigency.

It however needs to be appreciated that the pilgrimage circuit of the state has its peculiarities: (1) there is no single entry point and one can access the circuit from a number of entry points, and (2) there are habitations all along the route till the very end, and local people travel through the circuit in large numbers which makes differentiation of pilgrims and tourists difficult and registration challenging. It might therefore be hard to totally regulate the entire pilgrimage, but adequate registration and other measures have to be necessarily implemented in areas where people have to trek: Kedarnath, Gomukh, Hemkund Sahib, Yamunotri, and Chota Kailash-Kailash-Mansarovar. In these areas, only a specified number of persons should be allowed beyond the last roadhead at Gaurikund, Gangotri, Govindghat, Janki Chatti, and Tawaghat after duly registering their details.

People visiting these areas have often been facing health-related emergencies. Health check-up should therefore be made mandatory for all persons wishing to venture in these areas and only physically fit should be allowed. These persons should also be briefed on the terrain and weather conditions together with other associated hazards that the terrain might offer.

9.7.8 Support for Evacuees

Large number of disaster-affected persons had lost all their belongings and had no resources to manage their journey back home. Having come over from long distances, these people at the same time had no acquaintances around to look for help. Standard relief guidelines issued by the Ministry of Home Affairs, Government of India, have no mention of such situations.

In view of special circumstances, the state government made arrangement for the return journey of the evacuated persons and also provided special cash assistance for covering en route expenses. It is therefore required that provision for catering to such situations be standardized, particularly for areas that are routinely visited by people from far and wide in large numbers. This would ensure prompt dispatch of the disaster-affected persons to their destinations.

9.7.9 Surface Connectivity

Blockade of motor roads due to landslide and flash flood is common in the hills. Most tourist and pilgrim destinations in the Higher Himalaya have single road connectivity and blockade of the same often results in persons being stranded in large numbers. Providing logistics support to the stranded persons and ensuring their early evacuation thus becomes a major concern of the state.

After the disaster of June 2013, more than 150,000 persons were stranded at different places across the state due to blockade of roads due to landslides and washing of roads. Despite best efforts, major roads of the disaster-affected area could only be opened for light vehicles in September to October 2013. Large number of stranded persons had to be therefore evacuated by air. Constrains put forth by availability of airspace, helipads, refueling, and weather conditions thus delayed evacuation.

It is therefore required that surface connectivity in the region be improved, and alternative motor roads be planned and developed so as to ensure alternative connectivity during disaster incidences. In the disaster of 2013, most motor roads in the proximity of rivers and streams were washed off due to bank erosion. Wherever possible the alignment of new roads should therefore be kept sufficiently away from rivers and streams.

9.7.10 Aerial Evacuation

Due to prolonged disruption of roads, aerial evacuation of the stranded pilgrims and others was the only available option. A large number of helicopters were therefore mobilized for this purpose. These however could not be optimally utilized due to the constraints put forth by limitation of airspace and helipads. Moreover, adequate refueling facilities were not available in the disaster-affected area, and bowsers could not be moved in immediately due to road disruption. Valuable operational time was thus lost in refueling of the choppers.

In view of the terrain conditions, helipads with adequate stock of ATF should be developed in the hills at strategic locations. This would make disaster response prompt and effective.

9.7.11 Linguistic Issues

Apart from foreign nationals, the disaster involved people from more than 23 states of India. The disaster-affected persons thus exhibited distinct linguistic diversity, and large number of them could not communicate in a language that is commonly understood by the people of the state: Hindi and English. This put forth numerous problems for both rescue workers and relief providers. People responding to public queries also faced similar problems.

On the aftermath of a disaster involving people from different linguistic backgrounds, communication with disaster-affected persons could become a problem. It is therefore required that the relief camps be adequately staffed with persons who are conversant with language of the potential victims. For this volunteers could be registered and their services could be mustered as the requirement arises.

9.7.12 Missing Persons

Large number of persons went missing in the disaster of June 2013. Legal procedure in India for declaring a missing person dead requires the missing person to be unheard of for a minimum period of 7 years by the ones who would normally have information on his/her whereabouts. Moreover, ex gratia relief admissible out of SDRF could only be provided to the next of kin of those deceased in the disaster incidence. Besides claiming relief, the next of kin of the missing persons also required death certificate for settling various familial, societal, official, and legal issues.

At that time there were no guidelines in existence to declare missing persons as being dead. Due to this disbursement of relief and death certificates was delayed. This added to the trauma of the family members of the missing persons. This issue could only be resolved after the Registrar General of India issued fresh guidelines on 16 August 2013 and prescribed procedure to be followed for declaring missing persons as being dead and issuing death certificates.

This arrangement however was specifically for the disaster of 2013 and cannot be applied to other disasters in future. It is therefore required that standardized guidelines and procedures be put in place for declaring persons missing in disaster incidences as being dead so that the next of kin of the deceased persons are not unnecessarily traumatized.

9.7.13 Evacuation

Injured, ill, elderly, women, and children are generally accorded priority in evacuation, and this thumb rule was followed on the aftermath of this disaster as well. This went on well till there were injured and ill around but after that people declined to be dissociated from their group.

It was soon realized that the people had come over in close-knit groups to the disaster-affected area from far and wide for pilgrimage and had no acquaintance with the region. In many such groups, only one to two persons could speak and understand Hindi or English, and for the females, talking to strangers, particularly males, was taboo. Disassociation from the group that too after experiencing such a major disaster thus added to their trauma and infused sense of insecurity.

The ones evacuated alone in the initial phase thus denied to leave till others in the group were evacuated. This added to the burden of ones engaged in relief and rescue operations. Taking clue from the experience, it was later decided to evacuate people in groups rather than segregating them on the basis of age and sex. It is therefore necessary that the evacuation priorities be decided only after fully understanding the composition of the affected population.

9.7.14 Needs Assessment

Large volume of relief supplies that reached the disaster-affected area was not actually required, and handling of the same only added to the burden of the officials. Packaged water and old clothes were among such items; the former was not required and only added to trash, while the latter was not socially acceptable.

It is therefore a must after any major disaster to quickly undertake a needs assessment. The requirements so assessed should necessarily be widely publicized so that people do not send material that is not required. At the same time, items that are not acceptable to the people due to religious or cultural reasons should also be publicized.Though the impact has not been assessed but the supply of relief material has adversely affected local economy and business. It is therefore necessary that to the extent possible purchase of relief material should be done locally and only the items not available locally should be moved in from outside. This would boost economic recovery of the disaster-affected area.

9.7.15 Balanced Response

High media attention in Kedarnath-Hemkund Sahib area resulted in initial concentration of both rescue and relief efforts in this region. Ease of access to the center stage of disaster due to concentration of aerial rescue effort in this area further strengthened this trend. To the ones following the disaster through media, it seemed as if the disaster was restricted to Mandakini valley.

The assistance coming from civil society groups and corporate houses was thus concentrated in this region. This region therefore had excess of relief supplies as also other assistance, while disaster-affected areas of Pithoragarh and Bageshwar did not receive much attention. Moreover, even within this region, the stranded persons attained high attention, while the affected local population was not adequately catered during the initial response phase.

It is therefore necessary to plan response on the basis of assessed impact and needs, and all affected areas should be equally catered to. Mechanism also needs to be put in place for ensuring even distribution of the efforts put in by civil society groups and corporate houses.

9.7.16 Personal Effects

On the aftermath of the disaster, it was observed that large number of persons who had ventured to the high-altitude areas did not have adequate clothing and footwear. Persons visiting the region often do not have prior experience or knowledge of climatic conditions in the mountains where temperatures could drop drastically even in summers after rains. Moreover, coming from plains where temperatures go quite high in summers, it is hard for the visitors to contemplate need of carrying woolen clothing. Together with this, many people particularly women are not used to wearing shoes. Walking on the hilly track with slippers or sandals often becomes painful and cumbersome, particularly after rains.

Information on the weather-related peculiarities of the region together with clothing- and footwear-related advice should be made available to the potential visitors through various modes. The ones proceeding on high-altitude trekking route should necessarily be advised to carry woolen clothing, sturdy footwear, and raincoat/umbrella. These items should at the same time be made available en route on sale.

9.7.17 Logistic Support for Visiting Officials

The disaster involved people from more than 23 states of India and officials from many of these states were deputed to take care of specific requirements of the persons of their state. There were however no arrangements in place for the briefing of these officials that required information on geography, terrain, weather conditions, approach, efforts being made for search and rescue, and whereabouts of the evacuated persons. Arrangements for extending logistics and secretarial support to these officials were not in place. All this added to the workload of the persons manning the SEOC and carrying out other disaster management-related duties.

Based upon the composition of routine visitors, specific arrangements catering to the needs of the officials likely to come from different states/Embassies and High

Commissions on the aftermath of any major disaster have therefore to be incorporated in the concerning SOP.

On the aftermath of the disaster, there were visits of a large number of politicians and other high-ranking officials. Arrangements had therefore to be made for briefing of the visiting officials. Visits of this nature should be discouraged, if not curbed. Protocol-related formalities that are attendant to the visit of such officials should necessarily be formally waived off for disaster situations so that the officials engaged in post-disaster functions are not unnecessarily occupied with protocol-related arrangements that are hard to manage, particularly in disaster-affected and remote areas.

The national government should enact suitable legislation on this important issue and waive off all protocol-related formalities during disaster situations. Provisions pertaining to the same could be incorporated in DM Act, 2005.

9.7.18 Relief and Rescue Personnel

In view of the magnitude of the disaster, a large number of personnel from different state government departments were deputed to the disaster-affected and other areas for taking care of various emergency support functions. Unlike unformed services, the civilian departments do not have a culture of working in shifts as also that of rotation of persons undertaking stressful duties. These persons had therefore to carry out functions entrusted upon them continuously for long periods without any break. This had distinct adverse impact on the psychological and physical health of these persons that had distinct adverse impact on their performance. This was distinctly visible in the behavior and attitude of these persons.

It is therefore required that the working hours of the persons engaged in post-disaster functions, both in the disaster-affected area and at SEOC be fixed and the ones engaged in stressful tasks be rotated after a predetermined interval. This should necessarily be incorporated in the relevant SOPs. As the post-disaster relief and rescue period could extend for a long period, it is necessary to pay particular attention toward psychological, mental, and physical health of the relief and rescue personnel as deterioration in the same could adversely affect quality and effectiveness of the functions being discharged by them. In the stressful post-disaster conditions, particular care also needs to be taken to ensure that these personnel get proper rest and quality time for entertainment and exercise.

9.7.19 Demobilization of Resources

On the aftermath of the disaster of 2013, a large number of personnel and resources were requisitioned from uniformed forces and also various other departments of the government. These were often put under the control of the district administration of disaster-affected districts. As there were no protocols or SOPs in place for the demobilization of resources, these were not relieved even after they had discharged functions assigned to them, and there existed no specific assignment for them.

In the absence of protocols and SOPs, the district administration was reluctant to certify that there was no role whatsoever for these resources in the district, and therefore some of the resources were kept waiting for long periods. This was observed to be a cause of discontent for the organizations that had spared these resources at the time of exigency.

Moreover, it needs to be appreciated that in view of the emergency situation, resources provided by various organizations are often pulled out of deployment at some other location. Over the passage of time, these at the same time might well be required at some other location. This is particularly relevant for the resources of the uniformed forces as there are basically meant for catering to security and strategic requirements and should necessarily be demobilized at the earliest after these have served their purpose.

It is therefore required that protocols and SOPs be put in place for the demobilization of resources. Ceremonial or official demobilization with words of appreciation and thankfulness would further help in building better understanding and relations between the organizations.

9.7.20 Briefing and Debriefing

Unlike uniformed forces, the civilian departments do not have a culture of routine briefing and debriefing of personnel engaged in different tasks. After the disaster of 2013, large number of officials deputed by different departments were therefore inducted on duty without formal briefing on their roles and responsibilities. This often resulted in disrupted communication or chain of command. The officials were at the same time relieved without debriefing. This often resulted in vacuum and disrupted continuity of functions. Formal documentation of important lessons learned was also missed due to this.

It is therefore necessary to put in place SOPs for routine briefing and debriefing of the personnel. This should preferably be organized at the time of change of shift or duty so that the outgoing party could provide information on the ongoing tasks, problems being faced, plan of work, and targets set for the coming period. The party taking charge should at the same time be briefed on the chain of command together with reporting format and procedures.

9.7.21 Media Awareness

On the aftermath of the disaster, it was observed that the media personnel, in their bid to make the news sensational and salable, often resorted to reporting of personal routine and eating habits of the relief and rescue personnel. This often resulted in personal embarrassment besides demoralizing the relief and rescue personnel.

Media is therefore required to be educated that good mental, psychological, and physical health of relief and rescue personnel engaged in stressful and tiring tasks is inevitable for effective and smooth discharge of their functions, and for this they have to be provided healthy diet and recreational opportunity. To add to it, the media personnel should understand that the ones engaged in relief and rescue functions are in no way deprived of their right to personal liberty guaranteed by the Constitution of India, and the same needs to be respected and honored under all circumstances.

Though in limited number, unauthenticated, fabricated, and inflammatory stories were reported, while positive stories pertaining to acts of personal bravery, compassion, and empathy together with societal engagement were often not reported. Media persons have therefore to be educated on their role in encouraging and facilitating the masses to act and proceed in the right direction by maintaining right proportion of positive new items.

9.7.22 Legal Issues

Soon after the disaster of 2013, a number of public interest litigations were filed in the apex court alleging nonperformance on the part of the state and seeking immediate judicial intervention for providing relief to disaster-affected people. There were at the same time probes by the delegations of various quasi-judicial statutory bodies that included State and National Human Rights Commission, National Commission for Women, National Commission for the Protection of Child Rights, and National Commission for Scheduled Castes and Scheduled Tribes. These often sought specific information pertaining to the effect of disaster on their interest group and measures taken by the state to cater to their requirements.

No arrangements were in place for addressing legal issues, and the data was also not specifically segregated to cater to the queries of the various commissions. Therefore, the ones responsible for other disaster management-related functions and having familiarity with the sequence of events and progress of relief measures had to share the responsibility of preparing counter affidavits and replies.

It is therefore necessary to have an overview of the legal issues that could spring up on the aftermath of a major disaster incidence and accordingly manpower be put in place for addressing these. Apart from knowledge of legal issues, the person entrusted with this responsibility should have familiarity with ongoing post-disaster efforts and initiatives.

The national government has also to take a call on this important issue and enact suitable legislation to disallow judicial interference immediately after a major disaster when the state machinery is engaged in addressing other important, vital, and pressing issues. Provisions pertaining to the same could be incorporated in DM Act, 2005.

9.7.23 Warning Generation and Dissemination

Despite claims of advance warning of the incidence, the information received at SEOC from India Meteorological Department (IMD) was nothing more than a general forecast of particularly heavy rainfall all through the state. With location-specific and definitive warning of the impending disaster, many precious lives could have been definitely saved. The disaster of June 2013 thus reiterates the need of having a reliable and robust warning generation and dissemination infrastructure.

Reliable warning and its effective communication in a decipherable manner to the population likely to be affected by the incidence is the key to saving human lives and mitigating losses. To be effective, warning should however have sufficient lead time and be precise in space, time, and magnitude.

With the present state of scientific knowledge and technological advancement, it is possible to generate and disseminate warnings of hydrometeorological events well in advance. Sufficiently dense network of meteorological observatories with real-time data transmission capability is however a precondition for this, particularly in the Himalayan terrain where weather parameters are highly variable over short distances. Such a meteorological network could be integrated with rainfall threshold-based flood and landslide models to generate reliable warnings well in advance. A system capable of immediately communicating these warnings to the grassroots level, in a manner that suggests actions to be initiated by people at large, has to be an integral part of the warning infrastructure.

As the region is visited by tourists and pilgrims in large numbers, mobile messaging service with provision of automatic delivery of multilingual warning to all active mobile phones in the area likely to be affected by the said warning could be a viable option for this. At the same time, warnings have to be displayed at places where people gather in large numbers: bus/railway stations, taxi stands, transport registration offices, tourist information/registration centers, prominent road diversions, and the like. Warnings should also be aired through FM, community radio, and other radio networks and also telecasted through television channels. Provisions of the Disaster Management Act, 2005, should be utilized for ensuring overriding priority to these warning messages.

It however needs to be understood that it is not easy for the masses to understand the implications of the warning received in their specific context. Moreover, the warning received is of little use if the recipient is unaware of the course of action to be followed. For the effectiveness of the warnings, it is therefore necessary to undertake highly visible and aggressive mass awareness drive.

9.7.24 Settlement Pattern

Most losses caused by the disaster were generally observed to have restricted to the proximity of rivers and streams. Traditionally, the people of the area maintained safe distance from streams and settled down over middle or higher slopes of the hills. Alignment of motor roads along the rivers and associated commercial incentives has however induced people to settle down close to the streams and rivers.

On the aftermath of the disaster, it is required that land use zonation be undertaken and anthropogenic activities in close proximity of rivers and streams and also in identified hazard-prone areas be banned.

9.7.25 Aggradation

The issue of fast pace of aggradation in the disaster-affected area and ensuing enhanced vulnerability of many low-lying areas in the proximity of rivers and streams has been highlighted after the disaster of 2013. Uttarkashi, Bageshwar, and Sonprayag are among the settlements facing this problem. Reduced transport and erosion capacity of the rivers and streams due to arrested and regulated flow due to the construction of barrages and dams and increase in the debris production due to enhanced pace of developmental initiatives, landslides, and surface erosion in the hills are held responsible for this.

Debris production cannot be ruled out while undertaking developmental works in the hills, but it needs to be appreciated that the present practice of rolling down the debris aggravates mass movement and deforestation besides degrading soil and water quality. It often overruns productive land and other assets. All the loose material ultimately reaches the riverbed, enhances the pace of aggradation, and adversely affects productive life and storage capacity of the reservoirs.

It is therefore required that the practice of unscientific disposal of debris be discontinued and a policy be enacted for ensuring safe, systematic and scientific debris disposal with adequate punitive measures for non-compliance.

9.7.26 Risk Transfer

The disaster derailed the economy of the region that is largely dependent upon pilgrimage and tourism. Large number of persons engaged in hospitality, transport, and related sectors faced severe hardships due to sudden and unexpected downfall in the number of pilgrims, tourists, and others visiting the area.

In a bid to infuse new vigor into the economy, the state resorted to waiving off state dues besides enhancing the rate of relief admissible to disaster victims and bringing losses of almost all categories under the umbrella of relief. Thus the relief also covered losses incurred to commercial establishments. All this amounted to massive burden upon public exchequer.

It is therefore required that risk transfer (insurance) be made a precondition for operating any commercial establishment and the same be linked to their licensing. Risk transfer measures should also be promoted amongst general public, and financial institutions should be persuaded to ensure insurance of all assets created with their assistance. Besides reducing the burden upon public exchequer in case of a major disaster incidence, this would better compensate the disaster-affected population. Risk transfer at the same time has the potential of compensating indirect loss incurred to enterprises due to disaster incidences.

9.7.27 Risk-Informed Decision-Making

Post-disaster review of the situation brought forth the issue of risk assessment, risk communication, and risk-informed decision-making. All these were observed to be missing, and therefore risk of any of the hazards was not being taken note of even while taking planned organizational decisions. Unaware of the potential risk, masses were observed to do what suited them best. It was also observed that despite a strong tradition of disaster risk reduction, people were not following the age-old and time-tested principles that ensured safety of their community in this hazard-prone terrain all through. Lately, the people had started to settle down dangerously close to rivers and streams and over riverine terraces that were traditionally left for agriculture alone. The people were also observed to have discontinued traditional earthquake-safe construction practices (Rautela 2005, 2013, 2015; Rautela et al. 2008, 2009). All this is attributed to (i) status attached to modern infrastructure, (ii) social stigma attached to traditional practices that are considered backward, (iii) peer pressure, and (iv) emulation. These are held responsible for enhanced vulnerability of the masses in the hills.

It is therefore required that detailed risk assessment be undertaken and the results of the same be made available to the masses in an easily decipherable manner. Together with this, appropriate, site-specific, and simple risk reduction measures should be popularized. At the same time, it is required that the traditional disaster risk reduction practices of the people be researched, improvised, and amalgamated with modern science and technology so as to come up with socially acceptable, economically viable, innovative, and sustainable disaster risk reduction solutions.

References

Allen SK, Rastner IP, Arora IM, Huggel IC, Stoffel IM (2015) Lake outburst and debris flow disaster at Kedarnath, June 2013: hydrometeorological triggering and topographic predisposition. Landslides 1–15

Bandyopadhyay S, Sekhar KN (2014) The Kedarnath disaster: in search of scientific reasoning. Curr Sci 107(4):557
CAG (2010) Report of the comptroller and auditor general of India. Available online at http://cag.gov.in/sites/default/files/audit_report_files/Uttarakhand_Civil_2010.pdf
Chattoraj SL, Champatiray PK (2015) Simulation and modeling of debris flows using satellite derived data: a case study from Kedarnath area. Int J Geomatics Geosci 6(2):1498–1511
Chopra R (2014) Uttarakhand: development and ecological sustainability. Oxfam, India. 41 pp
Dobhal DP, Gupta Anil K, Manish M, Khandelwal DD (2013) Kedarnath disaster: facts and plausible causes. Curr Sci 105(2):171–174
Dube A, Raghavendra A, Amit A, Kuldeep S, Iyengar GR, Rajagopal EN, Swati B (2014) Forecasting the heavy rainfall during Himalayan flooding—June 2013. Weather Clim Extremes 4:22–34
Expert Committee Report (2014) Assessment of environmental degradation and impact of hydroelectric projects during the June 2013 disaster in Uttarakhand. The Ministry of Environment and Forests Government of India, New Delhi. 226 pp
IMD (2013) A preliminary report on heavy rainfall over Uttarakhand during 16–18 June 2013. Government of India, Ministry of Earth Science, Indian Meteorological Department, New Delhi. July 2013
Indian Standard (IS): 1893 Part I (2002) Criteria for earthquake resistant design of structures. Bureau of Indian Standards, New Delhi
Kotal SD, Soma SR, Roy Bhowmik SK (2014) Catastrophic heavy rainfall episode over Uttarakhand during 16–18 June 2013–observational aspects. Curr Sci 107(2):234–245
Rana N, Sunil S, Sundriyal YP, Navin J (2013) Recent and past floods in the Alaknanda valley: causes and consequences. Curr Sci 105(9):1209–1212
Rautela P (2005) Indigenous technical knowledge inputs for effective disaster management in the fragile Himalayan ecosystem. Disaster Prev Manag 14(2):233–241
Rautela P (2013) Lessons learnt from the deluge of Kedarnath, Uttarakhand, India. Asian J Environ Disaster Manag 5(2):1–9
Rautela P (2015) Traditional practices of the people of Uttarakhand Himalaya in India and relevance of these in disaster risk reduction in present times. Int J Disaster Risk Reduction 13:281–290
Rautela P, Joshi GC, Bhaisora B (2008) Earthquake – safe Koti Banal architecture of Uttarakhand, India. Curr Sci 95(4):475–481
Rautela P, Joshi GC, Bhaisora B (2009) Earthquake safety elements in traditional Koti Banal architecture of Uttarakhand, India. Disaster Prev Manag 18(3):299–316
Singh R, Singh D, Gokani SA, Buchunde PS, Singh RP, Singh AK (2015) Brief communication: climate, topographical and meteorological investigation of the 16–17 June 2013 Kedarnath (India) disaster causes. Nat Hazards Earth Syst Sci Discuss 3:941–953
Uttarakhand Flood Disaster (2013) Role of human actions. Dams, rivers and people. 11(5–6):2013, 1–32

Chapter 10
Uttarakhand Calamity: A Climate Revelation in the Bhagirathi River Valley Uttarakhand, India

Manish Mehta, Rakesh Bhambri, J. Perumal, Pradeep Srivastava, and Anil K. Gupta

Abstract On June 16 and 17, 2013, high-intensity rainfall (>400 mm) in different parts of the state of Uttarakhand caused devastating flash floods and triggered widespread landsides. In this event incurred heavy losses to the infrastructure, agricultural fields, human and animal lives, roads and widespread destruction of natural resources. Disaster of such a magnitude of disaster was perhaps not experienced by the region at least over the last 100 years. Thus, this disaster can be considered as an extreme climatic event of the century. The extent and intensity of the tragedy can easily be visualised by the fact that all the famous shrines of the Uttarakhand state are located in high mountainous, snowbound areas. These places, Badrinath (3133 m asl in Alaknanda valley), Kedarnath (3584 m asl in Mandakini valley), Gangotri (3140 m asl in Bhagirathi valley), Yamunotri (3291 m asl in Yamuna valley) and Hemkund Sahib (4433 m asl in Alaknanda valley), were badly affected by this extreme fury of the nature.

In this paper we cover the devastated area that lies in the Bhagirathi river valley, which forms a part of Uttarkashi, Tehri and Pauri districts in the Lesser and Higher Himalayas, Garhwal. During the field observation, we have collected a total number of 4955 points in the Bhagirathi valley; these points are points of interest (POI) (2084), damage to bridges and culverts (44), landslides (1034), eroded land cover and natural resource (92), riverbank erosion (170), damage to roads (494), damage to other infrastructures (320) and damage to buildings (717). These collected points can be used for rehabilitation and infrastructure development, implication in future plan.

Keywords Himalaya • Bhagirathi basin • Calamity • Rehabilitation

M. Mehta (✉) • J. Perumal • P. Srivastava • A.K. Gupta
Wadia Institute of Himalayan Geology, Dehradun, Uttarakhand, India
e-mail: msmehta75@gmail.com

R. Bhambri
Centre for Glaciology, Wadia Institute of Himalayan Geology, Dehradun, Uttarakhand, India

I. Pal, R. Shaw (eds.), *Disaster Risk Governance in India and Cross Cutting Issues*, Disaster Risk Reduction, DOI 10.1007/978-981-10-3310-0_10

10.1 Introduction

Extreme events in past few decades have drawn increased attention to the science seeking to understand their causes (Kerr 2013). The Intergovernmental Panel on Climate Change (IPCC) Fifth Assessment Report (Stocker et al. 2014) concluded that strong evidence exists for increases in some extremes worldwide since 1950, especially more frequent hot days and heavy precipitation events. But, with natural variability playing a substantial role in individual events and given the complexities of the weather and climate processes involved, many challenges still need to be overcome to authoritatively assess how climate change has affected the strength and likelihood of individual extremes. Goswami et al. (2006) suggested that there are year-to-year variability and significant increases in the frequency and intensity of extreme monsoonal rain events in India, which is 50% higher than the normal rainfall.

The Himalayan states are familiar with the adverse impacts of the extreme rainfall events, floods and related disaster. People dwelling under the shadow of mountain peaks in the Uttarakhand Himalaya live under constant threat from flash flood and landslides, particularly during the monsoon sessions. Due to these flash flood-induced landslides, Uttarakhand suffers great losses of human lives, livestock, infrastructures and natural wealth.

Recent climate changes have a significant impact on high-mountain glacial environment. Due to continuously rising air temperature all over the world, the precipitation pattern at higher altitude is changing from solid (snow) to liquid (rain) (Immerzeel et al. 2010). This high-altitude rain causes rapid melting of snow/ice resulting in formation and expansion of moraine-dammed, supraglacial and cirque lakes which will be potentially dangerous in downstream valley (Dobhal et al. 2013a, b).

10.2 Physiographical Setting of the Area

The area involved Bhagirathi river basin that additionally included Ganga river (downstream of Devprayag) and Nayar river. Geologically the area is composed of Palaeozoic-Mesozoic sedimentary succession resting upon the crystalline basement with fault/thrust contact. The area between Devprayag and Gangotri lies in metasedimentary Garhwal Group rock and Vaikrita Group (Valdiya et al. 1999) of Central Crystalline (Heim and Gansser 1936). In general rock types are sandstone, slate, phyllite, conglomerate, etc. in Garhwal Group and mica schist, quartzites, mylonite, gneiss, pegmatite, granite, etc. in Vaikrita Group. The sequence has been intruded by the Gangotri granite dated 465 ± 5 ma. Figure 10.1 shows the geological map of the study area in relation to its regional framework. Tectonically the southern front of the mountain is divided into three lithotectonic units that are separated by southward-younging thrusts. The southern and the youngest is the Himalayan Frontal

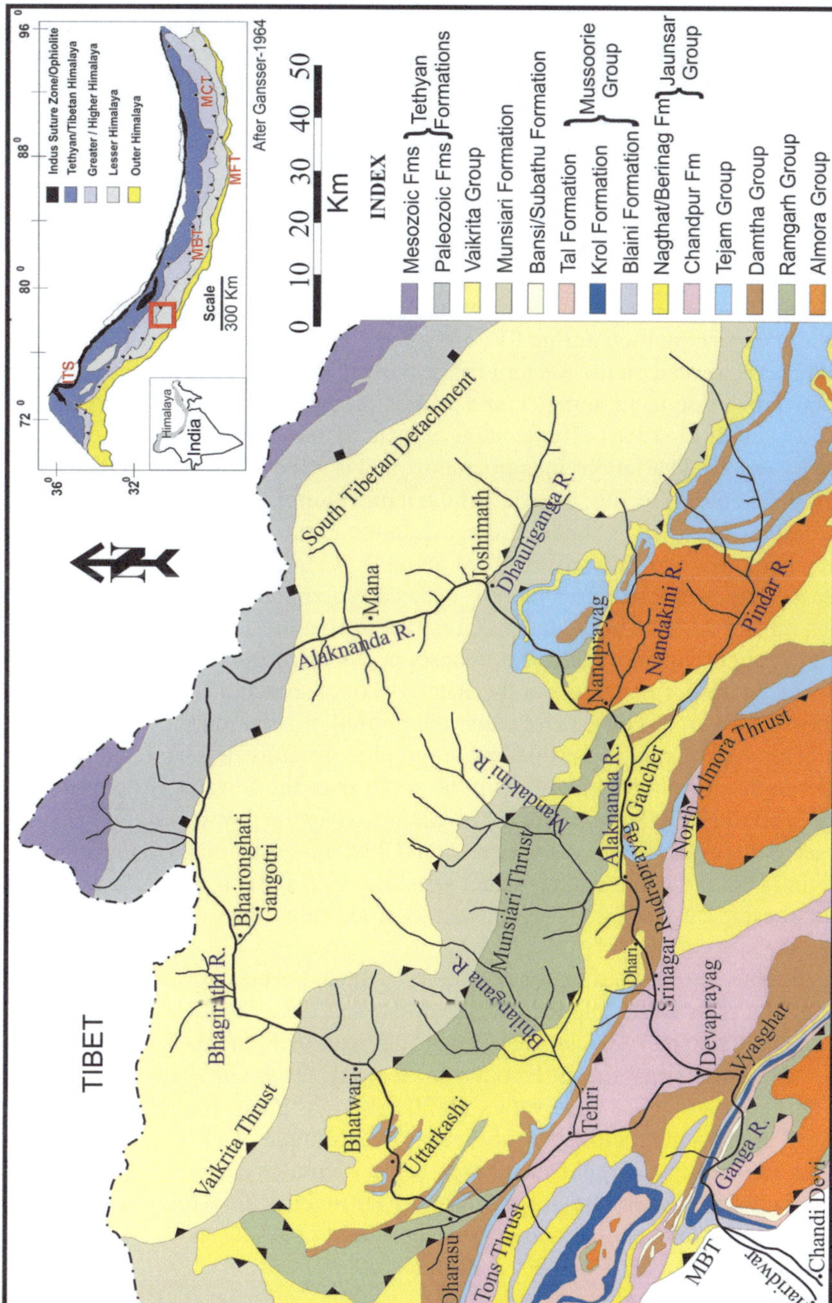

Fig. 10.1 Geology of the Bhagirathi river valley

Thrust (HFT) that brings Siwaliks on to Ganga foreland; the Main Boundary Thrust (MBT) brings Lesser Himalayan metasedimentaries over Siwaliks, and then Main Central Thrust (MCT) thrusts Higher Himalayan Crystallines over the Lesser Himalaya. All these thrusts zones are characterised by zones of weak and pulverised rocks that are prone to failures.

In Himalaya, the precipitation and basin runoff generally decrease from the east to west due to gradual weakening of westward-moving Indian summer monsoon (ISM) trough (Immerzeel et al. 2010; Ali et al. 2013). In the east, ISM precipitation dominates, while in the west, westerly circulation and cyclonic storms contribute about two-thirds of the total annual precipitation. Westerlies contribute two-thirds of the total snowfall in high altitudes during winter, the remaining one-third results from summer precipitation during SW monsoon circulation (Armstrong 2010). The Uttarakhand is located on the western fringe of the Central Himalaya, dominated by monsoon precipitation in summer and winter precipitation from western disturbances (Owen et al. 1996). The general climate of the Uttarakhand is temperate, marked by seasonal variations in temperature, and is affected by tropical monsoons. January is the coldest month, with daily high temperatures averaging below freezing in the north and near 21 °C in the southeast. In the north, July is the hottest month, with temperatures typically rising from about 7 °C to about 20 °C daily. In the southeast, May is the warmest month, with daily temperatures normally ranging between 38 and 27 °C. The annual precipitation of roughly 1500 mm in most of the states is brought by the southwest monsoon, which occurs from July through September. In the northern parts of the state, 3–5 m of snowfall is common between December and March. The general pattern of rainfall distribution shows two belts of high precipitation that are controlled by orographic structure of the mountain range. The first high-precipitation belt lies in the south over the Siwalik Hills where the average rainfall is 1200 mm/a, and the second lies over the physiographic transition of Lesser and Higher Himalayas where around 2000 mm of rainfall occurs annually. Remaining part of the mountain is rather semiarid. During June 16–17, 2013, places like Uttarkashi, Devprayag and Haridwar recorded more than 300 mm of rainfall within 2 days.

The climate of glaciated regions of the Himalaya, because of inaccessibility of the areas and poor meteorological and hydrological data, is rather not well documented. The Chorabari Glacier that lies at the head of Mandakini river has been monitored by Wadia Institute of Himalayan Geology, Dehradun, which began since 2003. This meteorological observatory (3820 m asl) installed monitors air temperature, wind speed and precipitation of glacier at an altitude of 3820 m asl. The data from this observatory as published recently by Mehta et al. (2014) suggests the average daily air temperature ranges from −13.5 to 11.6 °C, while the maximum and minimum air temperature ranges between −9.4 to 16.7 °C and −19.5 to 9.8 °C, respectively, between 2007 and 2012 (Mehta et al. 2014).

Summer precipitation in glaciated regions is highly influenced by the monsoon, and the average rainfall recorded between 2007 and 2012 was 1309 mm (June–October). Winter precipitation generally occurs between December and March, when the westerlies are dominant in the area as they move eastward over northern

India, and is the main source of snow accumulation. No instrumental data is available for winter snowfall; however, residual snow depth fluctuated between 25 and 50 cm in April and early May at 4000 m asl during the period from 2003 to 2010 (Dobhal et al. 2013); snow normally melts before the commencement of the monsoon in mid-June.

10.3 Fluvial Geomorphology of Bhagirathi River

All the major rivers in Uttarakhand like the Yamuna, the Bhagirathi, the Alaknanda, the Mandakini, the Pindar, etc. are origin from glaciers. On the basis of geomorphology and climate, Bhagirathi river valley can be divided into three major reaches from north to south: (1) The upper reach, arid-semiarid, glaciated with 'U'-shaped valleys lined with moraine ridges, largely falls north of the Main Central Thrust (MCT) in an altitude range of >6500–2600 m asl. The mapping of moraines in this reach indicates that the elevation of glacier snouts varied from 2604 m asl (~63 ka BP) to 3550 m asl (21–15 ka BP) to 3850 m asl at present (Sharma and Owen 1996; Mehta et al. 2012). (2) The middle reach is characterised by steep slopes, with 'V'-shaped river valleys forming deep gorges. It is located from the MCT zone to 60 km south in the Lesser Himalaya. This zone, between 2600 and 1200 m asl, forms an orographic barrier northward advancing Indian summer monsoon and is characterised by high rainfall and erosion. (3) Wider 'V'-shaped valleys with lower channel and slope gradients characterise the lower reach. This part of the river ranges from 1200 m asl in the Lesser Himalaya to ~250 m asl near the HFT at the mountain front. Impressive cut-and-fill staircases and terraces with underlying bedrock steps distinguish this reach (Ray and Srivastava 2010). In Bhagirathi valley Tehri, Uttarkashi, Netala, Latasera, Dunda, Gyansu and Atheli are good example of fill terraces.

10.4 The Event of June 2013 and Objectives of the Study

Exceptional early monsoon rain between June 15 and 17, 2013, combined with melting snow caused horrific flood in the rivers of Garhwal Himalaya (Bhagirathi, Alaknanda, Mandakini and Yamuna rivers) and subsequently triggered widespread landslides. Thousands of pilgrims got stranded at various pilgrim places and en route. Due to heavy downpour, the activation of landslide and flash flood in the region has caused huge damage to lives, infrastructure and property in Garhwal, Uttarakhand. In downstream of these rivers, places like Lambaghar, Govindghat, Bhyundar village and Pulna village in Alaknanda valley; Kedarnath, Rambara, Gaurikund and Sonprayag in Mandakini valley; and Uttarkashi, Gangori, Sangamchatti and Dharli in Bhagirathi basin were severely damaged or completely washed out. The catastrophic event of June 2013 took more than 6000 human lives with more than this number being untraceable and >100,000 people have been affected. The quick survey by an agency of the World Bank estimated the financial loss incurred due to the event was more than $250 million and over US$500 million damage appeared in the media. Due to this flash flood event in Uttarakhand, the landscape of the area has changed making the whole region more fragile and vulnerable. Such a magnitude of disaster was perhaps not witnessed by the region at least over the last 100 years.

The extent and intensity of the tragedy can easily be visualised by the fact that all the famous shrines of the Uttarakhand state, located in high-mountainous, snow-bound areas, such as Badrinath (3133 m asl on Alaknanda river), Kedarnath (3584 m asl on Mandakini river), Gangotri (3140 m asl on Bhagirathi river), Yamunotri (3291 m asl on Yamuna river) and Hemkund Sahib (4433 m asl on Alaknanda river) were badly affected by this extreme fury of the nature. In this report we cover the devastated area that lies within the Bhagirathi river basin, which forms the part of Uttarkashi, Tehri and Pauri districts in the Lesser and Higher Himalayas, Garhwal. The area is characterised by highly rugged topography and very high relief having sharp ridges with prominent dip slopes. Lithologically, the area constitutes medium- to high-grade metamorphic and sedimentary rocks classified as Garhwal formation and Central Crystalline.

The Department of Science and Technology (DST), Government of India, initiated a programme to map the disaster-wrecked area of Uttarakhand, and the programme was named as 'Map the Neighbourhood in Uttarakhand' (MANU). We shared this responsibility by working in Bhagirathi river valley and the DST project MANU (Bhagirathi) was taken up. To begin with the project work, 20 students were recruited in the first batch as interns in the project and were trained at IIRS, Dehradun. Subsequently, after the purchase of student field kits and mobile mappers (Android phone) and printing of forms, the task of mapping started on October 10, 2013. Five scientists and 20 students were divided into three groups, each deployed at Harsil, Uttarkashi and Tehri. This allowed us to cover Bhagirathi valley from three different fronts. The observations were first focused along the main channel of Bhagirathi and then fanned off into the tributary valleys. Attributes of all major and minor landslides, road damages and damaged buildings (i.e., GPS location, local geology, size, etc.) were directly uploaded on the Bhuvan portal. Location of point of interests (POIs) like hotels, police stations, hospitals, schools, panchayat bhavans, telephone towers and public conveniences were also located and data was duly uploaded. The quality of data sent to the portal was assured by coordinating scientists and confirmed by NRSC, Hyderabad, on a daily basis. Besides observations for uploading on Bhuvan portal, our team also visited all major villages and collected data on socio-economic structure and damage related to June 2013 event in the format supplied by the Department of Science and Technology, New Delhi. Data on tourist infrastructure and household damage was also collected.

10.5 Observations in Bhagirathi-Nayar River Valley (Up to Rishikesh)

Bhagirathi river, one of the major tributaries of Ganga river systems, originates from Gangotri glaciers at an elevation of ~4000 m amsl (snout) and drains southward for more than 170 km to meet Alaknanda river at Devprayag (458 m amsl). The longitudinal river profile of the river shows two major zones. *Zone I* that lies above Uttarkashi has channel gradient (~11.8°), characterised by the bedload dominant sediment load and steeper tributaries (Fig. 10.2). The lower part of this zone coincides with second physiographic transition with high rainfall and thick vegetation, and the headwaters are semiarid with thin vegetation cover. The size of the bedload in this zone ranges from more than a metre to couple of tens of centimetres. The upper reaches of the study area form deep and narrow gorges showing rapid vertical erosion. *Zone II* is characterised by gentler hill slopes and thick vegetation cover. The channel has lower channel gradient (~8°) and suspended load dominantly constitutes the sediment load.

The whole valley till Rishikesh including the Nayar basin was surveyed using software application provided by the Indian Institute of Remote Sensing and National Remote Sensing Centre. Twenty students selected from different colleges/universities of Uttarakhand were trained and then deployed in this survey. Table 10.1 provides a summary of the survey and Fig. 10.3 gives graphic distribution.

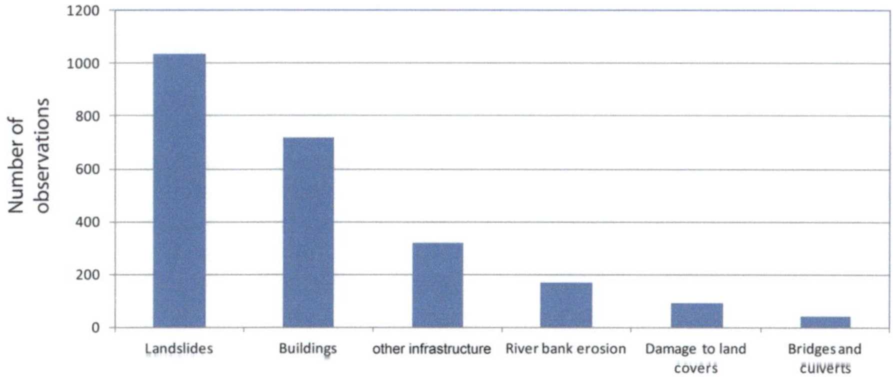

Fig. 10.2 Long profile along Bhagirathi valley. *Dotted rectangle* refers to predominance zone of damage related to surface process (Zone I). *Dotted arrows* indicate decreasing zone of surface process

Table 10.1 Total number of points collected in Bhagirathi river basin

S. no	Description	Number of points
1	Point of interest (POI)	2084
2	Damage to bridges and culverts	44
3	Landslide	1034
4	Land cover and natural resource	92
5	Riverbank erosion	170
6	Damage to roads	494
7	Damage to other infrastructures	320
8	Damage to buildings	717
Total		**4955**

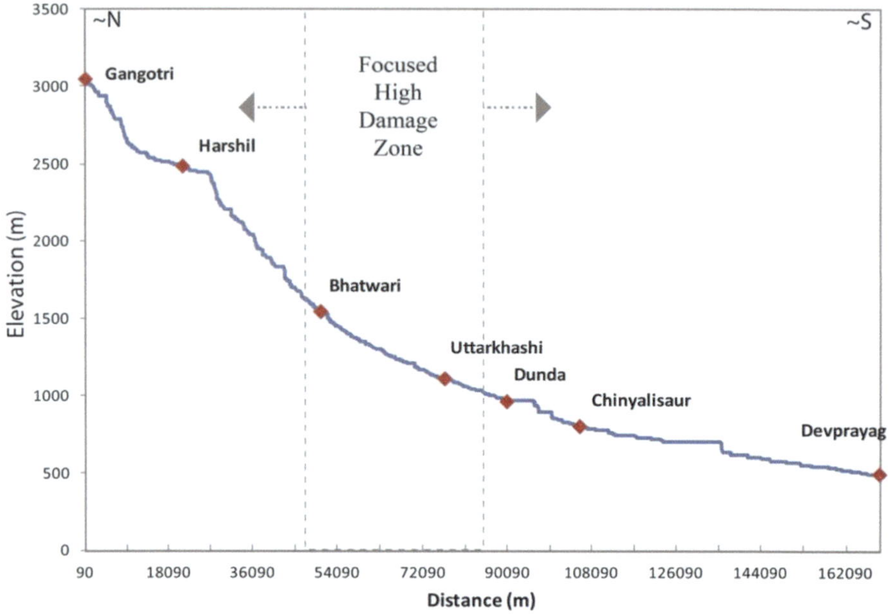

Fig. 10.3 Histogram showing extent of damages occurred along Bhagirathi valley

10.5.1 Result

The whole valley till Rishikesh including the Nayar basin was surveyed using software application provided by the Indian Institute of Remote Sensing and National Remote Sensing Centre. Twenty students selected from different colleges/universities of Uttarakhand were trained and then deployed in this survey. After completion of the survey, we have collected a *total of 4955* points in Bhagirathi valley; these points are points of interest (POI) (2084), damage to bridges and culverts (44), landslide (1034), eroded land cover and natural resource (92), riverbank erosion (170), damage to roads (494), damage to other infrastructures (320) and damage to buildings (717). The summarised results of the study are discussed in detail below.

10.5.2 Landslides

Several types of mass wasting that include landslides, debris flow and rock fall were mapped under this category. The survey shows that a total number of 1034 landslides got reactivated/initiated due to the high rainfall event of June 2013 (Fig. 10.4). When plotted on geological map, the mapped landslides appear to have occurred in two zones. Zone I falls in the part of catchment having phyllitic rocks. Phyllites are fine-grained low-grade metamorphic rocks. These rocks are fragile, and the hill

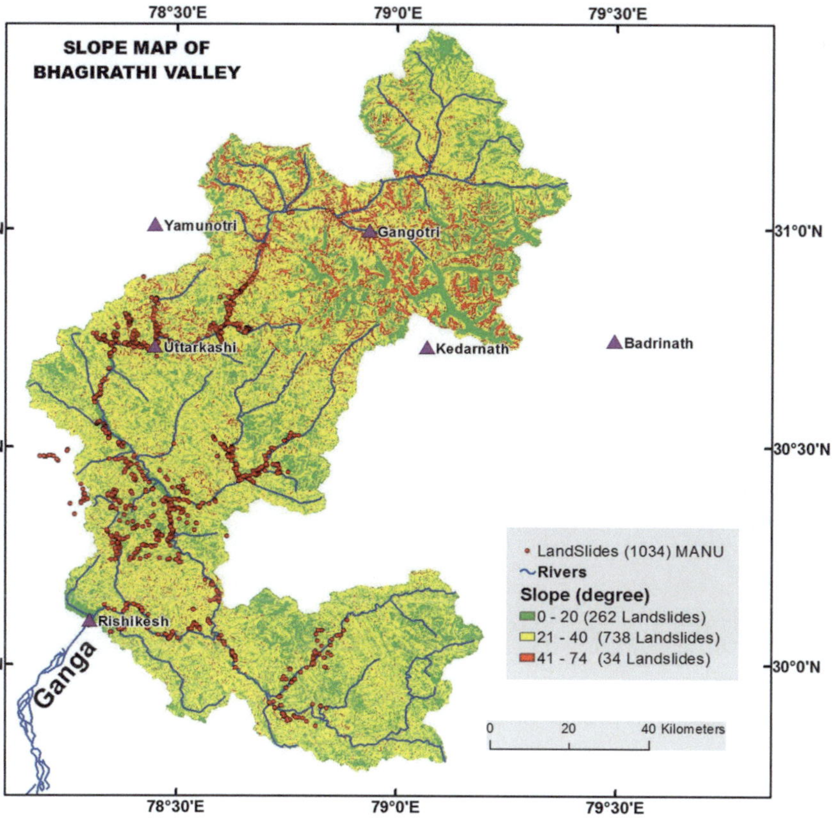

Fig. 10.4 Spatial distribution of landslides in different slope classes

slopes having thick mantel of weathered phyllitic rocks are vulnerable to failure during high rainfall events. Zone II largely coincides with physiographic transition between the Lesser and Higher Himalayas. This is a fragile zone which is characterised by a wide belt of crushed thrust belt of the Main Central Thrust (MCT) of Himalaya and high rainfall. The combination of steep hill slopes, poor rock strength and high rainfall leads to widespread mass wasting in this zone. Also the weaker rocks along the North Almora Thrust (NAT) and friable Siwalik belt compounded the causative factors. The maximum numbers of slope failure are observed between the floor thrust of MCT (Munsiari Thrust) and roof thrust of MCT (Vaikrita Thrust).

These landslides blocked and affected the national and state highways, damaging the arable land and forest cover.

Spatial analysis of landslides and geomorphic slope suggested that surfaces lying between 21° and 40° responded the most and created 738 landslides. The slopes incidentally fall in zone I (Fig. 10.4). Therefore during the extreme rainfall events, the zone along the MCT (zone I) in Garhwal Himalaya is most susceptible to landslide hazard.

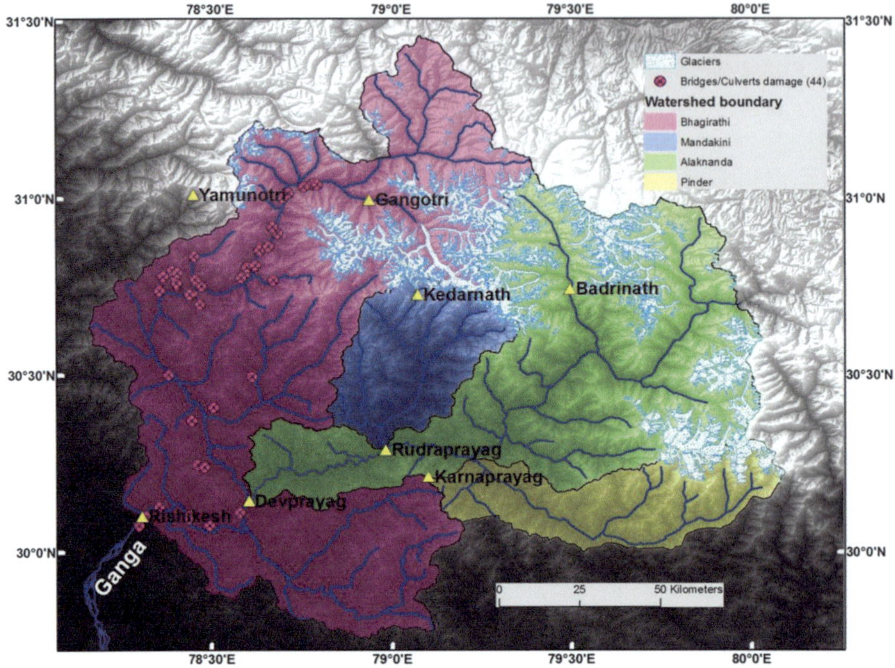

Fig. 10.5 Spatial distribution of damaged bridges and culverts in Bhagirathi-Nayar catchment

10.5.3 Bridges and Culverts

In Bhagirathi valley, bridges and culverts were found to be damaged at 44 locations (Fig. 10.5) where most sites of damage are present in the zone of river that has steeper channel gradients and larger bedload. It is inferred that a large amount of outsized bedload brought down by the trunk as well as the tributary channels is responsible for damages in bridges and culverts.

10.5.4 Damage to Buildings

A total of 717 buildings were found partially or totally damaged. This damage largely follows the two landslide clusters. The first cluster lies in the lower zones where damage is controlled by weaker rock type (phyllite) where most houses have their foundation in the weathered mantle of phyllitic rocks. The failure or creep in the weathered mantle induced by the increased pore pressure during the high rainfall event leads to the damage of several buildings in this zone, whereas the second

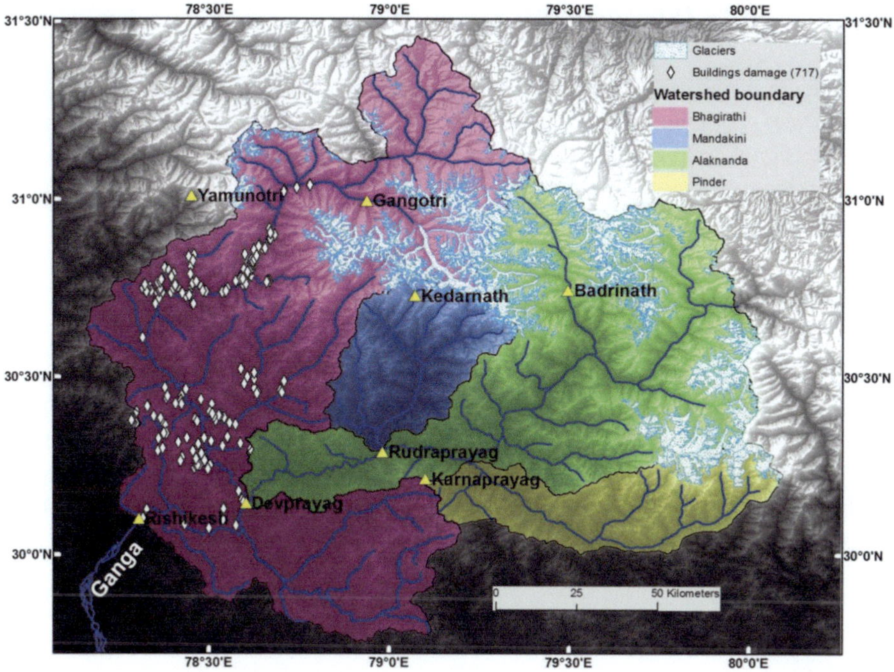

Fig. 10.6 Spatial distribution of damaged buildings and culverts in Bhagirathi-Nayar catchment

cluster that lies on the upper part of the catchment shows damage in the buildings due to steep gradients, thrust zones and high stream power. While in most houses minor cracks on the walls and roofs are seen, some have suffered the complete damage. The houses built on the riverbed or fill terraces are located adjacent to the river channel fell/completely damaged due to collapse of the terrace due to undercutting by the river. Figure 10.6 shows the general distribution of damaged building in the catchment.

10.5.5 Damage to Land Cover and Natural Resources

Damage to forest cover and agricultural land was mapped under this category. A total of 92 locations were found that showed such damage, where most are clustered in the lower reaches of the catchment that has phyllitic rocks. In most cases the agricultural fields were washed away due to high discharge in the small gadheras or toe cutting of sedimentary fill by major rivers. The forest cover, in most cases, is damaged due to landslides.

10.5.6 Other Infrastructures

Damages in the government buildings like panchayat bhavans, post offices, government offices, communication network, water supply lines, etc. were mapped under this category. A total of 320 numbers of sites were found damaged, and the damage seems to be clustered into two zones as landslides.

10.5.7 Riverbank Erosion

Riverbank erosion along Bhagirathi and Bhilangna occurred mainly above the town of Uttarkashi at 170 locations (Fig. 10.7). The catchment of Assi Ganga river was most affected. The damage has occurred along the course of the channel with relatively steeper gradients and where the terraces are of fill in nature. The fill terraces collapsed due to toe cutting by the channel. River ban erosion has mainly affected the agricultural terraces and has triggered landslides in the region.

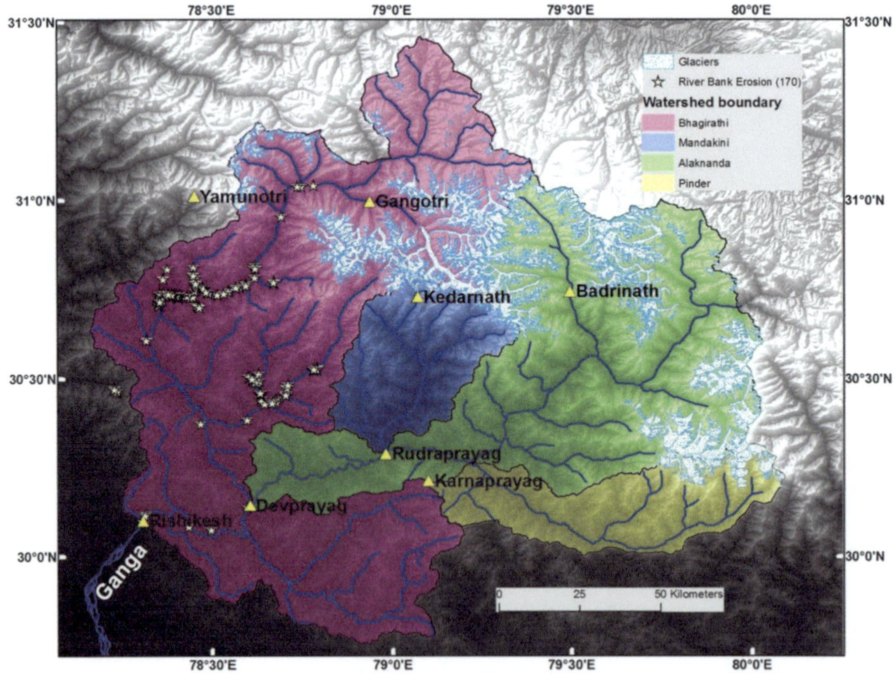

Fig. 10.7 Spatial distribution of riverbank erosion along the Bhagirathi-Nayar rivers

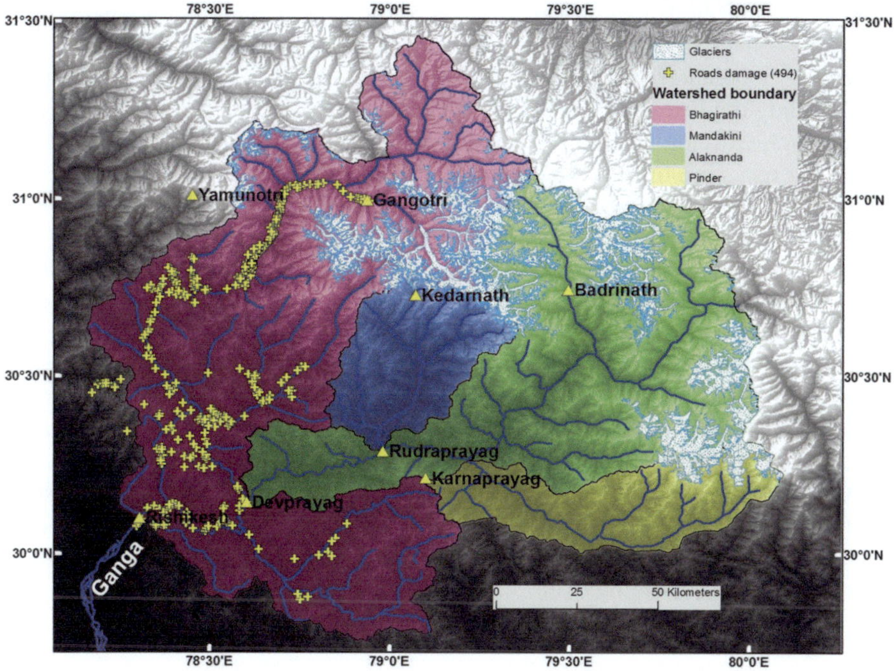

Fig. 10.8 Spatial distribution of damaged roads in Bhagirathi-Nayar catchment

10.5.8 Damage to Roads

Roads in Bhagirathi-Nayar river valleys mostly follow the course of the rivers and run 20–200 m above the channel. Roads are seen damaged at 494 places, and the damage is almost uniformly distributed and is controlled by landslides, riverbank erosion and overland flow. The topography of the mountain in general seems not to be controlling the road damage. Figure 10.8 shows the general distribution of damaged roads and the locations and extent of damage.

10.6 Conclusion

States of Himalaya frequently experience adverse impacts of extreme rainfall events, flood and related disasters. It is well known that the Himalaya has a fragile ecosystem because of ongoing tectonic activities as the Indian plate is underthrusting beneath the Tibetan plate and is being gradually consumed into the mantle, because of which thrusts and faults get seismically active (Valdiya et al. 1999; Yin 2006). Moreover, the climatic factors including the effect of global warming as indicated by the extreme hydrometeorological events are also responsible for the deterioration of the Himalayan environments and ecosystem. In addition, ongoing anthropogenic activities and poorly planned developmental work and rapidly increasing population have also affected the environment.

The study implied that there are two zones in Himalaya that are most vulnerable during such extreme events. Zone I lies above the Main Central Thrust (MCT) where the rocks are thrusted and tectonically deformed and hill slopes are steeper and where the mountains receive the highest rainfall. Such geologic and physiographic conditions make this region most reactive to heavy rainfall events. For example, this survey has shown that the area above Uttarkashi had the most number of landslides, where the bridges and culverts were also damaged. Zone II lies in the lower reaches where the hill slopes are rather gentler and rains are lesser intense, but the higher population density and anthropogenic interference combined with week rocks like phyllites made the few zones that failed during the event, and therefore the survey witnessed another cluster of damaged roads and high density of landslides in the Lesser Himalaya.

Another important information that this exercise brought out was the collapse of the fill terrace. The fill type of terraces that were within the reach of the flood or were densely populated with poor drainage and sewer network collapsed due to undercutting by the bulging river.

10.7 Recommendations

1. Dense network of river discharge measurement and automatic weather stations should be deployed. This network should be connected via satellite. This will help in understanding the changing pattern of rainfall and river response time.
2. Riverbeds and lowest level of fill terraces should not be allowed for any civil construction. However, these areas can be used for agricultural activities.
3. Complete ban on the construction of buildings or any forms of shelters over the highly unstable and active debris slopes like scree fans which are susceptible to change in angle of repose due to any alteration. The zones of river confluences should also be avoided.
4. It is a well-known fact that thick forests and vegetal cover minimise soil erosion and gulley formation due to protection of slope by leaves and branches and binding of the soil by root systems. Therefore a massive and well-adapted afforestation programme coupled with its protection leading to survival of plants should be taken up.
5. Roads should be aligned along the rivers but at higher elevation. There should be service roads linking villages and fields from the trunk roads.
6. In Bhagirathi valley there are two zones that are prone to landslides. Zone I lies above the MCT and the second in the periphery of the Tehri reservoir. Zone I which has lowest vegetation, steeper slopes is highly sensitive to such extreme events therefore should be avoided having permanent settlements. Since all shrines are located in these areas, we recommend that buffer community zones

should be created in the lower elevations and only controlled passage to pilgrims should be allowed. Several helipads and mobile hospitals with all emergency facilities should be developed during the Yatra season.

References

Ali SN, Biswas RH, Shukla AD, Juyal N (2013) Chronology and climatic implications of Late Quaternary glaciations in the Goriganga valley, central Himalaya, India. Quat Sci Rev 73:59–76

Armstrong RL (2010) The glaciers of the Hindu Kush-Himalayan region – a summary of the science regarding glacier melt/retreat in the Himalayan, Hindu Kush, Karakoram, Pamir, and Tien Shan mountain ranges. Kathmandu, ICIMOD, p 17

Goswami BN, Venugopal V, Sengupta D, Madhusoodanan MS, Xavier PK (2006) Increasing trend of extreme rain events over India in a warming environment. Science 314:1442–1445

Dobhal DP, Gupta AK, Mehta M, Khandelwal DD (2013) Kedarnath disaster: facts and plausible causes. Curr Sci 105(2):171–174

Heim GA (1939) "Central Himalaya" Geological observations of Swiss expedition 1936. Hindustan Publication Corporation, New Delhi, p 243

Immerzeel WW, Beek LPHV, Bierkens MFP (2010) Climate change will affected the Asian water tower. Science 324:1382–1385

Mehta M, Majeed Z, Dobhal DP, Srivastava P (2012) Geomorphological evidences of post LGM glacial advancements in the Himalaya: study from Chorabari Glacier, Garhwal Himalaya, India. J Earth Syst Sci 121(1):149–163

Kerr R (2013) In the hot seat. Science 342:688–689

Mehta M, Dobhal DP, Pratap B, Majid Z, Gupta AK, Srivastava P (2014) Late Quaternary glacial advances in the Tons River Valley, Garhwal Himalaya, India and regional synchronicity. The Holocene 24(10):1336–1350

Owen LA, Benn DI, Derbyshire E, Evans DJA, Mitchell WA, Richardson S (1996) Quaternary glacial history of the Lahul Himalaya, Northern India. J Quat Sci 11:25–42

Ray Y, Srivastava P (2010) Widespread aggradation in the mountainous catchment of the Alaknanda-Ganga River System: timescales and implications to Hinterland – foreland relationships. Quat Sci Rev 29:2238–2260

Sharma MC, Owen LA (1996) Quaternary glacial history of the NW Garhwal, central Himalayas, India. Quat Sci Rev 15:335–365

Stocker TF, Coauthors (eds) (2014) Climate change 2013: the physical science basis. Cambridge University Press, pp 1535

Valdiya KS, Paul SK, Tara C, Bhakuni SS, Upadhyay RC (1999) Tectonic and lithological characterization of Himadri (Great Himalaya) between Kali and Yamuna rivers, Central Himalaya. Himal Geol 20:1–17

Yin A (2006) Cenozoic tectonic evolution of the Himalayan orogen as constrained by along-strike variation of structural geometry, exhumation history, and foreland sedimentation. Earth Sci Rev 76:1–131

Part III
Disaster Risk Governance and Development

Chapter 11
Risk Governance for Improving Urban Disaster Risk Reduction

Chiranjiv Choudhary and Srinivasa Rajamani Neeli

Abstract Cities are the most complex dynamic systems in a continuous state of change. Urban conditions are very different from city to city, and their development is becoming a critical challenge. In developing countries, unplanned and unregulated urban growth is inviting risk in urban areas due to increased vulnerability. The urbanization is projected to increase multifold, and there would be a large-scale shift of rural population to urban areas and consequent growth of urban conglomerates. Therefore, urban disaster risk reduction will pose a challenge to the policy makers, administrators, and other stakeholders. Insufficient participation of key stakeholders in planning and management due to weak governance is one of the significant urban risk drivers. Planning, management, and regeneration based on CCA/DRR principles and governance are important aspects for resilient and sustainable development in the urban areas.

This chapter discusses in detail the present status of urban development, natural hazards, governance, disaster risk reduction, mainstreaming CCA/DRR, and risk governance, aimed at the creation of disaster-resilient cities.

Keywords Urbanization • Climate change • Urban governance • Sustainability • Disaster risk reduction • Risk governance • Resilience

11.1 Urbanization and Development

Urbanization is defined as a movement of people from rural to urban areas with population growth equal to urban migration and has been one of the most prominent trends of the twentieth and twenty-first centuries (Street 1997; United Nations 2010).

C. Choudhary (✉)
Commissioner of Horticulture & Sericulture, Agriculture & Cooperation Department, Government of Andhra Pradesh, Hyderabad, India
e-mail: chiranjivc@gmail.com

S.R. Neeli
City Project Coordinator, UNDP-GoI Project, Visakhapatnam, India
e-mail: rajamani74ns@gmail.com

I. Pal, R. Shaw (eds.), *Disaster Risk Governance in India and Cross Cutting Issues*, Disaster Risk Reduction, DOI 10.1007/978-981-10-3310-0_11

According to a report by the United Nations (2010), the ratio of urban population rose from 13% in 1900 to 29% in 1950, 50% in 2009, and it is projected to be 69% in 2050. Urbanization brings many benefits, such as diversity, market efficiency, jobs, education, and health improvement (Glaeser 1998; Christopher 2008). It is these benefits that attract a continuous flow of people from rural to urban areas. However, due to the rapid pace of urbanization, natural ecosystems are increasingly replaced by cities (Attwell 2000; United Nations Population Fund 2007).

The urban areas are recognized as one of the complex and highly dynamic landscapes on the earth's surface which supports more than half of the global human population, as well as hub of the worlds manufacturing and service industries (Kaplan et al. 2004). In spite of only 3% coverage of the earth's land surface, the urban areas are reported to exert marked effects on environmental conditions at both local and global scales (Grimm et al. 2000; Liu and Lathrop 2002; Herold et al. 2003). The driving factors (e.g., population or development), mediated by the socio-economic setting (e.g., market economy, resource institutions) and influenced by the existing environmental conditions, lead to changes in land use through the manipulation of the biophysical conditions of the land (Turner et al. 1995). Researchers emphasized that these changes of nonurban land uses however certainly provide many social and economic benefits but have adverse effects on natural environment (Tang et al. 2005) including global carbon cycle, climate, biodiversity, and landscape ecology (Houghton et al. 1999; Wickham et al. 2000) and lead to many problems, such as air and water pollution, depletion of cultivated land due to urban sprawl, global climate change, and others (Li et al. 2009; Yigitcanlar 2009). These problems present barriers to achieving sustainable development.

Cities are complex dynamic systems in a continuous state of change. They evolve in complex ways due to their size, social structures, economic systems, geopolitical settings, and the evolution of technology (Kennedy et al. 2007). Moreover, they require vast amount of resources to function, displaying diverse patterns, agglomeration, and intense competition for space with other land uses (Batty 2008). In the past, the depletion of the nearest and most accessible resources may have become a constraint on the growth of cities. However, technological and infrastructural innovations have driven the increments on urban inputs and outputs (Kennedy et al. 2007; Krausmann et al. 2008; Monstadt 2009). On a global scale and especially over the past two centuries, resource pressures have increased due to industrialization, rapid growth of the world population, urbanization, and technological development. Currently, cities are highly dependent on other cities and hinterlands to supply resources and dispose waste (Bai 2007). Hence, the environmental impacts are spread, thus enlarging the ecological (global) footprint of cities (Rees 1999; McNeill 2000; Monstadt 2009).

Cities have direct and indirect global impacts on the atmosphere, hydrosphere, geosphere, and biosphere by extracting large quantities of natural resources, in some cases leading to depletion and disposing of urban waste (Mills 2007). Diamond (2005) and Ponting (2007) described how the consequences of irreversible damage to the environment can cause the collapse of ecosystems and societies.

For the first time in history, more than half of the world population, which is 3.5 billion people, is living in urban areas. This urban fraction will increase to almost 60% by 2030 and 70% by 2050 (ESA-UN 2007). This large-scale urbanization requires large amount of resources, energy, and materials to build, feed, and fuel cities (Girardet 2003). Urban environmental problems (e.g., air pollution, waste disposal, open space fragmentation, and excessive fuel consumption) create an urgent need for urban sustainability. Such complexity poses a challenge to identify the causes of urban environmental problems and need to address them without causing greater deterioration. Environmental planning has traditionally addressed these problems with policies regulating the location and intensity of urban activities, often based on assumptions about urban and environmental dynamics that are rarely revisited (Ewing 1994, 1997; Alberti 1999; Chin 2002; Neuman 2005).

Urban growth all over the world is taking place which is unequal, but the rate of urbanization is very fast in the developing countries. Seventy nine million people were living in urban areas in India in 1961 but in 2001 about 285 million people resides in urban areas (Rahman 2009). In 1991, there were 23 metropolitan cities in India (Census of India 1991) which increased to 46 in 2011 (Census of India 2011). The highest rate of increase in the urban population has created many problems in the urban areas of Indian cities. The phenomena of accelerated urbanization without appropriate planning are the main culprit, wherein besides bringing a higher standard of living, it has also brought problems of growth of dense and unplanned residential areas, environmental pollution, nonavailability of services and amenities, and solid waste generation and growth of slums. Each urban center has a number of environmental problems with varying scales and scopes which are influenced by factors such as size of population and its density, climatic conditions, water resources, and the flora and fauna in and around the urban center (Rahman 2009). The state of urban environment all over India is deteriorating so fast that the sustainability of the cities is threatened.

While urbanization creates opportunities, it also exacerbates risks, and the speed at which it is happening challenges our capacity to plan and adapt. Inadequate urban planning and ineffective governance can bring significant economic, social, and environmental costs, threatening the sustainability of urban development (Renn and Walker 2008).

11.2 Disaster, Risk, and Vulnerability

"A 'disaster' is an event, be it man-made or natural, located in time and space, which produces the conditions whereby the continuity of structure and processes of social units becomes problematic. Disaster agents may differ in terms of their cause, frequency, controllability, speed of onset, length of forewarning, duration, scope of impact, and destructive potential" (Quarantelli 1980). Thus, "disasters" are the natural calamities, technological mishaps, and environmental extremes, while a "hazard" refers to the potential for damage. Traditionally, "risk" has been defined as "the product of the potential for hazard and the probability of that hazard being realized" (Murphy 1992). "Risk" is further described as a function of the nature of the hazard, accessibility or avenues of contact (exposure potential), characteristics of exposed populations (receptors), and the likelihood of occurrence of exposures and consequences (Kolluru 1996).

Risk can be defined as the magnitude of an unwanted consequence related to some activity multiplied by its probability (Morgan 1993), while Pencheon et al. (2001) in the *Oxford Handbook of Public Health Practice* define risk as the probability that a particular adverse event occurs during a stated period of time or results from a particular challenge. Risk is a measure of the probability and severity of adverse effects (Haimes 1991). Basically, three questions could be addressed in risk assessment including: (1) What can go wrong? (2) What is the likelihood of going wrong? (3) What are the consequences? Risk can be considered as the losses associated with a coastal disaster, and vulnerability refers to the effects of the event on the characteristics of the society or the environment. In recent years, a number of disaster events (natural, technological, and ecological) have made the global community aware of the immense loss of human lives and the productive resources that are caused regularly by such calamities. Equally important is the environmental damage provoked by these in the industrial areas, either as a direct or an indirect consequence; an increasing number of industrial complexes and hazardous materials concentrated in these units intensify the risks of secondary disasters, such as fires, explosions, radioactive radiation, etc., in the event of a natural hazard (IDNDR 1996).

Risk assessment is an analytical process based on scientific information, while risk management also involves the society's valuation of risks, benefits, costs, alternatives, consequences, etc. (Johannesson 1998). Site-specific risk assessment is the process of evaluating whether or not a site poses a risk to human health or to the environment (LaGoy 1994). "Risk management" and "risk assessment" are closely linked together. Risk assessment tries to answer the question "How big is the risk?" While the risk management process aims at finding out the best way to handle/reduce the risk. It is obvious that risk management must to a large extent be based on risk assessment (Anil et al. 2002).

11.3 Natural Hazards and Climate Change

A natural hazard is a potentially dangerous natural phenomenon that can cause injury or loss of life, property, infrastructure damage, and disruption of social and economic activities. Climate change will affect urban risks through changes in weather and climate hazards. The region is already experiencing changes in climate such as higher temperatures, changing rainfall patterns, varying frequencies of natural hazard events, and sea level rise. The consequences of natural hazard events and climate change on cities and towns are influenced not only by the natural hazard and climate change-related events themselves but also by a variety of physical, socioeconomic, and environmental factors. These include the geographic location of the urban center, poverty rates among urban residents, and levels of access to basic services, as well as the health of surrounding ecosystems. Climate change can be expected to result in changes in the nature and extent of natural hazards such as increased frequency, intensity, duration, and spatial extent of such events. Impacts of climate-related disasters are often high, dramatic, and, above all, still to some extent unpredictable.

There is a gap in translating the knowledge into practicees in the context of climate change and disaster risk, particularly in planning local, regional, and national urban development policies and practices. Natural hazards and climate change extremes pose a significant development challenge to the region. A single disastrous event has the potential to erode many years of economic development gains by damaging critical infrastructure and by diverting resources away from essential development spending, such as health and education services, toward disaster response and reconstruction efforts.

Vulnerability to risk is influenced by the coping and adaptive capacity of people, assets, and ecosystems that are exposed to natural hazards and climate extremes. Rapid urbanization has brought with it growing informal settlements, environmental degradation, unemployment, and increasing demands for the provision of basic services. Supply of infrastructure and urban services such as housing, roads, water, sanitation, drainage, and solid waste management has not kept pace with this growing demand, resulting in enormous pressure on existing infrastructure that was designed to cater to much smaller populations.

Urban poverty contributes to increased vulnerability to risk. Rapid urbanization has resulted in population settlement, economic activity, and infrastructure service provision in the cities and towns taking place in a largely unplanned manner. Weak urban governance arrangements and limited human and financial resources have contributed to the present situation. This low adaptive capacity has contributed change in urban centers.

11.4 Urban Governance and Sustainability

Cities are complex systems, weaving together thousands of economic, social, institutional, and environmental threats that affect individual and societal well-being. Cities experience a wide variety of shocks and stresses, such as industrial structural change, population inflow/outflow, disasters, disruption of the energy supply, and leadership change. Large urban systems are particularly vulnerable to foreseen and unforeseen threats; any sort of shock to complex systems such as these will have significant economic, social, environmental, and institutional repercussions.

Urban governance plays a critical role in enhancing resilience, mitigating climate change, resource efficiency, and thus ensuring sustainability. Institutionalizing an appropriate legislative, policy, and regulatory framework can help integrate climate change and disaster risk management (DRM) into all levels and sectors of government (all-of-government) decision-making. Governance can promote accountability, transparency, participation (all-of-society), and informed decision-making that actually implements risk reduction and climate action as a continuous process. This will facilitate an effective interface between government, communities, civil society, private sector, and other stakeholders, ensuring participation of different interest groups in decision-making.

Governance systems and mechanisms provide greater opportunities for an integrated development approach. Cities are composed of complex, interdependent systems that can be leveraged to support climate mitigation, adaptation, risk management, and sustainable development via effective local authorities supported by cooperative multilevel governance. This can enable synergies with infrastructure investment and maintenance, land use management, livelihood creation, and ecosystem service protection with resilience building as an overarching objective.

11.5 Urban Disaster Risk Reduction and Resilient Development

Disaster risk reduction cannot be achieved in isolation either by the government, NGOs, or the communities. Disaster risk reduction calls for collaborative and cooperative action and commitment by all important stakeholders – the government, NGOs, communities, media, and the academic community (UNISDR 2009). We have to make disaster prevention as inevitable, integral practice and approach to our development planning. Planning, management, and regeneration based on CCA/DRR principles and governance are important aspects to develop resilience and sustainable development in the urban areas.

Resilient cities are those able to absorb, adapt, transform, and prepare for past and future shocks and stresses in order to ensure sustainable development, well-being, and inclusive growth. Resilience is made up of seven building blocks:

- *Adaptive*: An adaptive urban system manages uncertainty by evolving – modifying the standards, norms, or past behavior – using evidence to identify solutions and applying the knowledge gained from past experience when making decisions about the future.
- *Robust*: A robust urban system can absorb shock and emerge without significant loss to its functionality. Robustness depends on a system that is well-designed, built, and managed to absorb the impact of a shock and continue to operate.
- *Redundant*: Redundant urban systems are able to meet the need for spare capacity when faced with unexpected demand, disruptive event, or extreme pressure. This entails intentionally developing or having access to more than one source of action, service, or service provider when necessary.
- *Flexible*: A flexible urban system allows individuals, households, businesses, communities, and government to adjust behavior or action in order to respond rapidly to change.

- *Resourceful*: A resourceful urban system can effectively and quickly restore the functionality of essential services and systems in a crisis or under highly constrained conditions, with the resources available.
- *Inclusive*: An inclusive urban system ensures that diverse actors and communities are fully consulted, engaged, and empowered in the policy process, including the policy design stage when possible.
- *Integrated*: An integrated urban system promotes a cooperative and, ideally, collaborative or participatory approach to policy and programming that transcends sectoral and administrative boundaries to ensure better coherent decisions and effective investment.

Disaster risk-resilient development is not only about protecting people's lives and livelihoods but also about social, economic, and environmental sustainability, aiming to reduce socioeconomic vulnerabilities to natural hazards. It is about building social, economic, and environmental resilience through the systematic integration of disaster risk reduction into overall development planning, policy, and programs. Disaster risk-resilient development requires a clear paradigm shift from a risk-insensitive development model to a risk-sensitive one, with disaster risk assessment recognized as a prerequisite for development planning. Disaster risk reduction must be an integral part of development planning, policy, and programs. By so doing, development will avoid generating new risks for people and communities by building instead their resilience to disasters.

Global implementation of the Hyogo Framework for Action has accumulated a wealth of evidence that reducing risks builds resilience to disasters and protects the gains made in sustainable development around the three pillars: social equity, economic growth, and environmental protection. Furthermore, it also contributes to good governance. Increasingly, research reveals that disasters lead to a variety of social issues, including those related to food security, housing, health, education, deepening poverty, and sometimes even political instability. Disaster risk-resilient development addresses social equity by reducing the social and economic vulnerabilities of people and communities to disasters. Disaster risk-resilient development protects economic growth by applying disaster risk assessment to development planning and programs. Risk-informed decision-making on development can protect lives and critical facilities from potential disasters, either by enhancing building codes or avoiding areas prone to natural hazards.

11.6 Mainstreaming Disaster Risk Reduction and Climate Change Adaptation

Mainstreaming as a process of governance has been variously defined and applied. According to UNDP (2010): "Mainstreaming of DRR is a governance process, enabling the systematic integration of DRR concerns into all relevant development spheres. In other words, responsive, accountable, transparent and efficient governance structures underwrite the environment where DRR can be institutionalized as an underlying principle of sustainable development."

Thus, the key to mainstreaming is to better understand and assess the implications of disasters and climate change on any development action and investment. Most importantly, it is an ex ante institutional process of development planning and decisions which take into account existing and future disaster and climate risks for a given system. The primary objective of mainstreaming DRR and CCA is to help reduce the potential impacts of disasters and climate change and strengthen the capacity of the system to recover and bounce back within a definite timeframe from any kind of shock or disruption. Investments in such mainstreaming planning and activities are cost effective and offer a high return.

According to the latest Global Assessment Report on Disaster Risk Reduction, 2015 (UNISDR 2015), India is one of the countries with high overall ratio of annual average loss (AAL) to its overall social expenditure, capital stock, and savings. India loses almost $9.8 billion every year due to natural disasters. Mainstreaming disaster risk reduction initiatives, among others, has been a key initiative to strengthen capacities of key governance systems, infrastructures, and development planning to better respond to and recover from natural disasters. Both the Disaster Management Act (2005) and the National Policy on Disaster Management (2009) of the Government of India categorically emphasize on mainstreaming disaster management into the development planning process by creating appropriate institution mechanism at various levels and the leadership of local authorities in this process.

Mainstreaming disaster risk reduction (DRR) and climate change adaptation (CCA) requires identifying the particular vulnerabilities and risk-building processes of the various sectors and levels of government and segments of the society. For instance, in looking at legal bases for strengthening democratic institutions, promoting human rights and integral development, special attention should be given to how these legal instruments may or may not reduce the vulnerability of women and children, elders and people with disabilities, indigenous people, and other marginalized segments of the society. One benefit is linked to the fact that many underlying causes of disasters are rooted in political and institutional structures that often are overlooked by DRR and CCA programs and projects. Strong democracies encourage good governance, which in the context of DRR refers to the process by which decisions are made and implemented to reduce the vulnerability of communities, increase their resilience; prepare, respond, and manage emergencies; and rehabilitate and reconstruct the damaged infrastructure as well as recuperate livelihoods. Thirdly, mainstreaming DRR and CCA will provide for the appropriate channels and mechanisms to address underlying causes of disasters.

The United Nations Development Program (UNDP) framework is designed to provide practical guidance to mainstream disaster risk reduction and climate change adaptation into development. The framework identifies the following entry points for mainstreaming:

- *Policy development* – integrating disaster risk reduction into development policies at the national and sector level, such as agriculture or education policies.

- *Organizational/institutional development* – identifying disaster risk reduction focal points across government agencies and strengthening cross-sectoral coordination mechanisms such as national platforms for disaster risk reduction.
- *Improving advocacy and knowledge* for disaster risk reduction – technical guidelines, training, and educational programs.
- Supporting the *implementation* of specific disaster risk reduction measures – conducting risk assessments and integrating risk reduction into recovery interventions.
- Supporting *broad participation* in disaster risk reduction – community-based disaster reduction plans and programs, as well as increasing the involvement of women in risk reduction plans.

Issues related to urban resilience:

- Resilience in cities runs across a spectrum. A city must necessarily absorb, adapt, transform, and prepare in the face of current and future shocks or stresses in order to maintain its core purpose. A city is either more or less resilient; but no city is entirely lacking in resilience. The challenge for policy makers is to identify where their city may lie along this spectrum and how far they wish to move it.
- The foundations of resilience include adaptive capacity, resourcefulness, robustness, redundancy, flexibility, inclusiveness, and integration. A system that cannot demonstrate these qualities in times of crisis will be less resilient.
- Enhancing resilience requires investment and investing to prepare for the unknown or unforeseeable circumstances.
- Building resilience is a multilevel governance exercise. Building resilience requires that city, regional, and national authorities work together as part of a broader urban policy and urban development agenda.

11.7 Risk Governance

An overview of the concept of risk governance, which promises to offer a comprehensive means of integrating risk identification, assessment, management, and communication, is given by Renn (2008). Risk governance denotes both the institutional structure and the policy process that guide and restrain collective activities of a group, society, or international community to regulate, reduce, or control risk problems (Klinke and Renn 2012). Foresight activities strengthen public awareness and risk perception among relevant stakeholders. Communication and participation is the key for successfully implementing risk governance (Renn and Klinke 2001).

Understanding the dynamics, structures, and functionality of risk governance processes requires a general and comprehensive conceptualization of procedural mechanisms and structural configurations. The literature on urban governance and pro-poor development indicates that there are both demand and supply constraints to achieving pro-poor governance. These can be applied effectively to governance issues related to adaptation and urban environmental management. On the supply side, municipal governments are constrained by a number of factors (Devas 2001; Satterthwaite 2001; Huq et al. 2007):

- Municipal boundaries often do not include areas where the poor reside, thus placing them outside of municipal jurisdiction.
- City governments are often not responsible for many public services including land allocation, housing, water, and other public services.
- Legal restrictions may prevent municipal governments from being able to act, for example, by not being able to supply those populations which do not pay property taxes.
- Weak managerial and technical capacities at the local government level.
- Lack of financial resources in order to increase service provision and build infrastructure. This is often exacerbated by international donors and development banks who reinforce the power of national governments and pay little attention to governance and needs at the local or municipal level.
- Lack of financial management capabilities and low financial incentives for staff resulting in corruption.
- Conflicts with national- or state-level governments.

Demand-side constraints commonly include low-frequency nature of high-impact events, low levels of awareness about the changing nature of climate-related risks due to climate change, and low levels of empowerment and mobilization by the groups of citizens most adversely impacted by climate shocks and stresses. In most cities at risk from floods, wealthier groups and formal enterprises do not face serious risk. For example, Mumbai has the resources capable of reducing risk from flooding, yet the costs are often borne by the low-income groups who live in more risk-prone settlements (Huq et al. 2007). Despite being at serious risk of increased flooding, Dhaka, Mumbai, and Shanghai have all attracted much private investment despite their vulnerability to storms and sea level rise (Sherbinin et al. 2007). The speed at which the economic "risk map" for cities will change is likely to be slower than the actual climate changes and risks (e.g., well-documented risk of New Orleans did not cause entrepreneurs and residents to move) (Huq et al. 2007). A so-called moral hazard may also play a role, as government decision-makers may not invest in a perceived low-probability event because they assume that the international relief community would come to their rescue in the event of a significant disaster (Sherbinin et al. 2007). Devas (2001) argues that the quality of pro-poor governance in urban settings is determined by the ability of the poor to influence political decision-makers. Broadly, these demand side variables include issues of institutional design whereby low-income citizens are able to participate meaningfully in elections and an active civil society, which is able to advocate for the rights of the poor. UN-Habitat (2003) highlights a number of principles for enhancing these mechanisms toward an "inclusive city":

- Responsibility for service delivery should be based on the principle of subsidiarity (taking action at the lowest appropriate scale).
- All urban citizens should have equal access to decision-making processes.
- Increased empowerment of low-income residents in order to enhance accountability and pro-poor decision-making.

11.8 Conclusions

Urbanization along with many benefits, because of its unequal growth pattern, brings adverse effects on environmental conditions at both local and global scales. These problems present barriers to achieving sustainable development. While urbanization creates opportunities, it also exacerbates risks, and the speed at which it is happening challenges our capacity to plan and adapt. Inadequate urban planning and ineffective governance can bring significant economic, social, and environmental costs, threatening the sustainability of urban development.

Upon the above situation, natural hazards and climate change extremes pose a significant development challenge in the urban areas. A single disastrous event has the potential to erode many years of economic development gains by damaging critical infrastructure and by diverting resources away from development spending, such as health and education services, toward disaster response and reconstruction efforts. Weak urban governance arrangements and limited human and financial resources have contributed to the present situation of low adaptive capacity that has contributed change in urban centers.

To address the shocks and stresses experienced by cities, urban governance can play a crucial role in enhancing resilience, mitigating climate change, resource efficiency, and thus ensuring sustainability. Governance can promote accountability, transparency, participation, and informed decision-making that actually implement risk reduction and climate action as a continuous process. This will facilitate an effective interface between government, communities, civil society, private sector, and other stakeholders, ensuring participation of different interest groups in decision-making.

Governance systems and mechanisms provide greater opportunities for an integrated development approach. Disaster risk reduction cannot be achieved in isolation either by the government, NGOs, or communities. Disaster risk reduction calls for collaborative action and commitment by all important stakeholders – the government, NGOs, communities, media, and the academic community. We have to make disaster prevention as inevitable, integral practice and approach to our development planning. Planning, management, and regeneration based on DRR/CCA principles and governance are important aspects to develop resilience and sustainable development in the urban areas.

Resilient cities can absorb, adapt, transform, and prepare for past and future shocks and stresses in order to ensure sustainable development, well-being, and inclusive growth. Disaster risk-resilient development is not only about protecting people's lives and livelihoods but also about social, economic, and environmental sustainability, aiming to reduce socioeconomic vulnerabilities to natural hazards. It is about building social, economic, and environmental resilience through the systematic integration of disaster risk reduction into overall development planning, policy, and programs.

Mainstreaming of DRR/CCA – a governance process can enable the systematic integration of DRR concerns into all relevant development spheres. The primary objective of DRR and CCA mainstreaming is to help reduce the potential impacts of disasters and climate change and strengthen the capacity of the system to recover and bounce back from any kind of shock or disruption. Mainstreaming DRR and CCA requires identifying the particular vulnerabilities and risk-building processes of the various sectors, levels of government, and segments of the society. Building resilience is a multilevel governance exercise, which requires that city, regional, and national authorities work together as part of a broader urban policy and urban development agenda.

Risk governance promises to offer a comprehensive means of integrating risk identification, assessment, management, and communication for effective implementation to minimize adverse impact. Risk governance denotes both the institutional structure and the policy process that guide and restrain collective activities of a group, society, or international community to regulate, reduce, or control risk problems. Foresight activities strengthen public awareness and risk perception among relevant stakeholders. Communication and participation is the key for successfully implementing risk governance.

References

Alberti M (1999) Urban patterns and environmental performance: What do we know? J Plan Educ Res 19:151–163

Anil KG, Inakollu VS, Jyoti M, Mohammad Y (2002) Environmental risk mapping approach: risk minimization tool for development of industrial growth centers in developing countries. J Clean Prod 10:271–281

Attwell K (2000) Urban land resource and urban planting – case studies from Denmark. Landsc Urban Plan 52:145–163

Bai X (2007) Industrial ecology and the global impacts of cities. J Ind Ecol 11:1–6

Batty M (2008) The size, scale, and shape of cities. Science 319:769–771

Census of India (1991) Provisional Population Totals, Office of Registrar, General of India. Government of India, New Delhi

Census of India (2011) Provisional Population Totals, Office of Registrar, General of India. Government of India, New Delhi

Chin N (2002) Unearthing the roots of urban sprawl: a critical analysis of form function and methodology. Centre for Advanced Spatial Analysis (CASA) Working Paper Series. University College London, London

Christopher D (2008) Health and urban living. Science 319:766–769

Devas N (ed) (2001) Urban governance and poverty: Lessons from a study of ten cities in the south. The School of Public Policy, University of Birmingham, Birmingham

Diamond JM (2005) Collapse: how societies choose to fail or succeed. Viking Penguin, New York

Disaster Management Act (2005) Govt. of India, 2005

ESA-UN (2007) World urbanisation prospects, Revised edn. United Nations, New York

Ewing RH (1994) Characteristics, causes and effects of sprawl: a literature review. Environmental and Urban Issues 21(2):1–15

Ewing RH (1997) Is Los Angeles-style sprawl desirable? J Am Plan Assoc 63(1):107–126

Girardet H (2003) Cities, people planet. In: Vertovec S, Posey DA (eds) Globalization, globalism, environment, and environmentalism: consciousness of connections. Oxford University Press, New York, pp 87–102

Glaeser EL (1998) Are cities dying? J Econ Perspect 12(2):139–160

Goy L (1994) Risk assessment: principles and applications for hazardous waste and related sites. Noyes publications, New Jersey

Grimm NB, Grove JM, Pickett STA, Redman CL (2000) Integrated approaches to long term studies of urban ecological systems. Bioscience 50(7):571–584
Haimes YY (1991) Total risk management. Risk Anal 11:1–8
Herold MN, Kent DB, Goldstein CK (2003) The spatiotemporal form of urban growth: measurement, analysis and modeling. Remote Sens Environ 86:286–302
Houghton RA, Hackler JL, Lawrence KT (1999) The U.S carbon budget: contributions from land-use change. Science 285(5427):574–578
Huq S, Kovats S, Reid H, Satterthwaite D (2007) Reducing risks to cities from disasters and climate change. Environ Urban 19(1):39–64
IDNDR (1996) Disaster reduction in urban systems, Volume 28 of Stop disasters. IDNDR, Geneva
Johannesson T (1998) Icelandic avalanche runout models compared with topographical models used in other countries. In: E. Hestnes (ed) 25 years of snow avalanche research. Voss. Norwegian Geotechnical Institute Publications; 2003:43–52
Kaplan DH, Wheeler JO, Holloway SR (2004) Urban geography. Wiley, NJ
Kennedy CA, Cuddihy J, Engel Yan J (2007) The changing metabolism of cities. J Ind Ecol 11:43–59
Klinke A, Renn O (2012) Adaptive and integrative governance on risk and uncertainty. J Risk Res 15(3):273–292
Kolluru RV (1996) Risk assessment and management handbook for environmental, health, and safety professionals. McGraw-Hill, New York
Krausmann F, Fischer-Kowalski M, Schandl H, Eisenmenger N (2008) The global sociometabolic transition: past and present metabolic profiles and their future trajectories. J Ind Ecol 12:637–656
Li F, Liu X, Hu D, Wang R, Yang W, Li D, Zhao D (2009) Measurement indicators and an evaluation approach for assessing urban sustainable development: a case study for China's Jining City. Landsc Urban Plan 90(3–4):134–142
Liu X, Lathrop RG (2002) Urban change detection based on an artificial neural network. Int J Remote Sens 23:2513–2518
McNeill JR (2000) Something new under the sun: an environmental history of the twentieth-century world. W.W. Norton, New York
Mills G (2007) Cities as agents of global change. Int J Climatol 27:1849–1857
Monstad J (2009) Conceptualizing the political ecology of urban infrastructures: insights from technology and urban studies. Environ Plan 41:1924–1942
Morgan MG (1993) Risk analysis and management. Scientific American, July, 24–30
Murphy P (1992) Citation in: researching health risks. U.S. Congress Office of Technology Assessment, Washington, DC
National Policy on Disaster Management (2009) National Disaster Management Authority, Government of India
Neuman M (2005) The compact city fallacy. J Plan Educ Res 25(1):11–26
Pencheon D, Guest C, Melzer D, Muir GJA (2001) Oxford handbook of public health practice. Oxford University Press, Oxford
Ponting C (2007) A new green history of the world: the environment and the collapse of great civilizations. Penguin Books, New York
Quarantelli EL (1980) Evacuation behaviour and problems: findings and implications from the research literature. Disaster Research Center, The Ohio State University, USA, p 214
Rahman A, Netzband M, Alka S, Javed M (2009) An assessment of urban environmental issues using remote sensing and GIS techniques an integrated approach: a case study: Delhi, India. In: de Sherbinin A, Rahman A, Barbieri A, Fosto JC, Zhu Y (eds) Urban population-environment dynamics in the developing world: case studies an Lessons learned. International Cooperation in National Research in Demography (CICRED), Paris, pp 181–211
Rees WE (1999) The built environment and the ecosphere: a global perspective. Build Res Inf 27:206–220
Renn O (2008) Risk governance. Coping with uncertainty in a complex world. Earthscan, London

Renn O, Klinke A (2001) Environmental risks—perception, evaluation and management: epilogue. In: Boehm G et al (eds) Environmental risks: perception, evaluation and management. Elsevier Science, Amsterdam, pp 275–299

Renn O, Walker K (2008) Lessons learned: a re-assessment of the IRGC framework on risk governance. The IRGC risk governance framework: concepts and practice. Springer, Heidelberg, pp 331–367

Satterthwaite D (2001) Environmental governance: a comparative analysis of nine city case studies. J Int Dev 13(7):1009–1014

Sherbinin A, Schiller A, Pulsipher A (2007) The vulnerability of global cities to climate hazards. Environ Urban 19(1):39–64

Street P (1997) Scenario workshops: a participatory approach to sustainable urban living? Futures 29(2):139–158

Tang J, Xue Y, Yuc T, Guan YN (2005) Aerosol optical thickness determination by exploiting the synergy of terra and aqua modis. Remote Sens Environ 94:327–334. 2

Turner II, B.L., Skole, D., Sanderson, S., Fischer, G., Fresco, L., R. Leemans., (1995). Land-use and land-cover change. Science/Research Plan. IGBP Report no. 35

Turner T, April C, Padma Dorothy K, Goutham J, Kalavati C, Rajanna B (2003) Health hazards due to pollution of waters along the coast of Visakhapatnam, east coast of India. Ecotoxicol Environ Saf 56:390–397

UNISDR (2009) UNISDR terminology on disaster risk reduction. UNISDR, Geneva

UNISDR (2015) Making development sustainable: the future of disaster risk management. Global assessment report on disaster risk reduction. United Nations Office for Disaster Risk Reduction (UNISDR), Geneva

United Nations (2010) World urbanization prospects, the 2009 version. United Nations, New York

United Nations Human Settlements Programme (UN-Habitat) (2003) The challenge of slums: global report on human settlements 2003. Earthscan, London

United Nations Population Fund (2007) State of world population: unleashing the potential of urban growth. United Nations Population Fund, New York

Wickham JD, Neill RVO, Jones KB (2000) A geography of ecosystem vulnerability. Landsc Ecol 15:495–504

Yigitcanlar T (2009) Planning for smart urban ecosystems: information technology applications for capacity building in environmental decision making. Theor Empir Res Urban Manag 3(12):5–21

Chapter 12
Risk Governance Measures and Actions in Sundarbans Delta (India): A Holistic Analysis of Post-disaster Situations of Cyclone Aila

Indrajit Pal and Tuhin Ghosh

Abstract The Sundarbans delta is a unique type of land marshes and tidal forests called 'mangroves', which have special characteristic to adapt themselves to the highly saline soil, strong winds and inundation of sea water twice a day during tides. The Sundarbans was declared as a 'world heritage site' in 1994 and as a 'biosphere reserve' in 1989. By virtue of its proximity to the Bay of Bengal, the land is very much prone to the wrath of bay cyclones and tidal disturbances during the monsoon. Apart from its vulnerable geo-climatic locations, the major part of the Sundarbans area is facing the problem of waterlogging because of its basinlike island geography.

In 2009, Cyclone Aila devastated the deltaic area of Sundarbans and impacted more than 3500 km of very old earthen embankments that are essential for the very survival of the islands. The natural levees were weakened by lashing sea waves during cyclones, often giving in to the tidal bores resulting in disastrous floods.

The impact of Cyclone Aila has led to heavy rainfall and damages in 16 districts in West Bengal. The worst affected districts are South 24 Parganas, North 24 Parganas, East Medinipur, West Medinipur, Howrah, Hooghly and Kolkata. The population affected in the state was more than five million. The damage impact assessment conducted by the Government of West Bengal and UNDP put the death toll at 96. Over 500,000 houses are damaged either fully or partially, and a similar amount of cropped area is affected. Over 60,000 people rescued have been accommodated in government relief camps. However, some areas in the Sundarbans river islands remain marooned and hard to reach.

I. Pal (✉)
Disaster Preparedness, Mitigation and Management (DPMM), Asian Institute of Technology, Khlong Luang, Pathum Thani, Thailand
e-mail: indrajit.pal@gmail.com

T. Ghosh
School of Oceanographic Studies, Jadavpur University, Kolkata, India
e-mail: tuhin_ghosh@yahoo.com

I. Pal, R. Shaw (eds.), *Disaster Risk Governance in India and Cross Cutting Issues*, Disaster Risk Reduction, DOI 10.1007/978-981-10-3310-0_12

This study will review the risk governance in post-Aila situations pertaining to the structural and nonstructural measures initiated in the region. This study will also analyse the mitigation and adaptation scenarios in line with the federal and state government disaster risk management frameworks and risk reduction plans. This study assesses the effectiveness of disaster management policy framework and existing institutional mechanisms in West Bengal, India, to cover natural disasters, particularly cyclonic disasters. This study is a preliminary attempt to identify existing gaps in the disaster management system and suggests upgradation to minimize effects of the impact in the future, which may lead to sustainable development in these cyclone-prone areas.

Keywords Cyclone • Coastal hazards • Sundarbans delta • Risk governance • Institutional mechanism

12.1 Introduction

Indian coastlines of 7516 km, is the longest in the world, and extremely predisposed to cyclones and its associated hazards like storm tide, high-velocity wind and heavy rains. The vulnerability of the long coastlines is again compounded with high population density and geophysical location in the North Indian Ocean (NIO) Basin. About 8% of the area and almost one-third of the country's population (32 crore) are susceptible to cyclone-related hazards (NDMA 2008). Climate change and its resultant sea-level rise can significantly increase the vulnerability of the coastal population (NDMA 2009). The current global average temperature is presumably higher now than in the last 2000 years (Jones 2004) mostly caused by anthropogenic influence (IPCC 2005). Recent measurements indicate a perceptible rise in sea surface temperatures, which are a main determinant of storm intensity. This suggests that a storm is likely to have more intense wind and generate more rainfall than without global warming (Trenberth 2005). Tropical cyclones cause significant damage, injury and loss of life and are the costliest natural calamities even in a developed country like the United States of America (Pielke and Landsea 1998). Though the frequency of tropical cyclones in the NIO region covering the Bay of Bengal and the Arabian Sea is the least in the world (7% of the global total), their impact on the east coast of India as well as the Bangladesh coast is relatively more devastating due to high density of vulnerable population (NDMA 2009).

On 25–26 May 2009, tropical cyclonic storm Aila created havoc in West Bengal and Bangladesh. A government report dated 9 July 2009 indicates that this multi-hazard emergency has taken 138 lives and affected nearly 6.8 million people. Cropped area damage reached 5 million hectares (WBGoV 2009). Tropical Cyclone Aila provides us the opportunity to consider this as a case study and investigate the effectiveness of existing institutional and coping mechanisms and policy framework at central, state, district, block and local levels in dealing with disaster of this magnitude. The study is based on secondary data analysis, personal interaction with representatives of different organizations involved to respond to Cyclone Aila. This study indicates that there is a scope of improvement in all aspects of disaster management. It is also important to identify the gaps that exist in implementing policies, assess the disaster preparedness levels of state and across agencies and degree of coordination across various government departments, measure efficiency of early warning mechanisms, evaluate access to emergency services and appraise NGOs and civil society's role in disaster preparedness, emergency services and relief activities in order to find out the current status of disaster management system in West Bengal. An assortment of sources has been utilized in this study such as reports and documents published by the Government of West Bengal, Government of India and United Nations on Cyclone Management and Disaster management and synthesis of reports published by Intergovernmental Panel on Climate Change (IPCC) and India's National communication (NATCOM) to United Nations Framework Convention on Climate Change (UNFCCC).

12.2 Cyclone Aila and Its Impact to Sundarbans Delta (India)

Cyclone Aila which hit West Bengal and Bangladesh on 25–26 May 2009 devastated the lives and livelihoods of nearly 6.8 million people in West Bengal alone. According to government records, the disaster hit 28,349 villages of 206 blocks in 18 districts in West Bengal killing 193 persons and leaving nearly half a million people homeless (Fig. 12.1). The cyclone that struck through the spine of West Bengal blew with devastating impact on the economy of the state with the total cost of damages placed at over 300 million USD; over 800 km long embankments were washed away in Sundarbans, salinating the farmlands and inundating villages. The intense rainfall and high wind over 100 km/h, triggered landslides in Darjeeling hills results in permanent displacement of hundreds. There is no accurate estimate of the long-term impact on the lives and livelihoods of people due to this cyclone.

The government and non-governmental organizations responded to this cyclone with positive impact on the people. From issuance of early warning, rescue efforts, coordinated relief and recovery programmes, the government of West Bengal, along with Government of India, has responded to the needs of affected people, and the partners and agencies of State Inter Agency Group (IAG) have supported these affected communities, to the tune of over 1600 million USD. (Rs. 647 crores from GoWB + Rs. 75 crores from INGOs/NGOs).

The West Bengal state government did a massive relief operation through various wings of the government, specially through the departments of disaster management, health, PHED, food and civil supplies, women and child development, civil defence, agriculture and irrigation. The entire cost of response by the state government's support alone is to the tune of Rs. 647 crores. With additional support from the union government, the total volume spent by the government by way of relief,

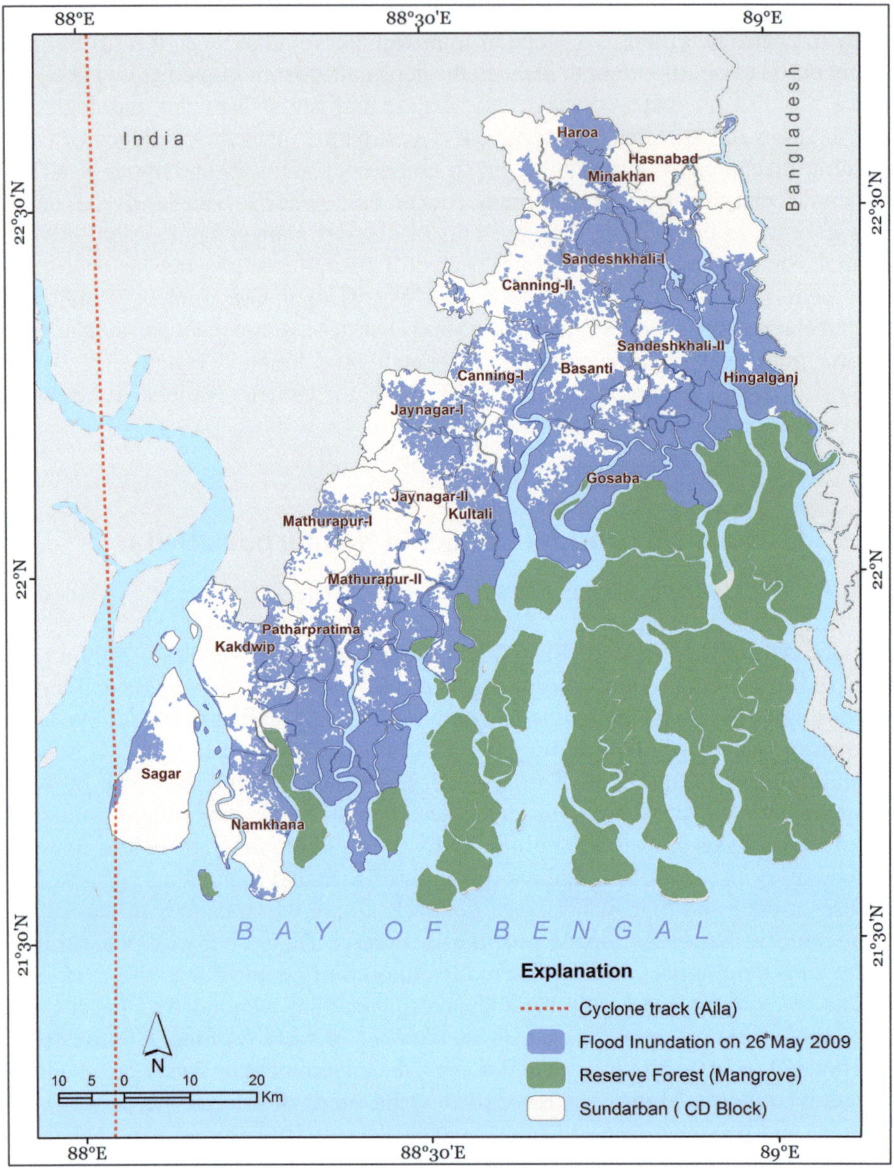

Fig. 12.1 Cyclone Aila 2009 and its impact to Sundarbans Delta (INDIA)

rehabilitation and recovery including embankment repairing would be over Rs. 5000 crores.

The State IAG-WB coordinated the efforts of the non-profit organizations, and its activities have been acclaimed and accepted as a great inspiration in coordinating humanitarian response. The support by the humanitarian agencies in the context of

Aila by way of response and recovery is about Rs. 60 crores. This assistance, though miniscule in comparison to the need of the people and support given by the government, has been well recognized by the government of West Bengal.

State Inter Agency Group State Inter Agency Group such as the forum of the Disaster Management Department and other international, national, state-level non-profit organizations and UN agencies has been closely working since 23 May 2010 to assist in the relief and recovery of the people affected by the cyclone and to avert the type of disasters unleashed by similar hazards. In this context, since it is nearly a year since cyclone Aila, the State IAG-WB along with all its partners plans to undertake a review of the Aila response and recovery programmes, to identify the learning, good practices and gaps in coordination, response and recovery and suggest way forward.

The impact of Cyclone Aila has led to heavy rainfall and damages in 16 districts in West Bengal. The worst affected districts are South 24 Parganas, North 24 Parganas, East Medinipur, West Medinipur, Howrah, Hooghly and Kolkata. On 28 May, the population affected in the state reached over 5.1 million. The damage impact assessment conducted by the Government of West Bengal and UNDP put the death toll at 96. Over 500,000 houses are damaged either fully or partially, and a similar amount of cropped area is affected. Over 60,000 people rescued have been accommodated in government relief camps. However, some areas in the Sundarbans river islands remain marooned and hard to reach. A sum of 15,000,000 has been announced for the state for contingency relief.

12.3 Response, Recovery, Reconstruction and Rehabilitations

Aila has impacted the metropolitan and other five adjoining districts of West Bengal on Monday, 25 May 2009. In Kolkata, the city became completely disrupted by the storm as most of the trees were uprooted and created havoc for the people and traffic along with electricity supply. International and domestic flights were delayed for several hours. All of a sudden, millions of people became stranded and paralysed from their normal life.

The compiled reporting of the incident says the following impacts:

- In India, at least 149 people were killed, two by electrocution, and hundreds others were left homeless as torrential rains led to flooding.
- High winds uprooted numerous trees, blocking roads throughout the region.
- More than 15,000 people in eight villages were reportedly isolated from relief crews by severe flooding.
- At least 18 of the 45 fatalities in West Bengal were in Kolkata, the region where Aila made landfall. All transit systems in the city of Kolkata were halted, and daily life was at a standstill due to the storm.

- In the West Bengal state, more than 100,000 people were left homeless as a result of Aila. At least 100 river embankments were breached by storm surge produced by the cyclone.
- Lakhs of mangrove trees uprooted and are also vulnerable due to the sudden change in the physico-chemical changes.
- The only mangrove tiger land in the globe, which houses around 100 odd numbers of the endangered Royal Bengal tigers, was inundated with 6.1 m (20 ft) of water. Dozens of the tigers were feared to have drowned in Aila's storm surge along with deer and other animals.
- As of 27 May 2009, one tiger has been found alive within a village which took shelter in a waterlogged cowshed following the cyclone's landfall. Additionally the forest remains under an estimated 2.4 m (7.9 ft) of water.
- On 27 May, conservationists have begun a search for the tigers throughout the forest. The search teams were supplied with fresh drinking water for the tigers as their natural water source was inundated with salt water from Aila's storm surge.

Problems

- Accessibility was the major problem faced to reach affected villages.
- Difficult to differentiate the waterways and land in waterlogged condition and of same blackish colour.
- Dead bodies of the cattle and decomposed biomaterials produced strong odour.
- Unavailability of safe drinking water and sanitation not to utter; food was not available.
- Breaching of embankments created flooding inside the villages, and homeless people were forced to take shelter on those weak embankments only.
- People were still searching their assets from their submerged/collapsed household.

Relief operation

- Lack of understanding of the situation was prominent.
- Absence of command was felt.
- Communication was weak.
- Weakness in planning was observed.
- More boats should have been deployed.
- More relief should have reached the interior villages. The nearby villages got excess relief.
- Managing dead bodies of the cattle and decomposed biomaterials was a total failure.
- Supply of drinking water was inadequate.
- Most of the NGOs were hyperactive but limited to the islands nearer to Kolkata.
- Relief supply was not proper as a lot of blankets were supplied in the month of May, in hot summer with rain.
- Staging ground of the relief was not designated and waited for even 2 days before handover.

12.3.1 Extent of Coverage by NGOs During Emergency Services and Relief Activities

Various national and international NGOs involved in West Bengal during the relief distribution formed a coalition group known as West Bengal State Inter Agency Group (IAG) for providing unified response to manage the situation and effectively provide distribution of relief material and reach out to masses. This group is a coalition of various NGOs working in West Bengal during relief phase of cyclone disaster and includes renowned agencies such as UNICEF, OXFAM and WWF. IAG helped to coordinate all efforts of various organizations and carry out all relief work in a very efficient way. To cope up with the situation emerging due to Aila, State IAG partners chose the worst affected block to help the worst affected communities (WBIAG 2009). Secondary data analysed in this study was obtained from this source to illustrate extent of relief measures in different sectors. Although this kind of data was not available from the state government to complete the analysis, at least a preliminary idea can be obtained regarding the gaps in relief measures.

The media reports (Thaindian 2009) corroborated the information that there was large section of population towards whom relief could not reach in a timely manner. Fresh water resources were polluted by saline water including groundwater in some areas. There were no proper sanitation facilities, and as houses were destroyed by cyclone and area submerged under flood water, people were forced to live on roads or on some higher grounds. Thus providing adequate shelter was also a challenge, and women were also forced to live in the same camp with men with no privacy. Health of people was affected due to lack of nutrition and also due to bad living conditions. Water-borne diseases were a major threat to the people affected areas.

12.4 Risk Governance and Institutional Mechanism

12.4.1 Cyclone Risk Mitigation and Preparedness Framework

A cyclone risk mitigation and preparedness framework is presented in Fig. 12.2. As illustrated in the figure, risk assessment is the most important and complex task in mitigation of cyclonic impact reduction. India too is in the process of devising appropriate strategies and framework for disaster mitigation in a holistic way as represented in the Fig. 12.2. The devastation caused by Cyclone Aila has clearly revealed the deficiencies in the mitigation process.

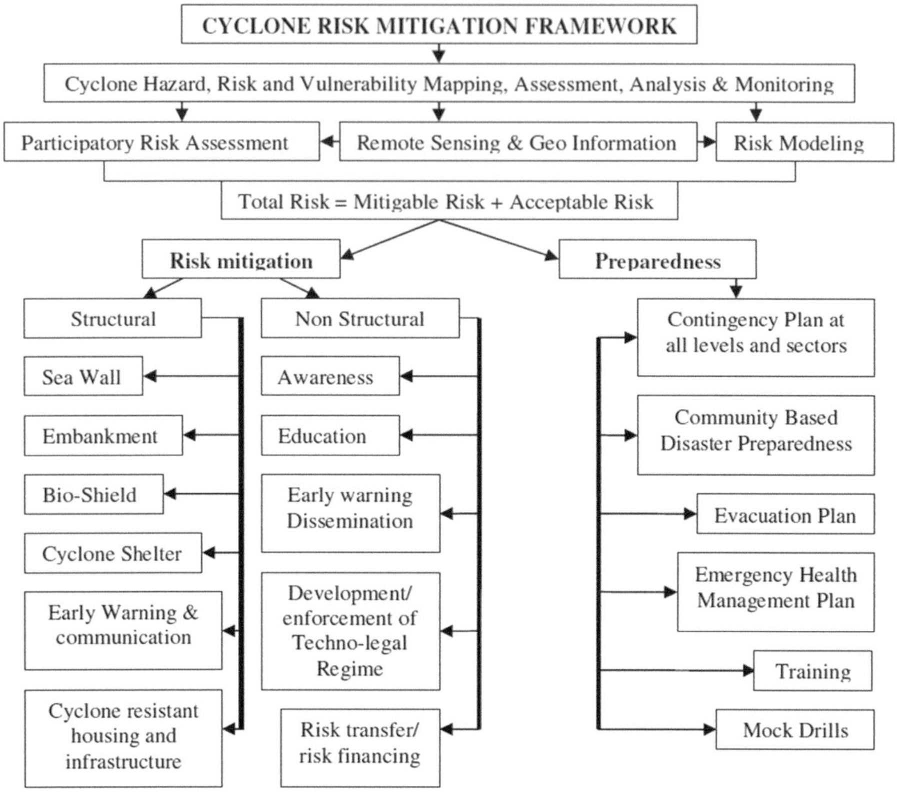

Fig. 12.2 Cyclone risk mitigation framework (SAARC 2009)

12.4.2 Preliminary Outline of Disaster Management in India

12.4.2.1 Organizational Structure

As illustrated in Fig. 12.2, the Disaster Management Act 2005 stipulates that a National Disaster Management Authority under the chairmanship of the prime minister of India shall be responsible for national policies, strategies and guidelines for disaster risk reduction and management. Similarly a State Disaster Management Authority under the chairmanship of the chief minister shall be responsible for the preparation of state plans of action. In each of the 600 plus districts, a District Disaster Management Authority under the co-chairmanship of the district magistrate and chairman of the elected Zilla Parishad shall be responsible for the preparation and implementation of District Disaster Management Plan (DM Act 2005).

12.4.3 Institutional Framework

The establishment of institutions following national guidelines issued by National Disaster Management Authority (Fig. 12.3) to cope with disaster management will enable enhanced coordination of stakeholders and concerned agencies and organizations at various levels to ensure an effectual and instantaneous response. The institutes also identify and provide resources including funds and trained personnel as well as establish effective disaster mitigation strategies for the communities (NDMA 2008) (Table 12.1).

12.4.4 Outline of Current Disaster Coping Mechanism at Place in West Bengal

12.4.4.1 Committee Framework

West Bengal government has formed the crisis management committees both at the state and district levels for the purpose of suggesting and taking advanced precautionary methods to mitigate the magnitude of disaster and also to commence measures during the post-calamity period to ensure quickest possible relief to the affected people (WBGoV 2009) (Table 12.2).

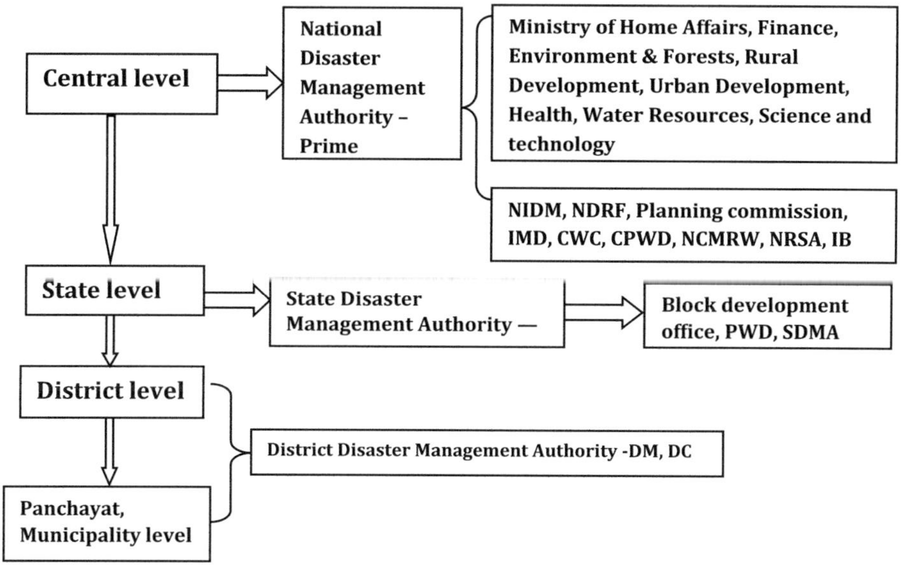

Fig. 12.3 Disaster management organizational structure (DM Act 2005)

Table 12.1 Institutions and responsible activities

Institutions	Activities
National Disaster Management Authority (NDMA)	Laying down policies, plans and guidelines for DM and coordinating their enforcement and implementation
National Executive Committee (NEC)	NEC is executive committee of NDMA, assists NDMA in the discharge of its functions, ensures compliance of the directions issued by central Government for the purpose of DM and coordinates immediate response in event of threatening disaster
National Disaster Response Force (NDRF)	Prompts response during any disaster and plays a pivotal role in community capacity building and public awareness; also has the responsibility of basic training of personnel of SDRF, police, civil defence, home guards and other stakeholders in disaster response
National Reserves	National reserves are intended to augment the resources of the states, collocated with NDRF battalions at different strategic locations in the country
National Institute of Disaster Management (NIDM)	Institutional capacity development, training, documentation of research, networking and development of a national level information base, synthesizing research activities
State Disaster Management Authority (SDMA)	Lays down policies and plans for DM in accordance with guidelines given by NDMA, coordinates the implementation of state plan, recommends provision of funds for mitigation and preparedness measures and reviews developmental plans of different departments of the state to ensure integration of prevention, preparedness and mitigation measures
District Disaster Management Authority (DDMA)	Planning, coordinating and implementing body for DM in district in accordance with the guidelines for prevention, mitigation, preparedness and response measures are followed by all departments of the government at the district level and the local authorities in the district
Local authorities	Will ensure capacity building of their officers and employees in DM; will carry out relief, rehabilitation and reconstruction activities in affected areas and will prepare DM plans in consonance with guidelines of NDMA, SDMAs, and DDMAs
Civil defence	Train the community for disaster response in the concerned districts, training modules for DM will cover awareness generation, first aid and rescue drills

12.4.5 Informal Community-Driven Responses

During any disaster, communities are always the first responders. Community partnership ensures local ownership, addresses local needs and promotes volunteerism and mutual help to prevent and minimize damage. Therefore, states should ensure all efforts to assist communities in understanding the extent of their vulnerability to enable them to effectively play the lead role in managing risks with less dependence on external entities. In West Bengal's cyclone-prone areas, the community has developed over the years certain mechanisms to combat cyclones. These initiatives include constructing the tube well at a safe height to ensure pumping of water and continuous supply of fresh drinking water. This is important since availability of

Table 12.2 Committee framework

Authority	Purpose
State Disaster Management Group	Overall supervision and management of natural calamities in the state to mitigate the sufferings of the distressed people and to formulate advance planning for their prevention, precaution and post-disaster restoration oversee the implementation of measures for DM and provide all directives for effective implementation of the relief operation
State level crisis management committee	Suggesting and taking advance precautionary measures to mitigate the sufferings of the distressed people, initiate measures to be taken in the post-calamity period for ensuring quickest possible relief to the affected people, responsible for formulation of policy and its implementation
State-level committee on the calamity relief fund	Administer the calamity relief fund; decide on all matters connected with financing of relief expenditure; assess the situation arising out of the flood, cyclone and major natural calamities; review the relief and restoration measures; take such action as may be necessary for proper and effective administration of calamity relief fund and matter connected herewith
District-level crisis management committee	Responsible for formulation of policy and its implementation at district level, initiate measures to be taken in the post-calamity period for ensuring quickest possible relief to the affected people
Subdivisional, block, gram panchayat, gram sansad level committees	These committees work at their respective levels as per instruction, guidelines given by state and district level crisis management committees, to ensure effective implementation of relief measures in the wake of natural calamity and at their respective levels for policy formulation and its implementation. The representatives of different governments as well as of the non-governmental agencies may be co-opted as members on these committees when necessary

fresh drinking water is major issue during calamity, as saline water may overflow into the tube well rendering the water non-drinkable. Also using their local techniques, communities make special type of structure to store food grains and other necessary things away from the reach of flood water. Thus, during calamity, people will develop the capacity to deal with the situation on their own till availability of external help. Thus, only immediate response can be initiated without expecting succour to arrive.

12.4.6 *National Cyclone Risk Management Programme (NCRMP) in West Bengal*

- World Bank-supported programme to undertake long-term cyclone mitigation measures in three coastal districts and in the city of Kolkata.
- Mitigation measures include construction of cyclone shelters, coastal shelter belt plantations, regeneration of mangrove plantation, construction of embankments to stop sea water inundation and construction of missing road/bridge links.

- Training and capacity building.
- Preparation of capital-intensive schemes by the Department of Irrigation and Waterways is completed.
- Proposals for construction of cyclone shelters.
- Proposal for augmenting the infrastructural facility of ATI for training and capacity building sent to NIDM.

12.4.7 Disaster Risk Management Programme in West Bengal

The emphasis of the programme is on sustainable disaster risk reduction in the state to build capacities at all levels to institutionalize the disaster risk management system in West Bengal. The programme aims to contribute to the social and economic development goals of the national and state governments, enable them to minimize losses to development gains and reduce the vulnerabilities to natural disasters. The programme was focused on awareness generation, education, training, management and recovery at community, district and state levels and strengthening of state and district disaster management information centres for effective and timely information dissemination. The multi-donor-funded programme started with initial support from the United Nations Development Programme which provided the ground for the establishment of the community-based disaster risk management framework and formalizing the collaboration between the stakeholders at various levels for implementation.

12.4.8 Community-Based Disaster Management in West Bengal

The government of West Bengal has introduced community-based disaster management programme with the assistance of UNICEF, state ATI and district administration in close collaboration with Panchayati Raj institutions. The main focus of this programme is to increase human resource development and involvement of communities by enlarging their capacities. Such involvements of local communities will ultimately benefit the development process and enlighten them for greater involvement in all spheres of activities.

12.5 Risk Reduction Measures and Challenges

The then chief minister of West Bengal alarmed the army in anticipation, while Border Security Force (BSF) and the police joined hands for the rescue operations in Sundarbans and other coastal areas of the South 24 Parganas where a number of thatched houses completely collapsed due to high level of tidal inundation within the islands. According to the official Government of West Bengal report till 1 June, the population affected in 18 districts in West Bengal due to Cyclone Aila rose to 6.6 million. In two districts – North and South 24 Parganas – the population affected crossed over one million in each district. The human death toll was reported at 126 and damaged houses were over one million in the state. The government has set up 765 relief camps and 84 medical camps in the most affected districts. The Inter Agency Group in West Bengal has been meeting every 72 h to coordinate response to the impact of Cyclone Aila. On 1 June, Sphere India held a meeting in New Delhi to discuss how humanitarian agencies at the national level could complement the efforts of the IAG in West Bengal. Official record has claimed that thousands of residents were evacuated from coastal areas ahead of Cyclone *Aila*. In addition, several warning alerts were issued before the cyclone hit Kolkata, though no official warning system worked in the remote coastal areas.

12.5.1 Gaps Identified in the Context of Cyclone Aila

West Bengal government has developed a State Disaster Management Plan in 2003, even before the Disaster Management Act of 2005. However, till date this plan has not been revised in accordance with Disaster Management Act 2005. Implementation of such a plan could have helped West Bengal state government to respond to Cyclone Aila in an effectual coordinated and efficient way at every level of administration. It is well known that investment in disaster preparedness will effectively minimize disaster-related economic losses.

12.5.2 Strengthening of Infrastructure

12.5.2.1 Embankments

Embankments play a key role in world's largest mangrove delta Sundarbans during any cyclone or flood. There is a strong need to replace weak mud embankments with strong ones made of concrete blocks. Many of these embankments were in a state of disrepair. Hence they could not withstand the intensity of tropical Cyclone Aila, and breaches occurred causing saline water to enter into the area and damage the fertility of land and pollute fresh water resources as well. A strong embankment also reduces the intensity of cyclone damage. But building strong concrete block embankments involves large sums of money. A preliminary estimate by the Government of West Bengal (personal interaction) indicates a sum of about Rs. 10,000 crore for the construction of permanent reinforced embankments in the coastal areas of the state, including the Sundarbans, to prevent flooding in the event of a natural disaster. The Government of West Bengal cited lack of funds in the construction of embankments. Preparedness was also hampered by the fact that it could not be predicted that the embankments would suffer such an extensive damage by cyclone of the type of Aila (Thaindian 2009; The Hindu 2009; Sinha and Bhattacharya 2009).

12.5.2.2 Bio-shields

Nature has provided biological mechanisms to protect coastal communities from the fury of cyclones, coastal storms, tidal waves and tsunamis. Mangrove forests constitute one such mechanism for safeguarding the ecological security of coastal areas and the livelihood security of fishing and farming communities in the coastal zone. The ecological, economic and social value of mangrove forests will further increase, if a rise in sea level takes place due to anticipated global warming. In the world's largest mangrove delta, Sundarbans in West Bengal, sea level is already rising. Hence special emphasis should be given to enhance growth and protection of mangrove near coastal areas. Mangroves along with embankments will slow down the intensity of storms/winds up to a greater extent, thus minimizing the losses. Due to increased population density, the mangrove forests are getting gradually depleted. A drive to plant mangroves on large scale is still lacking and needs special attention of community and government.

12.5.2.3 Efficiency of Early Warning System

India Meteorological Department (IMD) is the nodal agency in India for providing cyclone warnings through its Area Cyclone Warning Centres (ACWCs) at Kolkata, Chennai and Mumbai and the Cyclone Warning Centres (CWCs) at Visakhapatnam, Bhubaneshwar and Ahmedabad.

According to media reports, the predictions of tropical Cyclone Aila issued by Indian Meteorological Department (IMD) from time to time varied from its earlier prediction in terms of the location, time and exact direction.

In far remote villages in coastal area of West Bengal, communication system is very poor because of no electricity, and people generally did not have access to modes of communication like radios and telephones, and link through roads is also very poor. Therefore, it was a big challenge to communicate the information instantly regarding cyclones; hence, people have very little time to cope with emergency situations. Also warning issued in this case did not emphasize the seriousness of the cyclone, and the local people did not pay attention to the warning. Otherwise people would have moved to safer location like higher grounds or cyclone shelters thus minimizing the loss.

12.5.2.4 Degree of Coordination Across Various Government Departments

Coordination between different departments of state governments amongst themselves and with central governments departments, NGOs and international agencies plays a vital role in managing the situation in the wake of calamity. West Bengal has the following administrative and PRI structure (Fig. 12.4).

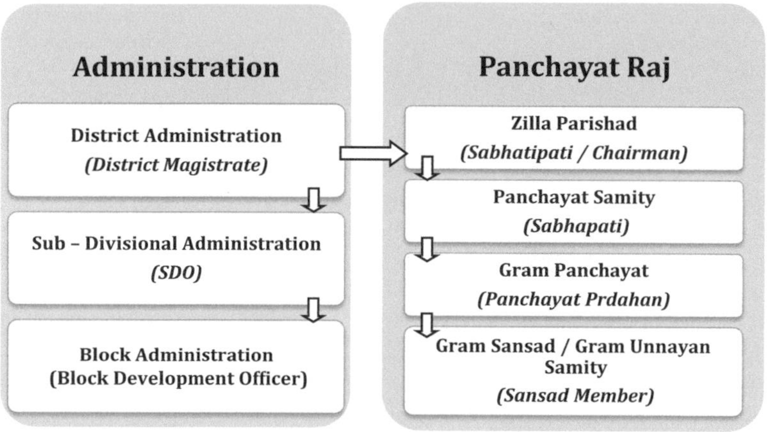

Fig. 12.4 Administrative and PRI structure in West Bengal (WBGoV 2009)

It was also noticed that although the government had started withdrawing from immediate relief and response phase claiming the situation was back to normal, the actual situation was quite otherwise. The absence of effective coordination between the administrations at block level and panchayat levels (Fig. 12.3) was visible at some places. This appeared to have resulted in duplication of relief materials provided and uneven distribution of relief (Sinha and Bhattacharya 2009). There was lack of role clarity amongst S.D.O., Deputy Magistrate and B.D.O in various situations. This created hindrances in taking decisions and efficient management of situation created by Cyclone Aila.

12.6 Community Vulnerability to Coastal Hazards

The increased vulnerability of coastal communities to potential hazards is partly due to the constant increase in coastal population (Adger et al. 2005). Currently, an estimated 23% of the world's population (1.2 billion people) lives within 100 km of a shoreline and 100 meters (m) of sea level (Small and Nicholls 2003). By the year 2030, an estimated 50% of the world's population will live in the coastal zone.

Most of the coastal population lives in relatively densely populated rural areas and small to medium cities, rather than in large cities. In these relatively rural communities, basic services and disaster warning and response mechanisms are limited (Fig. 12.5). Limited capacity of a community to plan for and respond to coastal hazards makes coastal populations increasingly vulnerable and increases disaster risk.

Economic development pressures along the coast, population density and distribution and human-induced vulnerabilities, coupled with increasing frequency and duration of storms, sea-level rise and other chronic coastal hazards, increase risk.

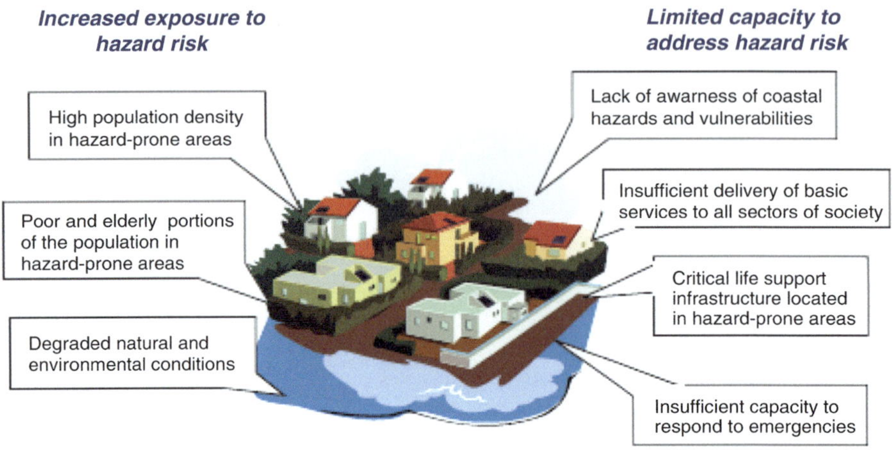

Fig. 12.5 Factors that contribute to vulnerability in coastal populations

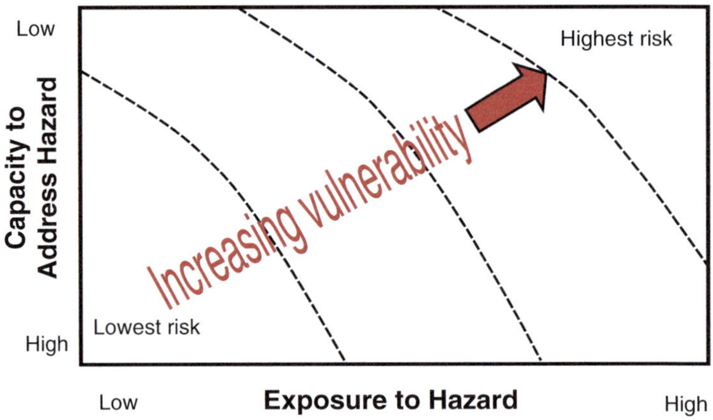

Fig. 12.6 Community vulnerability as a function of the degree of exposure and the capacity to address hazard risks

These conditions set the stage for more frequent and severe disasters and reduced time and capacity to recover. The time period between disaster event and recovery is becoming shorter, and some coastal communities find themselves in a state of perpetual response to and recovery from one disaster event after another. The assessment of risk is an important element of CCR. Communities must identify their exposure to hazard impacts to proactively address emergency planning, response and recovery and implement hazard mitigation measures (Figs. 12.5 and 12.6).

Disaster management consists of several components commencing with policy framework, preparedness, early warning system, relief distribution and access to various other emergency services. However initial investigations clearly indicate shortcomings in several of the aforesaid parameters. These always result in raising the level of disaster clearly indicating the lack of conjunctive appreciation and consequent action to effectively cover after effects of the consequences of the disaster. This study bears out this fact since a significant proportion of population failed to secure urgent and appropriate aid from government, NGOs and UN agencies. Further examination of primary data and role of government based on information gathered from the affected population through questionnaires is required to effectively conclude on the gaps associated with implementation of policies at ground level and subsequent alterations necessitated in the state's policy framework and institutional mechanism to render them more effective.

12.7 Conclusions

The preliminary analysis regarding tropical Cyclone Aila clearly indicates that there are substantial gaps in management of cyclonic disasters in all aspects from pre-cyclonic to post-cyclonic phases in West Bengal. This issue becomes very important in context of significant increase in frequency and intensity of cyclonic activities along with rising sea levels in coastal regions of India due to climate change. The present study finds that, although West Bengal Government already has a state disaster management plan, it is yet to be finalized according to the guidelines provided by the National Disaster Management Act 2005, which could have aided in enhanced coordination amongst various departments and other agencies minimizing losses and enhancing responses at every administrative stage from state level to village levels.

The study also reveals the inadequacy of existing infrastructure to combat cyclones of a stature such as Aila due to existence of weak embankments and depleted mangroves. There is also a strong need to properly maintain the embankment protection system. The central government should ensure adequate financial aid to West Bengal government to build and maintain reinforced embankments (Mukhopadhyay 2009).

The gaps in cyclone predictions issued by IMD can be rectified with sophisticated models and detailed meteorological observations. There were also problems to disseminate the cyclone information due to lack of proper communication facilities in these remote areas. Some villages did not get any information at all regarding the impeding cyclone. As communities in Sundarbans are already habituated to experience severe weather conditions, they did not feel the need to evacuate revealing a necessity to increase awareness amongst the affected communities regarding potential impact of cyclones through outreach programmes.

Lack of coordination, information sharing and role clarity was also visible within different departments of government which leads to uneven relief distribution. The government officials were not even aware of the situation in various remote areas leading to improper decision-making.

The data compiled by NGOs clearly revealed that a large section of affected population remains uncovered from the food, drinking water, health facilities, proper sanitation, education and shelter facilities provided by government and various other agencies. Therefore, there is a strong need to render strict bureaucratic procedures more flexible to expeditiously meet the requirements of the situation and enable quick decisions by government to respond to the situation.

The study reveals that challenges arising due to a tropical storm such as Aila are immense. The current study is fraught with uncertainties due to lack of collection of primary data and detailed investigations as mentioned earlier. There is also a necessity to conduct a long-term analysis of the morphological changes of the coastline through detailed analysis of satellite data and wave data collected by a wave recorder to properly assess the potential impact of cyclones. However despite the shortcomings, nevertheless this study provides a glimpse of the inadequacies of the current disaster risk governance mechanism in West Bengal, India.

Acknowledgement The authors are intended to convey their sincere thanks to Mr. Amit Ghosh, researcher in School of Oceanographic Studies, Jadavpur University, for his kind help in the map preparation.

References

Adger WN, Arnell NW, Tompkins EL (2005) Successful adaptation to climate change across scales. Glob Environ Chang 15:77–86

Disaster Management Act (2005) Govt. of India, 2005

Intergovernmental Panel on Climate Change (IPCC) (2005) Climate change: the scientific basis. Cambridge University Press, Cambridge

Jones M (2004) Life in the fast lane. Time Warner Books, New York

Mukhopadhyay A (2009) On the wrong side of the fence: embankment people and social justice in the Sundarbans. In: Bose PK, Das SK (eds) Social justice and enlightenment: West Bengal. Sage, New Delhi

National Disaster Management Authority (NDMA) Guidelines (NDMA), Management of Cyclones, Apr 2008- NDMA, Government of India

National Disaster Management Authority (NDMA). National cyclone risk mitigation project, Mar 3 2009

Pielke RAJ, Landsea CW (1998) Normalized U.S. hurricane damage, 1925–1995. Wealth Forecast 13:621–631

SAARC (2009) http://saarc-sdmc.nic.in/pdf/cyclone.pdf. Accessed on 24 Sept 2009

Sinha S, Bhattacharya P (2009) Cyclone Aila in West Bengal: a report by Jamsetji Tata Centre for Disaster Management, TISS, p 5

Small C, Nicholls RJ (2003) A global analysis of human settlement in coastal zones. J Coast Res 19(3):584–599. West Palm Beach (Florida), ISSN 0749-0208

The Hindu. Online edition of India's National Newspaper Monday, Jun 15, 2009. Accessed on 14 Jul 2009

Thaindian (2009) Government plans bailout package for Air India. [Online-Archived article, 24 June]. Available at: http://www.thaindian.com/newsportal/business/government-plans-bailout-package-for-air-india_100209081.html

Trenberth K (2005) Uncertainty in hurricanes and global warming. Science., 17June 2005:1753–1754

West Bengal Inter Agency Group, *iagwb*.forumotion.net/. Accessed on 10 July 2009

http://www.wbgov.com. Accessed on 15 July 2009

http://www.thaindian.com/newsportal/politics/no-steps-taken-to-maintain-sundarbans-embankments-jatua_100202030.html. Accessed on 14 July 2009

http://www.tiss.edu/Aila_report_webpage.pdf. Accessed on 14 July 2009

http://www.wbgov.com/portal/banglarMukh/Government/Departments/DepartmentListPortletWindow?action=e&windowstate=normal&mode=view. Accessed on 15 July 2009

Chapter 13
Good Governance to Achieve Resiliency and Sustainable Development

Chiranjiv Choudhary and Srinivasa Rajamani Neeli

Abstract Climate change has resulted in increased frequency, intensity, magnitude, and uncertainty of natural disasters. This has been clearly brought out in various assessment reports of Intergovernmental Panel on Climate Change (IPCC), in particular AR IV and AR V. As vulnerabilities cannot be predicted completely, good governance can play a critical role in management of the disaster risks, improve resilience, and meet the challenges of sustainable development. Effective disaster risk management to reduce vulnerability would require mainstreaming CCA/DRR into policy, planning, and implementation of development schemes as a multilevel approach which combines activities at various levels and uses top-down as well as bottom-up dynamics. For proactive planning and to identify flexible options for unknown and unpredictable future, disaster risk governance is to be strengthened both vertically and horizontally. Therefore, good governance is the need of the hour, and it would play a critical role in effective management of disaster risks.

This chapter highlights good governance as one of the important tools to effectively manage the disaster risks and achieve sustainability and resilient development. It presents a detailed discussion on good governance and disaster risk governance.

Keywords Climate change adaptation (CCA) • Disaster risk reduction (DRR) • Good governance • Disaster risk governance • Sustainable development

C. Choudhary (✉)
Commissioner of Horticulture & Sericulture, Agriculture & Cooperation Department, Government of Andhra Pradesh, Hyderabad, India
e-mail: chiranjivc@gmail.com

S.R. Neeli
City Project Coordinator, UNDP-GoI Project, Visakhapatnam, India
e-mail: rajamani74ns@gmail.com

I. Pal, R. Shaw (eds.), *Disaster Risk Governance in India and Cross Cutting Issues*, Disaster Risk Reduction, DOI 10.1007/978-981-10-3310-0_13

13.1 Introduction: Sustainable Development

The concept of sustainability emerged from the need for reducing environmental pressures stemming from human activities. Worldwide discourse on sustainability was launched with two documents: the club of Rome's Limits to Growth (Meadows et al. 1972) and the Brundtland Commission's Our Common Future (World Commission on Environment and Development 1987). The issue of sustainable development has been widely acknowledged and spread rapidly after United Nations conference on Environment and Development in Rio 1992 (United Nations 1992a, b).

Sustainable development was defined as "development that meets the needs of the present generation without compromising the ability of future generation to meet their own needs" (WCED 1987). Sustainable development is a development path that meets the major needs of the present without endangering subsequent needs and aspirations of future generation allowing for the conservation of nature (Gotlieb 1996; UNDP 2014). Sustainability is a process involving people, institution, natural resources and the environment. Sustainability is an attempt to provide the best outcomes for the human and natural environments, both now and into the indefinite future. Furthermore, sustainability is about intergenerational equity, of how nonreproducible resources should be shared between current and future generations (Bowers 1997).

The concept of sustainable development is charged with complexities as it involves and balances three different goals: the utility for economic development, the equity for social development, and the ecological integrity for environmental development. Therefore, an appropriate approach management system is essentially required (Munasinghe 1995). To be analyzed and carried out through a decision-making process, sustainability must be measured. Agenda 21 itself singles out, among main success factors for sustainable development projects, the availability of information, diagnosis activity, and monitoring capability (United Nations Agenda 21 1992a, b).

Climate change and disaster risks, together with population growth, represent a combined challenge to the achievement of sustainable development. An alarming rapid increase of exposure and vulnerability to natural hazards is taking place around the world. Unless the disaster risk and vulnerability are reduced and population growth and the right to access family planning services are addressed, sustainable development cannot be achieved as sustainability demands risk resilience and the ability of nations and communities to cope with shocks.

13.2 Governance

Governance, as defined by the UNDP (2004a, b), is the exercise of political, economic, and administrative authority in the management of a country's affairs at all levels. It comprises mechanisms, processes, and institutions through which citizens and groups articulate their interests, exercise their legal rights, meet their obligations, and mediate their differences. Governance encompasses, and also transcends, government (Ahrens and Rudolph 2006).

Governance refers generally to the set of instruments through which people living in a state, believing in common core values, govern themselves by the means of laws, rules, and regulations enforced by the state apparatus. It denotes a system of values, policies, and institutions by which society manages its economic, political, and social affairs through interaction among the state, civil society, and the private sector (Ahrens and Rudolph 2006; Bevir 2011).

It also denotes those processes and institutions through which citizens and groups articulate their interests, exercise their legal rights, meet their obligations, and mediate their differences. Governance has three components: economic, political, and administrative.

- Economic governance includes the decision-making processes that affect a country's economic activities and its relationship with other economies. This has major implications for equity, poverty, and quality of life.
- Political governance is the process of decision-making to formulate policies, including national disaster reduction and planning. The nature of this process and the way it brings together the state, on-state, and private sector actors determine the quality of the policy outcomes.
- Administrative governance is the system of policy implementation and requires the existence of well-functioning organizations at the central and local levels. In the case of disaster risk reduction, it requires functioning enforcement of building codes, land-use planning, environmental risk and human vulnerability monitoring, and safety standards.

Governance is about the processes by which public policy decisions are made and implemented. It is the result of interactions, relationships, and networks between the different sectors (government, public sector, private sector, and civil society) and involves decisions, negotiations, and different power relations between stakeholders to determine who gets what, when, and how. The relationships between government and different sectors of society determine how things are done and how services are provided. Governance is therefore much more than government or "good government" and shapes the way a service or set of services are planned, managed, and regulated within a set of political social and economic systems.

The very concept of "good governance" at local levels denotes quality, effectiveness, and efficiency of local administration and public service delivery; the quality of local public policy and decision-making procedures, their inclusiveness, their transparency, and their accountability; and the manner in which power and authority are exercised at the local level. While local government is the essential institutional building block for local governance, the wider governance sphere comprises a set of state and non-state institutions, mechanisms, and processes, through which public goods and services are delivered to citizens and through which citizens can articulate their interests and needs, mediate their differences, and exercise their rights and obligations.

The concepts of local governance and decentralization, at times used interchangeably, are related but different concepts. Decentralization is primarily a national, political, legislative, institutional, and fiscal process, while local governance can be affected by decentralization processes. UNDP has a holistic approach to defining the field of local governance and decentralization by using the concept of *decentralized governance for development.* "Decentralized governance is not a panacea or a quick fix. The key to human development-friendly decentralized governance is to ensure that the voices and concerns of the poor, especially women, help guide its design, implementation, and monitoring" (UNDP 2004a).

Decentralized governance for development is considered to be a key area of democratic governance which in turn is crucial to attaining human development and the Sustainable Development Goals (SDGs). For development and governance to be fully responsive and representational, people and institutions must be empowered at every level of society – national, provincial, district, city, town, and village.

There are four main *objectives* to undertaking an assessment of governance at the local level:

1. *Diagnostic:* An assessment will be done to identify a problem and its scope.
2. *Monitoring:* An assessment will be done at regular intervals to keep a check on the success or failure of an initiative, policy, or program.
3. *Evaluation:* An assessment will be done to assess whether an initiative, policy, or program has achieved its predefined results and outcomes.
4. *Dialogue:* An assessment will also serve to engage citizens and communities in informed discussions about shared goals and priorities.

Common stakeholders in most assessments of local governance will include:

1. *Local government representatives:* Local political and administrative leaders are crucial for launching, implementing, and using governance assessments. In many cases, local government will be in the driving seat of the assessment process. Local government representatives are especially active in the partnership, promotion, development, action planning/dissemination, and policy implementation phases.
2. *Central government representatives:* Central government (e.g., the ministry/department responsible for local government) is important in assessments as it has a significant role to play in capacity development of local authorities, including in the setting and maintaining of standards of performance, monitoring, and ensuring the establishment of mechanisms of accountability, and in the formulation and approval of local government policy frameworks. The central government may be especially active in the policy implementation phase, integrating the assessment results into its local government monitoring mandate.
3. *Local government associations*: these are comprised of local councils and express their collective voice in the national arena. Some illustrative examples of their roles and objectives include shaping public debate on local government issues, influencing policy at the national level, supporting capacity development that enables councils and their partnerships to deliver services, and working to

enhance democratic accountability and transparency in local government institutions. Where these exist, they represent an important stakeholder in assessments, especially in the partnership promotion, development, and policy implementation phases.
4. *Civil society organizations:* the existence of a vibrant and diverse civil society is an important indicator of good local governance. CSOs also need to play a role in the assessment process, including identifying and drawing attention to local governance deficits as well as using their expertise in data collection and analysis in the implementation of the assessment. CSOs are potentially active in every phase of the assessment process.
5. *Community-based organizations (CBOs):* These are a form of organized citizens and have a role in mobilizing local people around community development actions and to act as a watchdog. CBOs are also important for reflecting the views, rights, and interests of vulnerable or marginalized groups in communities.

Natural hazards on their own do not result in disaster. Rather, it is the vulnerability of populations in countries that has a direct bearing on levels of disaster. Supportive governance is necessary to ensure coping capacities in societies. Governance influences the way in which national and subnational actors (including governments, parliamentarians, public servants, the media, the private sector, and civil society organizations) are willing and able to coordinate their actions to manage and reduce disaster-related risk. Sufficient public awareness to recognize and address risk, coupled with the political will to set policy and allocate appropriate resources, is key. Equally critical is the need for processes and institutions with sufficient managerial and coordination capacity to manage and integrate the efforts of relevant sectors and account for vulnerable and poor communities. Such capacity is ultimately dependent upon the human, social, physical, economic, and environmental capital of a society (UN 2010; UNDP 2010).

Mainstreaming of DRR is a governance process, enabling the systematic integration of DRR concerns into all relevant development spheres. In other words, responsive, accountable, transparent, and efficient governance structures underwrite the environment where DRR can be institutionalized as an underlying principle of sustainable development. Therefore, building resilient communities in disaster-prone countries requires that (a) underlying risk factors are continuously considered in all relevant sectors and (b) risk reduction standards and measures are an integral part of the planning and delivery of core development services and processes, including education, environment, and health.

Disaster and climate change risks pose many difficult challenges for governance.

These include:

- *Distributional and equity linkages*: Disasters and the impacts of climate change affect different groups in different ways so there is no workable one-size-fits-all approach. This means that approaches need to take into consideration local contexts and that responses are designed and implemented accordingly.

- *Societal reach*: Climate change affects all sectors of society. This means that government, institutions, etc. need to understand a multitude of crosscutting issues and the interactions between these and climate change. This makes governance difficult and complex.
- *Scientific uncertainty*: Although we know more than ever about climate change, disaster and climate change risks still pose enormous uncertainties, making decision-making difficult.
- *Timescales*: The time it takes for positive impacts in disaster and climate change responses to take effect may exceed government tenure, which can make disaster risk reduction and climate change adaptation unattractive to decision-makers in government.
- *Global implications*: The causes and impact of climate change are international, so the response must be a collective one. Yet coordinating international efforts on such a scale remains a major challenge, leaving issues of global governance largely unanswered.

13.3 Good Governance

Governance is the exercise of political, economic, and administrative authority in the management of a country's affairs at all levels. It comprises formal and informal mechanisms, processes, and institutions through which citizens and groups articulate their interests, exercise their legal rights, meet their obligations, and mediate their differences. While governance encompasses government, it also includes all relevant groups in society, including private sector and civil society organizations, from household and local levels to provincial, national, and international levels (UNDP/EU 2004).

Good governance which drives the achievement of development results must also now rise to the challenge of achieving the equitable and sustainable development which will secure our common future (Doornbos 2001). Good governance for building disaster and climate resilience takes place when capable, accountable, transparent, inclusive, and responsive governments work together with civil society, the private sector, and at-risk populations to create an enabling environment to improve society's ability to prepare and respond to disasters and their capacity to adapt to changes in the climate.

Good governance is reflective of community empowerment and ecological wisdom; any concept of governance aiming for sustainability needs to start from two *observations*. One is the *accurate and honest realization* that the world we live in is on an unsustainable path and an understanding of why our path is unsustainable. The other is a *strong sense of ethics* to guide our search for a sustainable path.

13.4 The Eight Characteristics of Good Governance (UNDP)

Good governance has eight major characteristics. It is participatory, consensus oriented, accountable, transparent, responsive, effective and efficient, equitable, and inclusive and follows the rule of law. It assures that corruption is minimized, that the views of minorities are taken into account, and that the voices of the most vulnerable in society are heard in decision-making. It is also responsive to the present and future needs of society.

13.4.1 *Participation*

Participation by both men and women is a key cornerstone of good governance All men and women should have a voice in decision-making, either directly or through legitimate intermediate institutions that represent their interests. Such broad participation is built on freedom of association and speech, as well as capacities to participate constructively. Participation could be either direct or through legitimate intermediate institutions or representatives. It is important to point out that representative democracy does not necessarily mean that the concerns of the most vulnerable in society would be taken into consideration in decision-making. Participation needs to be informed and organized. This means freedom of association and expression on the one hand and an organized civil society on the other hand.

13.4.2 *Rule of Law*

Legal frameworks should be fair and enforced impartially, particularly the laws on human rights. Good governance requires fair legal frameworks that are enforced impartially. It also requires full protection of human rights, particularly those of minorities. Impartial enforcement of laws requires an independent judiciary and an impartial and incorruptible police force.

13.4.3 *Transparency*

Transparency is built on the free flow of information. Processes, institutions, and information are directly accessible to those concerned with them, and enough information is provided to understand and monitor them. Transparency means that decisions taken and their enforcement are done in a manner that follows rules and regulations. It also means that information is freely available and directly accessible to those who will be affected by such decisions and their enforcement. It also means that enough information is provided and that it is provided in easily understandable forms and media.

13.4.4 Responsiveness

Institutions and processes try to serve all stakeholders. Good governance requires that institutions and processes try to serve all stakeholders within a reasonable time frame.

13.4.5 Consensus Oriented

Good governance mediates differing interests to reach a broad consensus on what is in the best interests of the group and, where possible, on policies and procedures. There are several actors and as many viewpoints in a given society. Good governance requires mediation of the different interests in society to reach a broad consensus in society on what is in the best interest of the whole community and how this can be achieved. It also requires a broad and long-term perspective on what is needed for sustainable human development and how to achieve the goals of such development. This can only result from an understanding of the historical, cultural, and social contexts of a given society or community.

13.4.6 Equity and Inclusiveness

All men and women have opportunities to improve or maintain their well-being. A society's well-being depends on ensuring that all its members feel that they have a stake in it and do not feel excluded from the mainstream of society. This requires all groups, but particularly the most vulnerable, have opportunities to improve or maintain their well-being.

13.4.7 Effectiveness and Efficiency

Processes and institutions produce results that meet needs while making the best use of resources. Good governance means that processes and institutions produce results that meet the needs of society while making the best use of resources at their disposal. The concept of efficiency in the context of good governance also covers the sustainable use of natural resources and the protection of the environment.

13.4.8 Accountability

Decision-makers in government, the private sector, and civil society organizations are accountable to the public, as well as to institutional stakeholders. This accountability differs depending on the organization and whether the decision is internal or external to an organization. Accountability is a key requirement of good governance. Not only governmental institutions but also the private sector and civil society organizations must be accountable to the public and to their institutional stakeholders. Who is accountable to whom varies depending on whether decisions or actions taken are internal or external to an organization or institution. In general, an organization or an institution is accountable to those who will be affected by its decisions or actions. Accountability cannot be enforced without transparency and the rule of law.

According to UNDP (2002), good governance advances sustainable development for three reasons. First, enjoying political freedom and participating in the decisions that shape one's life are fundamental human rights. Second, good governance helps to protect people from economic and political catastrophes, such as famines, and other crises. There is a direct correlation between bad governance and famines. Third, good governance can promote sustainable development by empowering citizens to influence policies that promote growth and prosperity and reflect their priorities (UNDP/EU 2004).

13.5 Good Governance for DRR and CCA

Critical to DRR and CCA are the processes by which decisions are made and implemented – or not – to reduce the vulnerability of communities; increase their resilience; prepare, respond, and manage emergencies; rehabilitate and reconstruct the damaged infrastructure; as well as recuperate livelihoods. In the context of DRR, this is what is referred to as good governance. Furthermore, this process of decision-making affects all government sectors and levels and all segments of civil society. Thus building sound democratic institutions builds resilience and reduces vulnerability of societies to natural hazards.

Consequently, for the purposes of DRR and CCA, good governance must ensure that the process of decision-making is inclusive and transparent; is based on social consensus, achieved through active involvement of all members of the society; involves the use of sound information and communication systems and tools; provides for accountability of all – not only government institutions and officials but also all civil society organizations and community members; and follows the rule of law.

Good governance, including public participation and involvement, transparency, and accountability, is key to reducing risk. Active and responsible public participation results in self-awareness and sound risk assessment, while fostering transparency and accountability in the administration of emergency assistance, leading to more resilient and better prepared communities. Capacity building and institutional strengthening particularly at the local level, government, and organized civil society groups are also crucial, as these are the first to respond in a disaster and are in the best position to work with communities.

The information flow and the decision-making processes required to implement measures to reduce vulnerability, prevent and mitigate disasters, and prepare for and respond to disasters in a timely and effective fashion will make the difference between life and death and between the well-being of vulnerable communities and their suffering and marginalization.

Disaster Response and Emergency Management, to be effective, requires a great deal of preparedness, which can only be obtained with sound interinstitutional coordination, able to capitalize on comparative capabilities of all government sectors and all levels of government – from national to local. It also requires the active involvement of all segments of civil society – from private enterprises, NGOs, and religious groups to advocacy groups and other organized groups of the civil society.

The sustainability and successful implementation of community-centered early warning systems, for example, will depend on the formulation and execution of sound public policy and good governance that ensures accountability across all government sectors and levels, as well as all segments of civil society, and effective coordination.

Increasing attention to climate change and the need to advance strategies and measures for adaptation add a layer of complexity to the integration of DRR into development policies and plans. Interinstitutional coordination at all levels, but particularly at the national level, becomes even more critical if financial resources to reduce vulnerability, mitigate disasters, and adapt to climate change are to be effectively used.

Where disaster prevention and mitigation are concerned, interinstitutional coordination and cooperation become even more critical. Many of the organizations involved in disaster prevention and mitigation are not as strong as their counterparts in disaster response/emergency management in the area of coordination and logistic capabilities. So, thus, strong democratic institutions, sound institutional arrangements, and coordination and cooperation mechanisms for DRR and CCA are critical to good governance in DRR and CCA.

13.6 Disaster Risk Governance

> Disaster risk governance refers to the way in which the public authorities, civil servants, media, private sector, and civil society coordinate at community, national and regional levels in order to manage and reduce disaster and climate related risks. (UNDP 2013)

Disaster risk governance may be considered a subset of the concept and practice of governance in general. For the purposes of the present review, its terms of reference establish that disaster risk governance refers to *"the way in which public authorities, civil servants, media, private sector and civil society coordinate at community, national and regional levels in order to manage and reduce disaster- and climate-related risks. This means ensuring that sufficient levels of capacity and resources are made available to prevent, prepare for, manage and recover from disasters. It also entails mechanisms, institutions and processes for citizens to articulate their interests, exercise their legal rights and obligations and mediate their differences"* (UNDP 2013). Despite the fact that the use of the term disaster risk governance is now common currency in disaster risk reduction and management circles.

13.7 Principles of Good Disaster Risk Governance (DRG)

Good governance most generally refers to a list of positive characteristics or principles of how government decision-making and policy implementation ought to be carried out. For the World Bank, the characteristics of good governance comprise voice and accountability, political stability and absence of violence, government effectiveness, regulatory quality, rule of law, and control of corruption (World Bank 1994). The UNDP singles out characteristics such as participation, responsiveness, transparency, accountability, effectiveness, efficiency, rule of law, and equity as its most important characteristics (UNDP 2013). "Disaster risk governance" can be defined as the way actors at all levels manage and reduce disaster and climate-related risks (UNDP 2013).

A number of key points arise from the literature on disaster risk governance. From the general guidance literature, notable points are:

- There are certain *key entry points* for mainstreaming disaster risk governance. These include policy development, institutional development, advocacy and knowledge, supporting implementation of measures, and supporting broad participation.
- *Development and disaster risk are closely related* and impact each other in several ways. For example, development can lead to urbanization which can present new disaster risks. Development initiatives should incorporate disaster risk considerations.
- *Climate change adaptation and disaster risk reduction/management are closely related,* as would be the governance processes and mechanisms in these areas.
- *Disaster risk governance relates to many levels and actors*. It involves the governance mechanisms and processes of national decision-makers, local communities, as well as inter- and intragovernmental organizations.

From the literature focused on national-level governance, key points are:

- *Parliamentarians can and should play an important role* in improving governance for disaster risk.
- *More guidance is needed that delineates responsibility* between global, regional, national, and local actors.
- There are *various monitoring mechanisms for identifying the degree of mainstreaming of disaster risk management* in governance arrangements. Examples include tracking budgets for disaster risk and evaluating national policies, planning processes, and decision-making.

From the literature focused on local governance, key points are:

- *Decentralization and capacity building* of local governments, communities, and networks are important to manage disaster risk.
- *Decentralization, by itself, does not guarantee* greater efficiency, social participation, or accountability in relation to disaster risk management, but can create conditions conducive to these.
- *Political commitment* from local and national actors is important to institutionalize effective disaster risk governance.

Finally, from the literature focused on institutional arrangements, key points are:

- A number of *institutional arrangements affect disaster risk management decision-making*. These include incentive structures, information gaps, and intra-governmental relations.
- Both *formal and informal institutions* help shape exposure, sensitivity, and capacities of individuals, social groups, and social-ecological systems to respond to disaster risk.
- *New legislation* on disaster risk management can be a key enabler of disaster risk reduction and management.
- *Other governance reforms* such as "New Public Management" reforms can adversely impact disaster risk governance.

13.8 Conclusion

Sustainable development is a development path that meets the major needs of the present without endangering subsequent needs and aspirations of future generation allowing for the conservation of nature; it is a process involving people, institution, natural resources, and the environment (United Nations Development Program 1997). It is an attempt to provide the best outcomes for the human and natural environments, both now and into the indefinite future. Climate change and disaster risks, together with population growth, represent a combined challenge to the achievement of sustainable development. An alarming rapid increase of exposure and vulnerability to natural hazards is taking place around the world. Unless the disaster risk and vulnerability are reduced and population growth and the right to access family planning services are addressed, sustainable development cannot be achieved as sustainability demands risk resilience and the ability of nations and communities to cope with shocks.

Governance is the exercise of political, economic, and administrative authority in the management of a country's affairs at all levels. It comprises mechanisms, processes, and institutions through which citizens and groups articulate their interests, exercise their legal rights, meet their obligations, and mediate their differences. Governance encompasses, but also transcends, government. It denotes a system of values, policies, and institutions by which society manages its economic, political, and social affairs through interaction among the state, the civil society, and the private sector. It has three components: economic, political, and administrative. It is the processes by which public policy decisions are made and implemented. The "good governance" at local levels denotes quality, effectiveness, and efficiency of local administration and public service delivery; the quality of local public policy and decision-making procedures, their inclusiveness, their transparency, and their accountability; and the manner in which power and authority are exercised at the local level (World Bank 1992).

The concepts of local governance and decentralization, at times used interchangeably, are related but different concepts. Decentralization is primarily a national, political, legislative, institutional, and fiscal process, while local governance can be affected by decentralization processes (UNISDR 2015). UNDP has a holistic approach to defining the field of local governance and decentralization by using the concept of decentralized governance for development. Decentralized governance for development is considered to be a key area of democratic governance which in turn is crucial to attaining human development and the Sustainable Development Goals (SDGs). For development and governance to be fully responsive and representational, people and institutions must be empowered at every level of society – national, provincial, district, city, town, and village.

Natural hazards on their own do not result in disaster. Rather, it is the vulnerability of populations in countries that has a direct bearing on levels of disaster. Supportive governance is necessary to ensure coping capacities in societies. Mainstreaming of DRR is a governance process, enabling the systematic integration of DRR concerns into all relevant development spheres. In other words, responsive, accountable, transparent, and efficient governance structures underwrite the environment where DRR can be institutionalized as an underlying principle of sustainable development. Therefore, building resilient communities in disaster-prone countries requires that (a) underlying risk factors are continuously considered in all relevant sectors and (b) risk reduction standards and measures are an integral part of the planning and delivery of core development services and processes.

Good governance is reflective of community empowerment and ecological wisdom. Any concept of governance aiming for sustainability needs to start from two observations. One is the accurate and honest realization that the world we live in is on an unsustainable path and an understanding of why our path is unsustainable. The other is a strong sense of ethics to guide our search for a sustainable path.

Good governance has eight major characteristics. It is participatory, consensus oriented, accountable, transparent, responsive, effective and efficient, equitable, and inclusive and follows the rule of law.

Good governance advances sustainable development for three reasons. First, enjoying political freedom and participating in the decisions that shape one's life are fundamental human rights. Second, good governance helps to protect people from economic and political catastrophes, such as famines, and other crises. There is a direct correlation between bad governance and famines. Third, good governance can promote sustainable development by empowering citizens to influence policies that promote growth and prosperity and reflect their priorities. Disaster risk governance is the way in which the public authorities, civil servants, media, private sector, and civil society coordinate at community, national, and regional levels in order to manage and reduce disaster and climate-related risks. It is the way actors at all levels manage and reduce disaster and climate-related risks.

References

Ahrens J, Rudolph PM (2006) The importance of governance in risk reduction and disaster management. J Conting Crisis Manag 14(4):207–220
Bevir M (ed) (2011) The SAGE handbook of governance. Sage, Thousands Oaks
Bowers J (1997) Sustainability and environmental economics – an alternative text. Longman, London
Doornbos M (2001) "Good governance": the rise and decline of a policy metaphor? J Dev Stud 37(6):93–108
Gotlieb Y (1996) Development, environment and global dysfunction, towards sustainable recovery. St. Lucie Press, Florida
Meadows DH, Meadows DL, Randers J, Behrens WW III (1972) The limits to growth: a report for the Club of Rome Project on the predicament of mankind. Universe Books, New York
Munasinghe M (1995) Making growth more sustainable. Ecol Econ 15:121–124
UNDP (2004a) Reducing disaster risk: a challenge for development a global report. Bureau for Crisis Prevention and Recovery, United Nations Development Program, New York
UNDP (2004b) Reducing disaster risk: a challenge for development. United Nations Development Programme, Geneva
UNDP (2010). Disaster risk reduction, governance & mainstreaming. Bureau for Crisis Prevention and Recovery – UNDP (BCPR-UNDP)
UNDP (2013) Issue brief: disaster risk governance. Bureau for Crisis Prevention and Recovery, UNDP, New York
UNDP (2014). Disaster risk reduction makes development sustainable.
UNDP/EU (2004) Governance indicators: a users' guide. UNDP and Eurostat, Oslo
UNISDR (United Nations International Strategy for Disaster Reduction) (2015) Sendai framework for disaster risk reduction 2015–2030. UNISDR, Geneva
United Nations (1992a). Agenda 21. Rio de Janeiro, Brazil: United Nations conference on environment and development (UNCED), UN General Assembly
United Nations (1992b). Rio declaration on environment and development. Rio de Janeiro, Brazil:United nations conference on environment and development (UNCED), UN General Assembly; 1992

United Nations (2010) Local governments and disaster risk reduction: good practices and lessons learned. UNISDR, Geneva

United Nations Development Program (1997). Governance and sustainable development. New York

United Nations Development Program (2002) Human development report 2002: strengthening democracy in a fragmented world. Oxford University Press, New York

WCED (1987) Our common future. Oxford University Press, Oxford

World Bank (1992) Governance and development. World Bank, Washington, DC

World Bank (1994) Governance – the world bank's experience. The World Bank, Washington, DC

Chapter 14
A Participatory Approach to Enhance Disaster Risk Governance: The Case of Delhi, India

Sunil Prashar and Rajib Shaw

Abstract The climate-related disasters are increasing in urban areas, and every year, they affect millions of people through loss of life and massive damage to basic infrastructure, livelihoods, and personal assets. The risk in urban centers is growing faster not only due to urban growth in hazardous locations, lacking infrastructure, vulnerability of housing and infrastructure, increasing poverty, and inequality but also due to weak disaster risk governance, which also reflects how risk is valued against socioeconomic development concerns. Thus, disaster risk governance is utmost important and one of the main components of building disaster resilience of urban areas. This chapter deals with a participatory research on prioritizing community actions for climate-related disaster resilience in Delhi and showcases how the research attempted to enhance disaster risk governance in Delhi, India. The research describes how critical factors such as stakeholder involvement, cooperation and collaboration between stakeholders, partnership, and communication can effectively contribute to enhance disaster risk governance. In addition, the chapter also discusses the critical challenges of disaster risk governance in general and specific to Delhi. Finally, the way forward provides some suggestions to address the key challenges of disaster risk governance.

Keywords Disaster risk governance • Participatory approach • Climate-related hazards • Community actions • Resilience • Delhi

S. Prashar (✉)
Asian Disaster Preparedness Center, Bangkok, Thailand
e-mail: sunilparashar111@gmail.com

R. Shaw
Graduate School of Media and Governance, Keio University, Shonan Fujisawa Campus, Fujisawa, Kanagawa, Japan
e-mail: shaw@sfc.keio.ac.jp

I. Pal, R. Shaw (eds.), *Disaster Risk Governance in India and Cross Cutting Issues*, Disaster Risk Reduction, DOI 10.1007/978-981-10-3310-0_14

14.1 Introduction

Asia-Pacific region is one of the most disaster-prone regions in the world and is hit by both large and small disasters. Some of the world's natural disasters occur in this region from 2005 to 2015 including Japan, earthquake and tsunami (2011); Myanmar, tropical cyclone (2008); China, earthquake (2008); Pakistan, earthquake (2005); Nepal, earthquake (2015); the Philippines, typhoon (2013); India, flood (2013); and Indonesia, earthquake (2006) (UN-ESCAP 2016). Moreover, the most frequent disasters that occurred are climate related, which mainly comprise of meteorological, hydrological, and climatological subgroups of natural disasters. As per the Emergency Event Database (EM-DAT) of the Center for Research on the Epidemiology of Disasters (CRED), the human impacts (i.e., lives lost and people affected) recorded from climate-related disasters are higher than any other disasters (UN-ESCAP 2016). Moreover, these disasters are also influenced by change in climate, such as changes in the means (i.e., temperature, precipitation, sea-level risk), changes in extremes (i.e., extreme rainfall/tropical cyclones, droughts, heat or cold waves, abrupt climate change), and changes in exposure (i.e., population movements, biological changes, etc.). The twenty-first century noted that the world's growing population is concentrated in the urban centers and most of them are facing an increase in natural disasters (Brown 2012). Moreover, people affected by such disasters are also growing.

Typically, in urban areas, disaster occurs as a result of disturbance created by shocks and stresses. Shocks refer to sudden events or hazards that impact on the vulnerability of the system, and its components and stresses are the long-term trends, such as rapid urbanization, natural resource degradation, and climate change, among others, which undermine the potential of a given system or process and increase the vulnerability of the actors within it. The shocks and stresses have also influenced the system of disaster risk governance, which tends to focus more on concepts such as disaster risk, disaster risk reduction (DRR), and disaster management. "Disaster risk governance refers to the specific arrangements that societies put in place to manage their disaster risk (Gall et al. 2014; UNDP 2013; UNISDR 2011) within a broader context of risk governance" (Renn 2008). In recent decades, several countries have invested in strengthening disaster risk governance due to the fact that their governance system, including institutional system failed to address the major disasters. In less developed countries, disaster risk governance has failed due to fundamental and systematic limitations such as inequality, ineffective governments, a lack of resources at local levels, and lack of trust, among others (Djalante et al. 2011; Gall et al. 2014; Highfield and Brody 2013; Nolte and Boenigk 2011; Zanotti 2010). Thus, lessons learned by such countries call for a better stakeholder involvement, capacity building, decentralization, and transfer of powers to the local levels (Dahiya 2012; Gall et al. 2014; Samaratunge et al. 2012; Wilkinson 2012). Moreover, cases of weak civil societies have been noted which limit the effectiveness of disaster risk governance in both poor and developed countries. For example, such observations were found in Japan after the Kobe earthquake (Shaw and Goda 2004) and Haiti after the 2010 disaster (Pelling 2011).

The issue of weak civil societies in limiting the disaster risk governance was also noted in several empirical studies, and in order to address such issue, several governments promoted community participation in DRR through decentralization (UNISDR 2015a). Studies have also shown that the participation of the communities, including beneficiaries in the design and implementation of projects and policies, can mobilize their commitment and cooperation to a greater extent (Ahrens and Rudolph 2006; Brown 2012). Moreover, the model proposed by the Hyogo Framework for Action for DRR considers "community participation" as one of the key activities relating to disaster risk governance (UNISDR 2015a). Unfortunately, in reality, the progress toward community participation or engagement in disaster risk governance has been restricted to specific short-term projects or programs supported by the nongovernment organizations (NGOs) (UNISDR 2015a).

Given this background, this chapter deals with enhancing disaster risk governance through a participatory approach, which aims to enhance community participation in DRR. The research focuses on prioritizing community actions for climate-related disaster resilience in Delhi. The chapter is structured in the following order. First, it gives a theoretical overview of governance, risk governance, and disaster risk governance. Second, it describes disaster risk and disaster risk governance in Delhi and its key challenges. The third part deals with the study on prioritizing community actions for climate-related disaster resilience in Delhi. It is followed by a discussion, which analyzes the study through the lens of disaster risk governance. Finally, the way forward discusses issues and means to achieve effective disaster risk governance.

14.2 Governance, Risk Governance, and Disaster Risk Governance: An Overview

The term "governance" is defined in many theories as an organizing concept that guides administrators as administrative practices shift from the bureaucratic state to what is called the "hollow state" or what Osborne and Gaebler (1993) call "third-party government." The shift from a bureaucratic state is also noted in the definition proposed by the Frederickson and Smith (2003), "Governance refers to the lateral and inter-institutional relations in administration in the context of the decline of sovereignty, the decreasing importance of jurisdictional boarders and a general institutional fragmentation." Later, at the end of the twentieth century, the term gained more attention of international agencies, donor agencies, social scientists, and civil society. For example, in 1997, the United Nations Development Programme (UNDP) in its policy paper defined governance as "the exercise of economic, political and administrative authority to manage a country's affairs at all levels. It comprises the mechanisms, processes and institutions through which citizens and groups articulate their interests, exercise their legal rights, meet their obligations and mediate their differences" (UNDP 1997). The definition clearly focuses on the changes in the functions or roles that may formerly have been carried out by public entities that are now frequently dispersed among a diverse set of actors, such as private sector, and civil society entities.

Moreover, the movement from government to governance is also due to the social changes, namely, the rise of practices related to contracting and outsourcing, new forms of public–private partnerships, and replacement of hierarchical, bureaucratic systems of control with more decentralized network forms of organizations. Similarly, the rise of complex problems, such as climate change, natural hazards, and disasters, cannot be addressed by individual organizations and institutions. Risk governance provides means to address these problems. It is a wide-ranging and inherently multidisciplinary activity that "requires consideration of the legal, institutional, social and economic contexts in which risk is evaluated, and involvement of the actors and stakeholders who represent them" (Renn 2008). "Risk governance looks at the complex web of actors, rules, conventions, processes and mechanisms concerned with how relevant risk information is collected, analyzed and communication, how management decisions are taken" (Renn 2008).

Renn (2008) provides a comprehensive framework for understanding risk governance, and the framework involves four consecutive phases of pre-assessment, appraisal, characterization/evaluation, and management (Fig. 14.1). The fifth element, risk communication, is at the center of the framework and important in all phases and providing flows of information and dialogue between them (Fig. 14.1). The framework is not linear, but it is open, cyclical, and iterative and interlines as shown in Fig. 14.1, and it also means risk is never fully controlled or eradicated because risk itself and the activities within each element are constantly evolving (Renn 2008; Walker et al. 2010). Later this framework was also adopted by international institutions, for instance, in 2014, the International Risk Governance Council proposed the similar framework for risk governance.

Pre-assessment (Fig. 14.1) deals with the determination of the context, the goal, and the purpose, as well as the boundaries of the analysis (Renn 2008; Walker et al. 2010). For instance, in the case of natural hazards, the stakeholders whose actions can trigger the disasters or those who can be affected by the disasters can together define the boundaries of the risk analysis. Appraisal (Fig. 14.1) on the other hand deals not only with the conventional scientific risk assessment but also involves assessment of different stakeholders' potential concerns about the risk (Renn 2008; Walker et al. 2010). It also includes data collection and sharing, interdisciplinary scientific and concern assessment, and capability assessment. Characterization/evaluation (Fig. 14.1) deals with judgment about the overall severity of risk, for example, whether or not the risk is acceptable or not (Renn 2008; Walker et al. 2010). In addition, if risk is not acceptable, whether risk reduction is considered necessary to make it more tolerable. The need to take management measures is based upon both risk governance elements, such as pre-assessment and appraisal, and evaluation of other factors, such as social, economic, and political (Renn 2008; Walker et al. 2010). Management (Fig. 14.1) deals with the generation, assessment,

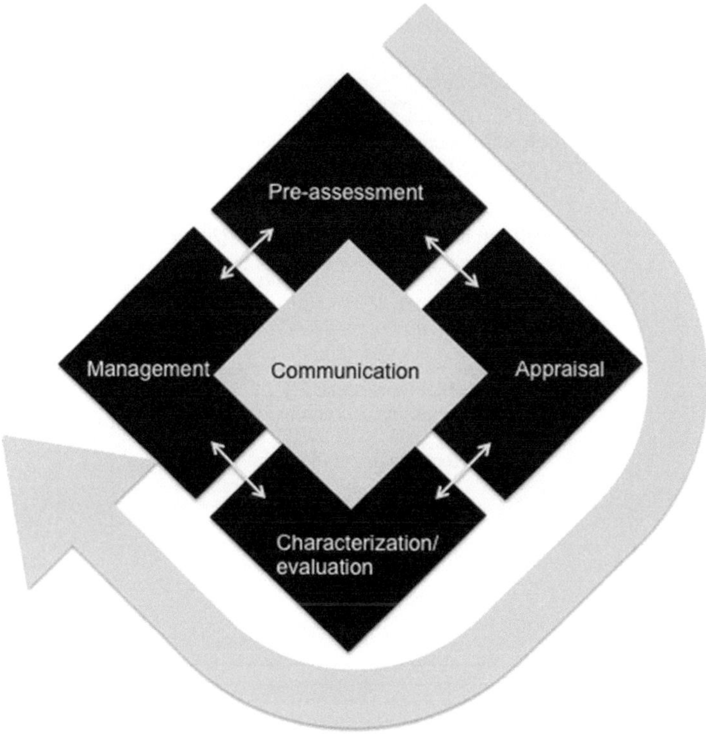

Fig. 14.1 Five elements of risk governance (Adopted from Renn 2008, p. 48)

evaluation, and selection of appropriate risk-reduction options, implementing the selected measures, monitoring their effectiveness, and reviewing the decision (Renn 2008; Walker et al. 2010). Finally, communication (Fig. 14.1) is the most important aspect of risk governance as it enables risk assessors and risk managers to develop a common understanding of their tasks and responsibilities (Renn 2008; Walker et al. 2010). It also empowers stakeholders to understand the risk and the rationale for risk management. For example, once the risk management decision is made, communication should explain the rationale for the policy decisions and allow people to make informed choices about the risk and its management. Two-way communication dialogue creates trust in risk management (Renn 2008; Walker et al. 2010).

In order for risk governance framework (Fig. 14.1) to operate effectively, it is vital to place this framework in a wider social and political context in which decisions are being made (Fig. 14.2; Renn 2008; Walker et al. 2010).

As Renn (2008) emphasizes, risk governance is not something that can be applied in a standard way in all locations, political cultures, organizations, and risk situations, but the entire process of risk governance must be open to adaptation in order to reflect the specific context of each risk (Renn 2008; Walker et al. 2010). Figure 14.2 showcases how risk governance can be placed in a wider context. In the

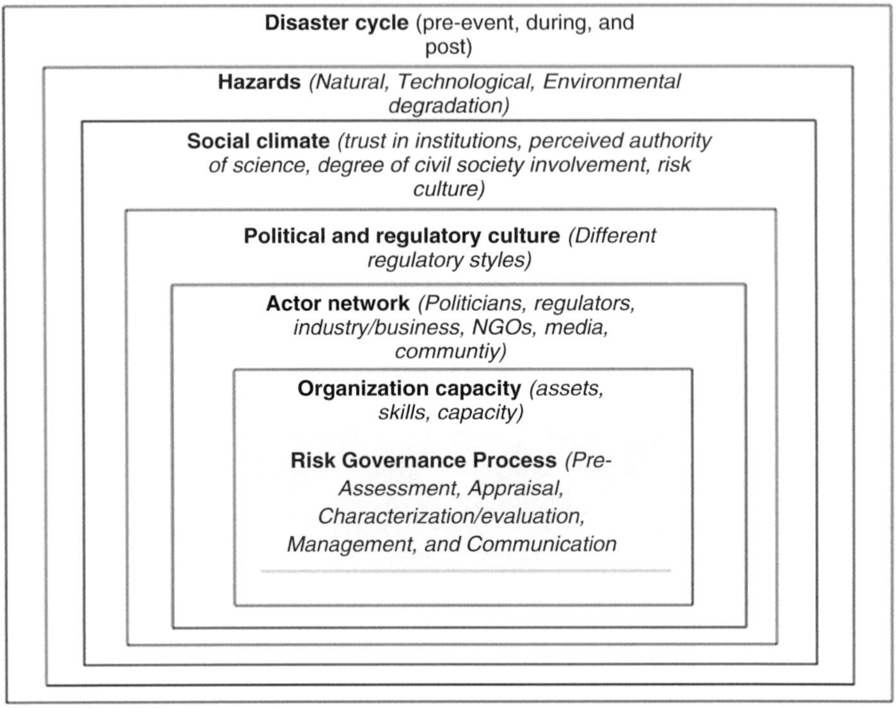

Fig. 14.2 Placing risk governance process in the wider context

context of natural hazards, the model risk governance can be placed to manage the disaster risk.

14.2.1 Disaster Risk Governance

Disaster governance – sometimes also called disaster risk governance or adaptive (disaster) governance – is related to management of complex social–environmental problems and associated risks (Gall et al. 2014). Disaster risk governance is defined as "the specific arrangements that societies put in place to manage their disaster risk (UNISDR, 2011; UNDP, 2013; Gall et al. 2014) within a broader context of risk governance (Renn 2008)" as highlighted in Fig. 14.1. It deals with institutions, organizations, laws, regulations, and contributions from civil society and private sector actors that influence risk management (Brunner et al. 2005; Gall et al. 2014; Holley et al. 2011). Several research studies have been carried out to characterize disaster risk governance and identified its key elements, namely, stakeholder involvement, cooperation and collaboration, flexibility, accountability, and transparency (Table 14.1; Gall et al. 2014).

Table 14.1 Key elements of disaster risk governance

Element	Description	Indicator (examples)
Stakeholder involvement	Advocates for disaster risk reduction and provide coordination, analysis, and advice on areas of priority requiring concerted action through a coordinated and participatory process	Platforms for DRR – regional, national, local, city level, etc.
Cooperation and collaboration	Distribution of government functions across variety of state and non-state actors facilitating horizontal and vertical disaster risk management. It creates local capacity, establishes trust, and enhances cooperation	Community participation and decentralization
Flexibility	Adjustment of policies, regulation, etc., creation of ad hoc groups and networks, and community self-organization	Policy framework, policy instrument, tool, legal, regulatory system for DRR
Accountability and transparency	Ability to monitor and measure the beneficial (or adverse) effects of disaster governance	Resilience-specific data, census data, etc.
	Monitoring the effectiveness of the governance systems in reducing disaster risk requires data on the state of society and the environment and the human actions	

Text adopted and modified from Gall et al. (2014); UNISDR (2005)

Since 1990, stakeholder involvement (Table 14.1) was recognized and considered as important to disaster risk governance (Gall et al. 2014; IPCC 2012; UNISDR 2005, 2011). Stakeholders are those "who have an interest in a particular decision, either as individuals or representatives of a group. This includes people who influence a decision, or can influence it, as well as those affected by it" (Hemmati 2001). In the multi-stakeholder involvement, processes aim to bring together all major stakeholders in a new form of communication, decision finding (and possibly decision-making) on a particular issue. Providing platforms for multi-stakeholder dialogue is not a new area of development, for example, since 1990, several international institutions and companies promoted platforms for multi-stakeholder dialogues on the contentious issues, including disaster and climate risks. For example, based on the gains made in DRR during 1990–1999 International Decade on Natural Disaster Reduction (IDNDR), the UN/ISDR secretariat was established and promoted together with other UN agencies in the identification of needs for establishment or further development of national platforms for DRR (Gall et al. 2014). In 2005, at the World Conference on DRR, 168 governments adopted the Hyogo Framework for Action (HFA) 2005–2015. One of the strategic goals of HFA was "making DRR a priority" which aimed to promote disaster risk governance. The increase in the national platforms to promote multi-stakeholder dialogues across the world was noted during HFA review (UNISDR 2015a).

Cooperation and collaboration (Table 14.1) are also important elements of disaster risk governance. The key for effective cooperation and collaboration is the involvement and participation of a diversity of local actors who influences the decisions or in other words the distribution of decisional power. It also includes distribution of government functions, for example, administrative, managerial, regulatory, etc., across a variety of state and non-state actors, facilitates vertical as well as horizontal disaster risk management and creates local capacities, establishes trust, and enhances corporation (Boyer-Villemaire et al. 2014; Djalante et al. 2011; Tompkins et al. 2008). For effective decentralization and community participation, cooperation and collaboration are most critical. Most theories of participation by local actors are rooted in the Arnstein's ladder of participation with different steps reflecting an increasing share of power with civil society. Limited information sharing and manipulation of the opinions characterize the lowest level of participation, whereas the highest involves citizens making decisions (Arnstein 1969). The HFA promoted community participation in DRR through the adoption of specific policies, the promotion of networking, the strategic arrangement of volunteer resources, the attribution of roles and responsibilities, and the delegation and provision of the necessary authority and resources (UNISDR 2015a).

Flexibility (Table 14.1) deals with adjustments of policies and regulations, creation of ad hoc groups and networks, and community self-organization (Gall et al. 2014). These are widely perceived as essential and important components of disaster governance. Cosens (2013) argues that in order to make a system resilient, the restructuring of the current system is not an easy task. The governance systems require careful attention to procedural elements to assure legitimacy in implementation. Two areas may help in reducing the barriers to implementation of an adaptive approach to management of risk: (i) procedural elements to assure legitimacy in governance and (ii) coordination across the scale of governance to assure that legitimacy carries through various scales of decision-making (Cosens 2013). Thus, such kind of change or reform will require authorization for greater flexibility in decision-making. In addition, several literature studies have also documented the benefit of flexible governance from disaster response and recovery experience. Shaw and Goda (2004) noted that the Kobe earthquake caused major changes to the governance, for instance, increased cooperation between local government and resident's association and enhanced participation of people in the decision-making process. In addition, community governance, administrative reforms, and cooperation and partnership were the major priorities of the Kobe Action Plan to develop a sustainable society. These actions provided strength for equal partnerships under the new autonomy system (Shaw and Goda 2004).

Accountability and transparency (Table 14.1) elements are also important for disaster risk governance to be more effective. The accountability ensures that politicians can be held responsible for their actions, and it is reinforced through participation, i.e., formal or informal channels for citizens to influence policymakers (Ahrens and Rudolph 2006), whereas transparency deals with publication of reliable information to other actors in a timely manner, reduces opportunities for corrupt behavior, improves the analysis and articulation of public policy choices, and enhances their acceptance (Ahrens 2002; Ahrens and Rudolph 2006; Piciotto 1997). But in order to assess the system's transparency and accountability, it is important to monitor the effectiveness of the governance system. Monitoring further requires data on the state of society, the environment, and human actions as well as the development of benchmarks and measures.

14.2.1.1 Linking Disaster Risk Governance with Resilience

Djalante et al. (2011) emphasizes that effective disaster risk governance produces resilience. In other words, resilience can be the outcome of the effective disaster risk governance due to the fact that there are close conceptual ties between the two frameworks. The concept of resilience is developed, adopted, and interpreted differently in different fields of study. In disaster field, resilience is the ultimate goal for reducing the disaster risks, and it is defined as the ability of a community or society exposed to hazards to resist, absorb, accommodate, and recover from hazards timely and efficiently (UNISDR 2009). Moreover, the elements of disaster risk governance help in building resilience. Participation and collaboration, for instance, favorably influence the capacity to manage resilience, and pooling of knowledge from various participants can lead to effective process and better outcomes (Djalante et al. 2011; Pahl-Wostl 2009). Similarly, engaging stakeholders in various roles and assuming that they take administrative, regulatory, managerial, and mediating functions, which is previously undertaken by a central government, can have high potential to influence the capacity to manage resilience. As Ostrom 2010 emphasizes, those opportunities for learning and innovation through utilization of local and multiple sources of knowledge can also lead to better adaptation strategies and contribute to resilience (Djalante et al. 2011).

14.2.1.2 Key Challenges

Several empirical studies have also noted the challenges faced by international institutions as well as governments in implementing the disaster risk governance. For example, lack of resources, responsibilities, political legitimacy/recognition, adequate stakeholder representation; linkages to establish network; adequate technical competence; policies, norms, and standards for DRR; and community participation is the core challenge of disaster risk governance (Aysan and Lavell 2014; Djalante 2012; Coskun 2013; UNISDR 2009, 2015a). Moreover, the effectiveness of multilayered institutions reduces and can create unnecessary overlaps making coordination difficult. In addition, nontraditional stakeholders such as private sector and academia are frequently absent, and NGOs are not adequately represented at the local levels (Djalante 2012). Due to lack of resources, a long-term arrangement is challenging for volunteers as well as nongovernmental actors. Moreover, governance system functions effectively during and after disaster, but not when there is an urgent need to generate and implement transformative, systematic changes that reduce disaster risk or adapt to climate change over the long run.

The model proposed by the HFA on disaster risk governance has also faced similar challenges. For instance, most resources invested in strengthening capacity for disaster management, but limited success in applying policies, norms, standards, and regulations to manage and reduce risk across development sectors (UNISDR 2015a). Countries made commitments to integrate DRR into poverty reduction and other development strategies, but such commitments in law and policy have not been translated into real priorities and investments due to weak political authority, governance arrangements, and technical competencies to do so (UNISDR 2015a). Moreover, the political support to the importance of disaster risk management sector in most countries is transient, and it is further reflected in inadequate financing of disaster risk management and insufficient investment in weak human and institutional capacities (UNISDR 2015a). Even political support to decentralization has recorded uneven, for instance, there is agreement on the critical role of local government in disaster risk management, but the institutional capacity for that has remained restricted particularly in smaller governments and rural areas (UNISDR 2015a). The community participation in several countries has become another hyperreality, and progress in the community-based or local level DRR has been limited to short-term projects or programs only (UNISDR 2015a). The corruption has been also noted in hazard-prone areas, and as a result the vulnerability of peoples and their assets increased (UNISDR 2015a). In such areas, the underlying drivers of disaster risk include the absence or limited voice of the citizens and accountability of decision-makers to the people they represent (UNISDR 2009a).

Low- and middle-income countries lack the necessary regulatory quality of norms and standards to be applied due to weak accountability of local to central government, government to citizens, and across government sectors that has undermined the effectiveness of norms, standards, laws, and policies (Coskun 2013; UNISDR 2015a). The weak governance has also led to the adoption of inappropriate strict codes and standards, which derived more development into the informal development.

Finally, the disaster risk governance adopted in many countries relied mainly on the specialized emergency management organizations, which means that such agencies have assigned with two different tasks that are same by nature. Emergency management has evolved as a stand-alone sector addressing the challenges of responding to accidents, but governance arrangement required to manage disaster risks need – by definition – to interweave with and flow through the broader governance arrangements used by countries to manage economic and social development (Aysan and Lavell 2014; UNISDR 2015a).

14.3 Disaster Risk in Delhi

Delhi, National Capital Territory (NCT), the capital of India, is sprawled over the west bank of Yamuna River. It is surrounded on three sides by the states of Haryana and to the east across to the river of Yamuna by the state of Uttar Pradesh. The total area of NCT of Delhi is 1483 km^2, the largest city in India in terms of the geographical area (Govt. of NCT of Delhi n.d.). It is divided into nine districts: North West, North, North East, East, New Delhi, Central, West, South West, and South (Prashar et al. 2012; Parvin et al. 2011). The total population of Delhi was 16 million in 2011, which had grown rapidly in the recent past. The city has experienced very fast population growth between 1991 and 2001, expected to reach 28.41 million by 2026 (Census of India 2001). With the rapid pace of urbanization and growth of urban population, the rural population and rural areas have been continuously decreased as noted by the reports of Census of India. Rapid development process and ever-increasing population pressure have posed serious disaster risks to the city. Delhi is prone to different kinds of natural and man-made hazards, namely, earthquake; flood; bomb blasts; other acts of terrorism; fires; industrial and nuclear, biological, and chemical hazards; flash floods; building collapses; road accidents; and waterlogging (Prashar et al. 2012). This chapter particularly focuses on the natural hazards that have caused massive impacts to the city in the form of disasters in the past. Some of the major disasters include nine major floods since 1900 in the years 1924, 1947, 1976, 1978, 1988, 1995, 1998, 2010, and 2013 (DDMA 2009). Most floods occurred in Delhi due to the overflow of river of Yamuna. Yamuna catchment up to Delhi is divided in two areas: upper catchment areas (from source in Himalayas to *Kalanaur* in Haryana) and lower catchment areas (from *Kalanaur* to old Delhi rail bridge) (DDMA 2009). Based on flood atlas and maps prepared by the Central Water Commission, it shows that most flood-prone areas include North Delhi on the west bank of the Yamuna and entire Trans Yamuna area on the east bank. The underlying risk factor climate change is also posing significant risk in the form of heat waves and cold waves. For example, the temperatures fall sharply for a few weeks during January and February (DDMA 2009). During summer, tens of thousands of people suffer from heat stress. Cloudburst is not a regular hazard, but it occurred recently in 2011 (DDMA 2009). Thunderstorm and squall often occur during summer and lead to broken tree branches, overturned cars, traffic jams and power outages, etc. In addition, epidemic disease outbreak after any major natural disaster is also common and affects thousands of people in the past. Delhi is also prone to hydrological droughts and water scarcities. The groundwater level has decreased in recent years between 0 and 29 m (max.), and as a result, there is a gap of water demand and supply, which has increased rapidly in the last few years (DDMA 2009). Finally, environmental degradation in the form of air pollution, water pollution, and land pollution is also happening and affecting the whole city. In order to address the risks from environmental degradation, it is most important that sectors, namely, water, forestry, buildings, lighting, energy, renewable, transportation, and waste management, should be targeted with a systematic approach to address future risk.

14.3.1 Disaster Risk Governance in Delhi

The Disaster Management (DM) Act, 2005, serves as the foundation for disaster risk governance in Delhi, India. As per the act, National Disaster Management Authority (NDMA), State Disaster Management Authority (SDMA), and District

Disaster Management Authority (DDMA) are constituted, and they serve as a platform for DRR (DDMA n.d.). Moreover, as per the notification under subsection (I) of section 3, a National Executive Committee (NEC) to assist the national authority is constituted, and under subsection (I) of section 14, a State Executive Committee (SEC) is established to assist state authorities (Ministry of Law and Justice 2005). In order to address the disaster risk in Delhi and for undertaking a holistic, coordinated, and prompt response to any disaster situation, the Delhi Disaster Management Authority (DDMA) was constituted in 2008 and reconstituted in 2015. Similarly, the SEC was constituted in 2008 to assist DDMA (DDMA n.d.). SEC in Delhi is composed of stakeholders or officials (i.e., ex officio) from the Home Ministry, Public Work Department, Revenue Department, and Urban Development, whereas the NEC is composed of stakeholders or officials (i.e., secretaries of Government of India) from agriculture, atomic energy, defense, drinking water supply, environment and forests, finance, health, power, rural development, science and technology, space, telecommunication, urban development, water resources, and the chief of the Integrated Defense Staff (Ministry of Law and Justice 2005). Finally, the DDMA merely acts as a district planning, coordinating, and implementing body for disaster management work according to the guidelines laid down by the national and state authorities.

The DM Act, 2005, also provides a mechanism for the coordination and cooperation, but the approach is more top-down in nature in Delhi, India. The central and state governments have been assigned with more power to make decisions on disaster management, and local authorities have limited power and cannot make decisions on disaster management. For example, the local authority is only responsible for ensuring that its officers and employers are trained in disaster management; resources related to disaster management are maintained; ensuring all construction projects under it conform to the standards laid down by the national, state, and district authorities; and carry out relief, rehabilitation, reconstruction activities in affected areas. In Delhi, the DDMA emphasizes on the community awareness and community preparedness planning for disaster response. The government has developed large-scale information and communication and education materials that are distributed in the offices, schools, institutions, and residential colonies. Community disaster management planning is an important component of community preparedness and combat disaster in predefined areas. The main roles of community (i.e., NGOs, residential and welfare associations) and other community-based organizations in Delhi during disaster are to be the first responders when disaster strikes, assist in basic relief and rescue, and help in maintaining peaceful coordination between administration and general public. DMA of all districts maintains a list of NGOs and CBOs of the district, along with details of their functional specialization/capacity and geographical coverage.

The act also provides directions to the ministry or the department of the Government of India to prepare a disaster management plan that will include necessary measures, such as the specifications regarding integration of mitigation measures in its development plan and its roles and responsibilities in relation to preparedness and capacity building to deal with any disaster situation, among others

(Ministry of Law and Justice 2005). The Government of India has also formulated the National Disaster Management Policy, and a number of states have also prepared their state disaster management policy, but Delhi is still in the process of preparation. The 11th 5-year national plan has emphasized the need and importance of mainstreaming of DRR into development planning process and programs. Chapter 5 of the Delhi Disaster Management Plan (2015) also emphasizes on "Mainstreaming disaster management concerns into development plans/ programs/ projects." It is one of the key priorities of central government as well as Delhi government, and in addition, the plan also provides a checklist for the Expenditure Finance Committee and Detailed Project Report formats for mainstreaming. In addition, section 5.4 of the plan provides sector-specific guidelines for mainstreaming disaster management concerns in housing, infrastructure, health, education, and financial service sectors.

The act also promotes accountability and transparency, especially in the area of disaster response. For example, in the compensation and relief to the victims of disaster, there shall be no discrimination on the ground of sex, caste, community, descent, or religion (Ministry of Law and Justice 2005). Similarly, both national and state authorities shall prepare once in every year an annual report giving a true and full account of their activities during the previous year (Ministry of Law and Justice 2005). In the case of Delhi, the DDMA prepares such report and submit to the Delhi government.

14.3.1.1 Key Challenges of Disaster Risk Governance in Delhi

The government of Delhi has made significant process by establishing a well-defined legislative and institutional arrangement for disaster management, but the focus is still heavily weighted on disaster response and not ex ante risk management. The major challenge is "how to" make DRR a major policy issue when it is difficult to engage policymakers in dialogue on reducing disasters and also when it is difficult to show its social and economic benefits immediately.

The participation of local communities or civil society or local authorities is only limited to disaster response, but not in the systematic approach of DRR (see Sect. 14.3.1). For example, communities' role in public sector DRR – policy consultation, reviewing standards, and implementation – is not promoted and also not clearly mentioned in the DM Act. Moreover, local communities in Delhi often do not have a direct relationship to national authority, but they have a vital role in disaster risk management because local communities are often more conversant with both disaster risks experienced and the necessary resources and existing opportunities to identify and manage disaster risks. Thus, coordination and cooperation between local communities, local authorities, DDMA, NDMA, SEC, and NEC need to be further enhanced in the future. As emphasized by UNISDR (2015a), while disaster management response coordination requires centralized command, there is a need to decentralized DRR, which means that DRR at the local level needs to be encouraged and supported. Thus, dedicated human resource support and funding arrangements are needed for strengthening District DMA and DDMA in Delhi.

There are several gaps need to be plugged especially for mainstreaming DRR into development policies, plans, and projects. Merely checklists for mainstreaming DRR and sector-specific guidelines are not sufficient to implement mainstreaming, but a comprehensive, detailed risk assessment data and information must be collected and provided to each department in Delhi to incorporate risk information into development projects and plans. All ministries can use risk information to revise their national policies and plans. Mainstreaming is also difficult if land use planning standards are inadequate, construction standards are poor, and policies to enforce standard code for all types of construction are weak. In Delhi, for example, there is no legal framework that ensures all constructions in the city follow or implement appropriate standard code (e.g., seismic code), the standard code for different infrastructures is not regularly updated, and systematic studies are missing on vulnerability of different types of construction in the city. Thus, legal and regulatory frameworks need to further strengthen to address disaster risk in Delhi. Disaster risks often exacerbate due to underlying risk factors, such as climate change, rapid urbanization and migration, and poverty, among others. In Delhi, there is no systematic provision on how underlying factors are contributing to increasing disaster risk in the city. Without addressing the underlying risk factors, it will be difficult for DDMA to achieve its vision of effective disaster management.

In Delhi, there is a clear absence of accountability and transparency in monitoring the effectiveness of the governance system in reducing the disaster risk. Monitoring the governance system requires data related to state of society, the environment, and human actions as well as development of benchmarks and measures such as indicators. Since the effectiveness of disaster governance produces resilience, it is essential that resilience-specific data should be collected on a regular basis.

Finally, resource allocation poses a real challenge in a situation where so many demands compete for limited resources. In Delhi, the budget allocation for DRR (i.e., disaster mitigation) fund is allocated to the SEC and DDMA, but no budget is allocated to local government or authorities. The devolution of power and financial resources to local authorities has been a major challenge to ensure decentralized planning and development. DDMA needs to allocate more power and resources to the local authorities.

14.4 Prioritizing Community Actions for DRR in Delhi: A Participatory Approach

The study aimed to address the climate-related disaster risk of the urban area in Delhi, India, through a participatory approach, where the process of community-prioritized actions was developed for DRR (Prashar 2012). It was built on the assumption that community-prioritized actions reduce disaster risk of residents living in urban areas. The study answered three research questions, namely, how to understand and analyze the disaster resilience of an urban area, how to identify community actions that reduce disaster risk in urban area, and how to analyze the

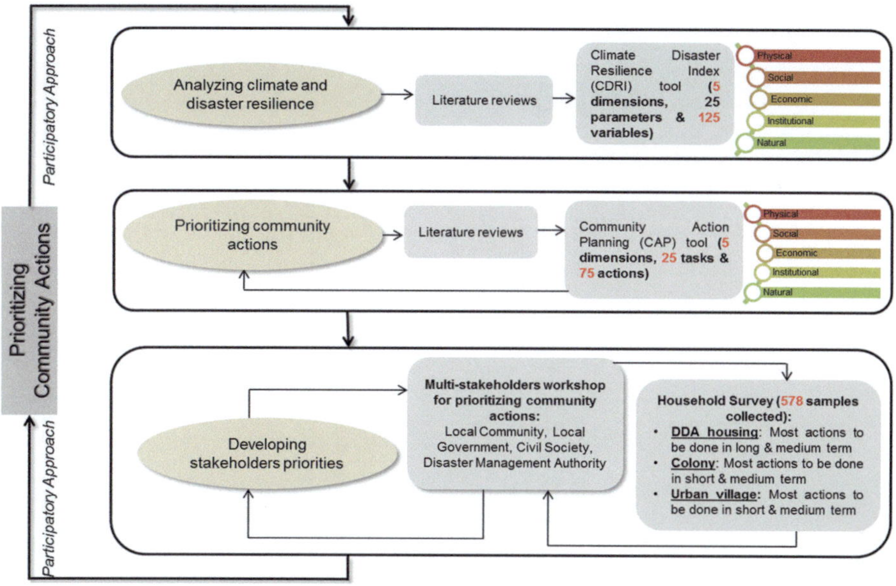

Fig. 14.3 Conceptual framework for prioritizing community actions in the context of Delhi (Prashar 2012)

roles of different stakeholders in identifying community actions for DRR (Prashar 2012). To address these research questions, the study developed a participatory approach, for instance, tool such as community action planning (CAP) that was utilized for prioritizing community actions for DRR. Figure 14.3 shows the conceptual framework for prioritizing community actions in the context of Delhi.

The following section briefly discusses three major components of the conceptual framework, namely, analyzing climate and disaster resilience, prioritizing community actions, and developing stakeholders' priorities (Fig. 14.3), and also describes the key findings from each component.

14.4.1 *Analyzing Climate and Disaster Resilience*

The first research question (see Sect. 14.4) was addressed by utilizing the Climate Disaster Resilience Index (CDRI) tool, which assesses the disaster resilience from five dimensions: physical, social, economic, institutional, and natural. Each dimension consists of five parameters, and each parameter carries five variables (Joerin et al. 2012; Table 14.2). All in all, CDRI analyzes disaster resilience on 125 (5 × 5 × 5) key quantifiable variables (Table 14.2). Each parameter is evaluated on five choices between (1 = poor) and (5 = best). The respondent is requested to provide a choice between poor and best. In addition, each variable under a specific parameter

Table 14.2 Dimensions and variables of CDRI

Dimensions	Physical	Social	Economic	Institutional	Natural
Parameters	Electricity	Population	Income	Mainstreaming of DRR and CCA	Intensity/severity of natural hazards
	Water	Health	Employment	Crisis management	Frequency of natural hazards
	Sanitation and solid waste disposal	Education and awareness	Household assets	Knowledge dissemination and management	Ecosystem services
	Accessibility of roads	Social capital	Finance and savings	Institutional collaboration	Land use
	Housing and land use	Community preparedness	Budget and subsidy	Good governance	Environmental policies

is required to be ranked or weighted on a scale of (1 = not important) and (5 = very important). All collected data are analyzed using Weighted Mean Index (WMI) and Aggregate Weighted Mean Index (AWMI) formula. WMI provides the resilience score for all parameters. Similarly, AWMI provides resilience score for all dimensions. The calculated value of AWMI of one dimension is the CDRI of that dimension. The CDRI scores vary from 1 to 5, where (1 = very poor), (2 = poor), (3 = moderate), (4 = good), and (5 = best), and presented through a spider diagram (Fig. 14.4). Higher CDRI values are equivalent to higher preparedness to cope with climate-related disasters. The quality of results depends on the knowledge of respondents. Needless to say, the results presented are not absolute values, but can serve as policy guidance.

Key Findings of Analyzing Climate and Disaster Resilience For the purpose of data collection, a CDRI questionnaire survey was developed and utilized in the study area. All data were mainly collected in nine administrative districts of Delhi (see Sect. 14.3) from district project officers (DPOs) to project officers (POs) of the DDMA, Government of NCT of Delhi, who were the key respondents to this survey.

The study revealed that East Delhi was least resilient, and New Delhi was most resilient among all of the nine districts of Delhi due to all five dimensions where physical and natural dimensions were the crucial factor responsible for the low overall resilience of East Delhi (Fig. 14.4). Moreover, the results showed interesting correlation between population density and overall resilience which means districts with high overall resilience scores had a low population density and vice versa such as New Delhi and East Delhi. In addition, water supply, income and employment, community preparedness, and land use were the highly prioritized sectors, where community preparedness could further enhance through developing community-prioritized actions. The institutional aspect showed high resilience in all nine districts, and since the chapter focuses on disaster risk governance, the following section briefly discusses institutional findings from CDRI.

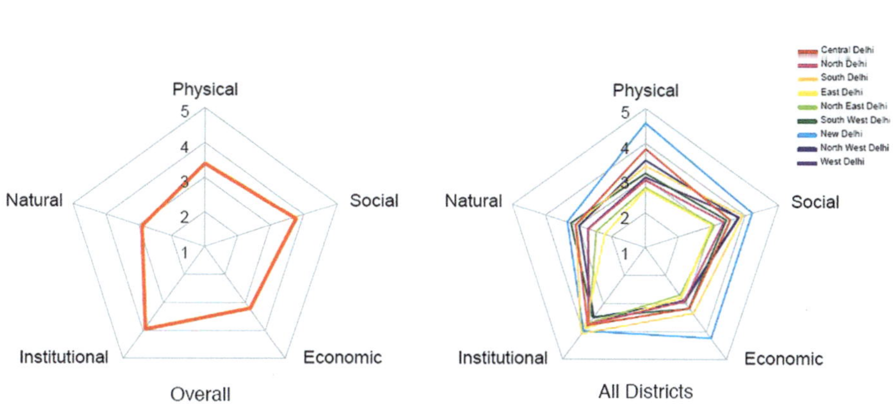

Fig. 14.4 Spider diagram showing overall resilience of Delhi

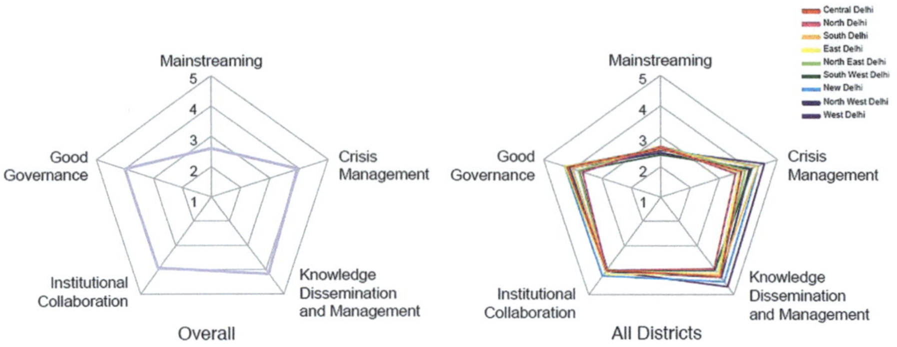

Fig. 14.5 Spider diagram showing institutional resilience

CDRI for institutional and its parameters in Table 14.2 captured some of the key aspects of disaster risk governance, for instance, institutional collaboration, good governance, and mainstreaming of DRR and CCA (Fig. 14.5). Overall the disaster resilience on institutional dimension showed moderate to good score in Delhi due to the fact that organizational capacities have improved on preparedness for disaster response, but not on mainstreaming DRR and CCA (Fig. 14.5). Mainstreaming DRR and CCA could be seen at some extent in education, but not in other sectors, including land use, housing, transport, and environment policies or plans. The score of crisis management framework was good (approx. 4.0; Fig. 14.5) due to well-established crisis management plan that includes institutional framework, search and rescue team, and communication mechanism. Similarly, the score of knowledge dissemination and management was also better (more than 4.0; Fig. 14.5) due to existing knowledge management platforms (i.e. websites of DDMA, Delhi eatherquake safety intiatives, India disaster resource network, etc.); training programmes on disaster risk management for masons, town planners, NGOs, policy makers, maisons, architects, etc.). In order to strenghten the score, such intiatives need to be expanded so that all government departments, statutory agencies, research organizations/institutions, and humanitarian organizations share their knowledge and technical expertise collectively. The documentation of best practices could have further strengthened knowledge management. The institutional collaboration score was also good (approx. 4.0; Fig. 14.5) due to well-established DDMA and District Disaster Management Group. In addition, the districts are well supported by the institutions established at the state level, such as DDMA, SEC, State Crisis Management Group, Delhi Disaster Response Force, Control Room of DDMA, and Emergency Support Functions. Finally, the score of good governance was also good (approx. 4.0; Fig. 14.5) due to established institutions for early warning system, such as Indian Meteorological Department, EOC, etc. Disaster drills were regularly conducted, but there were some lacunas in the areas of incident site, hospitals, and relief camps, which need to be rectified for conducting effective drills.

14.4.2 *Prioritizing Community Actions*

The second research question (see Sect. 14.4) was addressed through identifying and prioritizing community actions for DRR through CAP. "CAP is a participatory approach that aims at community development through problem solving" (Hamdi and Goethert 1997; Prashar et al. 2011; Prashar 2012). The key essence of CAP is the development of community-led action plans. It mainly targets risk reduction by focusing on specific themes including physical improvement, strengthening of community structures, and identification of community-led environmental improvement initiatives (Bhatt et al. 1999). The concept of action planning derives from traditional urban planning, which was orthodox and less proficient at delivering its benefit on the ground (Hamdi and Goethert 1997). Consequently, very little benefit of urban planning was reaching the poor. Thus, by involving the local community, urban planning can be further improved. CAP mainly consists of three stages: the problem and opportunity identification, prioritizing solutions, and implementing and monitoring actions. The key aspects of CAP are community participation and stakeholder involvement. For example, the CAP often includes stakeholders such as community, government authorities, NGOs, and other civil society organizations.

The study first identified the problem and opportunities through analyzing the findings of CDRI and a literature review of past community development activities. Moreover, several brochures and pamphlets describing communities and their actions in community development were examined. As a result, the study developed CAP tool to identify and prioritized community actions for DRR. The tool consists of 5 dimensions, 25 parameters, and 75 actions (see Chap. 14 Appendix).

To identify prioritized community actions for DRR, a questionnaire survey was developed. Similar to CAP tool, it was developed from five resilience-based dimensions: physical, social, economic, institutional, and natural of CDRI study (Joerin and Shaw 2011). Each dimension has some specific parameters and each parameter in turn has three actions. For example, in the physical dimension, parameter P1 is "Task for reducing interruption in electricity," and the three actions are (A1) promoting the use of generator/inverter/emergency light, (A2) promoting the use of low energy/eco-friendly appliances, and (A3) registering complaints with Delhi Electricity Board (see Chap. 14 Appendix). In addition to three actions, each parameter also carries an additional choice A4 for respondent's own action. The A4 choice is not a compulsory option for the respondent. Therefore, all in all, the questionnaire has five dimensions, 25 parameters, and 75 actions (without A4) as shown in Chap. 14 "Appendix." In the survey, respondents were requested to rank orderly these actions on a 1–3 scale (i.e., 1, most important; 2, important; and 3, least important). If a respondent chooses to provide additional action A4 for any parameter, then he/she also needs to rank all four actions based upon his/her priority. Using the data collected from questionnaire survey, the results were analyzed.

Key Findings of Prioritizing Community Actions All data were collected from the President or Vice-President of Residential and Welfare Association (RWAs) residing in East Delhi. The district was selected based upon CDRI findings and the long history of experiencing major disasters in the past. RWAs are neighborhood associations which have as their main objective to work for the welfare of their residents or people through engaging in public services such as electricity, water, road maintenance, street lighting, solid waste management, and parking (Chakrabarti 2007; Government of National Capital Territory of Delhi 2008). Eighty-nine (43.84%) RWAs were interviewed out of 203 in total. They are situated in three types of settlements in East Delhi: Planned Delhi Development Authority (DDA) housing and societies, colonies, and urban villages. In Delhi, as per the housing data of 2000, the percentage of population living in the planned housing and societies accounts for 23.7%, the colonies around 30.7% (include unauthorized colonies, regularized colonies, and resettlement colonies), and the urban village is 6.4% (Mehta 2009; Planning Department 2001).

All in all, 25 tasks and 75 community-prioritized actions were identified by RWAs where out of 75, 32 actions were prioritized as most important in all three settlements (Fig. 14.6; Chap. 14 "Appendix"). In addition to these actions, the respondents also suggested some additional actions (A4), for example, in respose to P3 (i.e. task to improve the collection of solid waste), the respondents from the colonies suggested "door to door collection of solid waste" as A4. Most prioritized actions were community level, but require support from stakeholders for implementation, and such actions broadly highlighted the social aspects where community can make decisions effectively. In addition, the results also highlighted the importance of involving multi-stakeholders in implementing community-prioritized

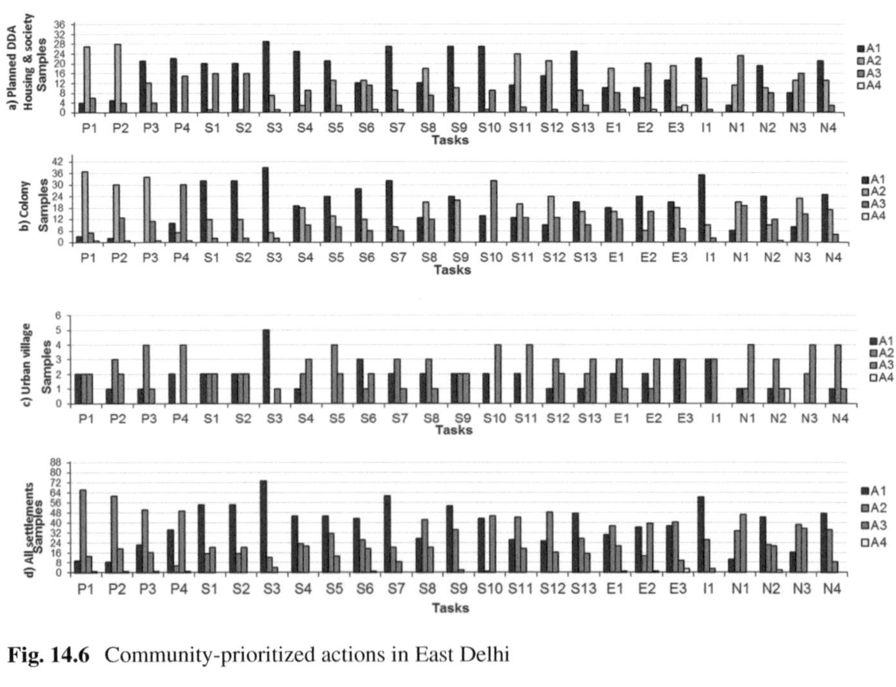

Fig. 14.6 Community-prioritized actions in East Delhi

actions, which means that based on the nature of actions, NGOs, district disaster management authority, academia, other civil society organizations, local government, and related departments can be engaged as key stakeholders in the implementation of prioritized actions.

To validate community-prioritized actions, the study also conducted interviews with various stakeholders in Delhi, namely, Center for Science and Environment (area of work: environment consecration), Center for Urban and Regional Experience (urban civic services), DDMA – East (disaster management), *Shaheed Bhagat Singh Sewa Dal* (local NGO) – East Delhi (community development), *Bhagidhari* (citizen partnership), Cell (community development), and some RWAs – East (resident welfare). All in all, 11 respondents (six RWAs and five state/local organizations) were interviewed. The validation of community-prioritized actions was carried out for two purposes, namely, to understand the challenges in implementing the prioritized actions and to understand the opportunities in implementing the prioritized actions. Some of the challenges identified by RWAs include lack of information about local area development projects, knowledge gap, and lack of financial resources. Some of the opportunities identified were RWAs experience of past activities with local people, local knowledge, and regular interaction with the people. Some of the opportunities identified by the state/local institutions include past experience of working with the community or people, their expert knowledge, and similar working areas. Most of the interviewed organizations felt that "partnership" is very important for the effective implementation of prioritized actions as it can help in addressing the main challenges. Moreover, a strong partnership between RWAs and relevant organization would imply corporation among the partners (Fig. 14.7), which would further limit the challenge of information gaps and lack of resources with RWAs.

14.4.3 Developing Stakeholders Priorities on Risk Reduction Actions

The process of developing stakeholders' priorities for risk reduction actions was conducted to answer the third research question (see Sect. 14.4). The objective of developing stakeholders' priority was to develop a systematic time frame for implementing community-prioritized actions. The study developed three different time frames: short term (between now and the next 2 years), medium term (for the next 2–5 years), and long term (beyond 5 years), which provided community with long-term vision to implement prioritized actions. To develop time frames, the study utilized stakeholder workshop and conducted household questionnaire survey in East Delhi. A half-day multi-stakeholder consultation workshop was organized by the Alliance for Adaptation and DRR, and in total 34 stakeholders participated (Table 14.3).

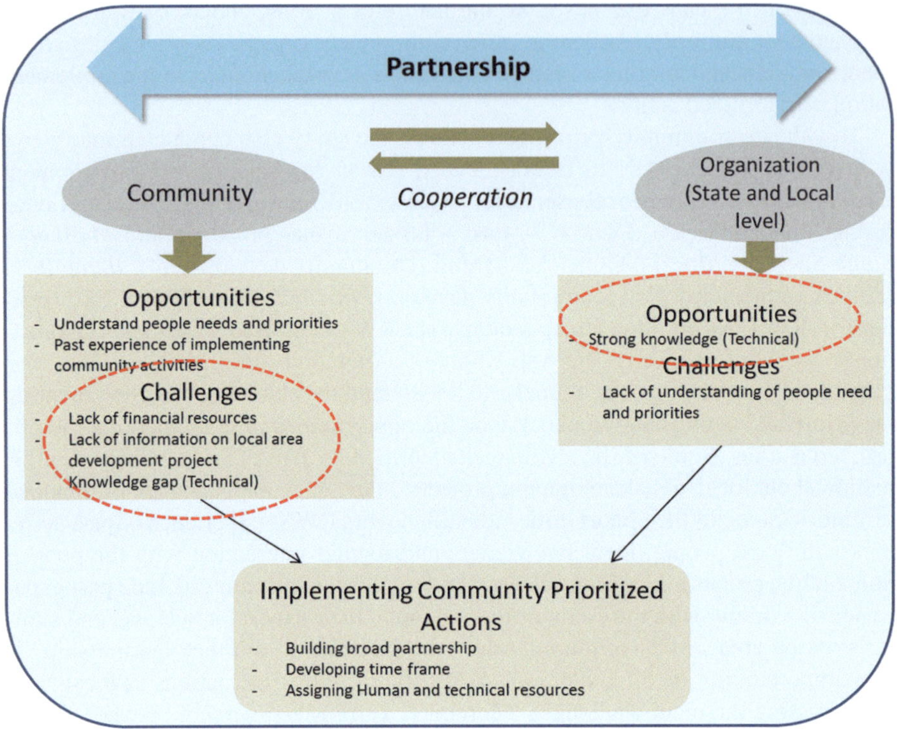

Fig. 14.7 Framework showing challenges and opportunities by the respondents for implementing community-prioritized actions for DRR

Table 14.3 Stakeholders participated in developing time frame for community-prioritized in East Delhi

Group name and no. of participants	Physical (8)	Social (8)	Economic and institutional (8)	Natural (8)
Participants	RWAs	RWAs	RWAs	RWAs
	District Resources Center	Red Cross Society	District Authority (East Delhi)	District Pollution Authority
	District Management Unit	*Shaheed Bhagat Singh* (NGO)	Indian Institute of Public Administration (IIPA)	Knowledge Link (NGO)
		District Disaster Management Authority		
		Civil Defense		

Four sets of template were prepared for four different groups (i.e., physical, social, economic and institutional, and natural) (Table 14.3). Each group was given 30 min for discussion and to classify actions into three time frames (as discussed earlier). The groups classified all community-prioritized actions into three time frames based upon nature of action, need of action, and availability of resources with community. Thus, outcomes of this exercise classified all community-prioritized actions equally into three time frames.

Key Findings of Developing Stakeholders' Priorities for Risk Reduction Actions The actions prioritized in the short term were P1A3, P2A3, P3A2, P4A3, S1A1, S2A2, S3A1, S4A2, S5A2, S6A1, S7A2, S8A1, S9A3, S10A2, S11A2, S12A1, S13A1, E1A3, E2A3, E3A1, I1A3, N1A3, N2A2, N3A1, and N4A2 (see Appendix); the actions prioritized in the medium term were P1A1, P2A1, P3A3, P4A2, S1A1, S2A3, S3A2, S4A3, S5A3, S6A3, S7A3, S8A2, S9A2, S10A3, S11A1, S12A2, S13A2, E1A2, E2A2, E3A2, I1A1, N1A2, N2A1, N3A2, and N4A1 (see Appendix), and the actions prioritized in the long term were P1A2, P2A2, P3A1, P4A1, S1A3, S2A1, S3A3, S4A1, S5A1, S6A2, S7A1, S8A3, S9A1, S10A1, S11A3, S12A3, S13A3, E1A1, E2A1, E3A3, I1A2, N1A1, N2A3, N3A3, and N4A3 (see Chap. 14 Appendix). From the classification, it was noted that actions that were new and rarely implemented in the past by the community had been classified into medium- or long-term actions, which imply that community requires more knowledge, resources, guidance, and information before they can be implemented.

Since the household priorities for risk reduction actions were difficult to capture in the workshop due to limited space and time, the study also carried out a household questionnaire survey to obtain the perception of households about the community-prioritized actions by RWAs and to understand households' classification priorities. The criteria selected to classify actions in different frames were based upon the need and urgency of specific action. In addition, the questionnaire survey also captured the disaster experience of households in the recent past.

Key Findings of Developing Household Priorities for Risk Reduction Actions In order to collect data, three sets of questionnaire were developed for three different types of settlements: planned DDA housing and societies, colonies, and urban villages in East Delhi. In total 578 household samples (approx. 40.65% female) were collected from all three settlements. Using the data collected from questionnaire survey, the results were analyzed. In all three settlements, the actions prioritized were S1A1, S2A1, and S8A2 (as short term); P2A2, S3A1, S9A1, and I1A1 (as medium term); and P1A2 and I1A1 (as long term). The classification at the household level reflects the urgent needs of three settlements, which can also be understood as planning and management issues, such as rising unemployment, demand and supply gaps of civic services, infrastructure pressure, etc. For example, task to reduce unemployment among the youth was prioritized by 60% household in urban villages, which reflects the unemployment issue. Based upon the needs, in DDA housing, most actions prioritized were classified into medium and long term. In the colony, most actions prioritized were classified into short and medium term. Finally, in urban villages also, most actions prioritized were classified into short-term frame. Moreover, household survey results revealed that most of the respondents experienced disasters in the past are from colonies and urban villages, and they had not taken any measures or actions to reduce their disaster risk. The findings from the household survey also supported the CDRI results particularly of East Delhi, which says that physical, social, and natural dimensions are key factors responsible for low overall resilience.

14.5 Discussion

The section discusses linkages between the studies on prioritizing community actions for disaster risk in Delhi: a participatory approach and disaster risk governance. They provide evidence of some factors that contributes toward enhancing disaster risk governance. For instance, the CDRI study and its findings mobilize the creation of communication of strengths and weakness of the current system and promote appropriate risk culture, which is one of the critical factors for successful DRR as well as for enhancing disaster risk governance. The CDRI findings about the institutional resilience pinpoints to the urgent need for future investment in mainstreaming DRR and CCA by the government (see Sect. 14.4.1; Fig. 14.5). This implies designing and implementing integrated risk management strategies and setting priorities, including resource allocation for mainstreaming DRR and CCA. Moreover, CDRI findings also provide evidence-based information, which can be very useful for policy decision and actions (Fig. 14.4). All of these are important for enhancing disaster risk governance.

The process to prioritize community actions was conducted through a participatory approach, which involved various actors from different agencies such as community, government authorities, NGOs, and other civil society organizations (see Sect. 14.4.3; Table 14.3). Similarly, validation of prioritized actions through interviews with other stakeholders promoted the partnership approach for implementing the prioritized actions (Fig. 14.7). The CAP process brings together most of the actors who have a stake in risk issue, and an interest in risk reduction, and commitment in the implementation of actions that robust in the long term. In addition, developing stakeholders' priorities for risk reduction actions (see Sect. 14.4.3) engages several stakeholders (Table 14.3) in order to develop time frames for implementing the prioritized actions for DRR. Such kind of engagement between public authorities, nongovernmental actors, civil society, and private sector contributes toward strengthening disaster risk governance. Thus, the process of CAP in Delhi contributes in strengthening engagement and partnership, which are also very crucial for disaster risk governance.

The study also contributes toward disaster risk governance by involving and engaging diverse local communities in prioritizing actions for DRR. For example, the household survey targeted 538 households in all three settlements (see Sect. 14.4.3), which included male, female, old, young, adult, migrants, etc. It implies that all actors who have a stake in disaster risk and its management support in developing the time frames for prioritized actions by the RWAs. In addition, the households also identified their issues, which are more related to the urban planning and management, which means that such issues are the priorities of the local communities. In other words, such priorities help government authorities in better understanding the local context and resource allocation. Inclusiveness of diverse local communities and setting priorities for resource allocation are important for disaster risk governance.

Finally, the study also links the whole process of prioritizing community actions for DRR with DDMA as shown in Fig. 14.8. For example, in the pre-disaster phase, one of the roles of DDMA is to develop risk assessment and emergency plans that focus on disaster preparedness and mitigation. The authority can utilize the CDRI results to understand the situation of each district as well as areas of improvement, such as community preparedness during disasters, effectiveness of crisis management framework, knowledge dissemination, and management, among others. Similarly, DDMA should utilize prioritized actions by RWAs and associated stakeholders while developing the capacity building programs, which imply that the programs would be more targeted toward the need of the community. The main reason for developing these linkages is to promote sharing of information through proper communication between different actors. It may trigger changes in institutional and individual behaviors. Moreover, effective communication is vital in the risk governance (Figs. 14.1 and 14.8).

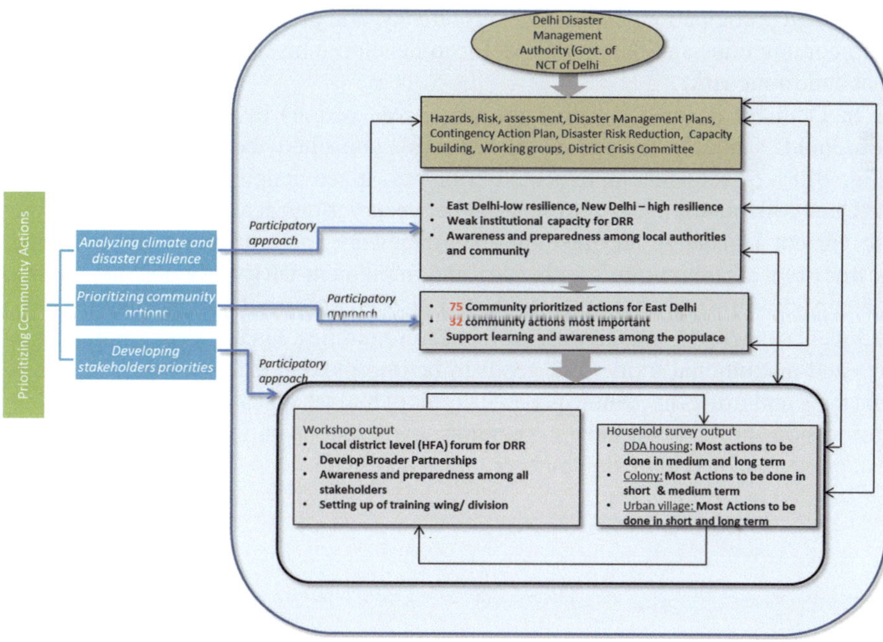

Fig. 14.8 Conceptual framework showing the output of prioritizing community actions and linkages (applicable to the study area)

14.6 Way Forward

The approach developed in the study seeks to enhance the disaster resilience of the community through a combination of bottom-up and top-down approach, where the community can become effective partners of local government by forming a broader partnership and mutual learning as shown in Fig. 14.8. Thus, coordination and cooperation are very important not only across different actors but also across sectors, ministries, departments, and all levels of government, as well as across borders, for effective disaster risk governance. The study also stresses on the notion that communities, citizen, and civil society organization are the key actors and have a core contribution to make to reduce existing and future disaster risk.

In order for the community to make a core contribution toward DRR in the future, the core challenges of disaster risk governance need to be addressed. As discussed in Sect. 14.2.1.2, the challenges are more related to the implementation. For instance, regulatory mechanism, framework, accountability, compliance measures, and incentives are still weak in several low- and middle-income countries (see Sect. 14.2.1.2) and require a robust review. In order to have robust review, it will be necessary to strengthen cooperation with civil society, the private sector, and other key stakeholders, review legal and regulatory frameworks for DRR, measure the efficacy disaster risk management, promote accountability, expand outreach to communities most at risk, and promote more inclusive and engaged approach to build resilience (UNISDR 2015b). Governance approach that promotes more innovation and participation, facilitates transparency, and promotes data availability can help communities and the private sector to develop innovative solution to the current and future risk.

In Delhi, it is important that the prioritized actions by the stakeholders and households should be communicated and consulted before implementation (Fig. 14.8). Successful implementation of prioritized actions would require more efficient collaboration and coordination between various actors due to the fact that the current legislative and institutional framework in Delhi gives no or limited power to local communities to develop and implement DRR activities. In addition, the capacity of local governments should also be enhanced so that they have the technical capacity to engage with the local communities. Finally, the current legislative and institutional framework needs to be improved to make it more relevant so that local authorities have more power in terms of budget allocation, decision-making, and responsibility to promote community participation in the process of disaster risk prevention and reduction on every level and scale.

Appendix: CAP Tool for Prioritizing Community Actions for Disaster Risk Reduction in Delhi

Resilience dimensions	Tasks	Community-prioritized actions
Physical	P1. Task for reducing the interruption in electricity	A1. Promoting use of generator/inverter/ emergency light, etc.
		A2. Promoting use of low energy/eco-friendly appliances
		A3. Registering complaint with Delhi Electricity Board
	P2. Tasks for reducing the interruption in water	A1. Promoting use of backup like tube wells, hand pumps, purchasing water
		A2. Promoting to reduce water consumption
		A3. Registering complaint with Delhi Jal Board (water leakage, theft, etc.)
	P3. Tasks to improve the collection of solid waste	A1. Promoting to incorporate 3R (reduce, reuse, and recycle)
		A2. Promoting community collective actions for waste collection
		A3. Promoting creation and maintenance of waste collection point
	P4. Tasks to improve treatment of solid waste before dumping	A1. Promoting practices of waste segregation at household level
		A2. Promoting "say no to plastic bags"
		A3. To appoint outsider for waste segregation

Resilience dimensions	Tasks	Community-prioritized actions
Social	S1. Tasks to reduce number of people suffer from waterborne diseases every year	A1. Promoting community awareness on waterborne diseases
		A2. Community drills
		A3. Promoting community to monitor sanitation condition
	S2. Tasks to reduce number of people suffer from vector-borne diseases every year	A1. Promoting community awareness on vector-borne diseases
		A2. Community drills
		A3. Promoting community to monitor sanitation condition
	S3. Tasks to reduce number of people suffer from waterborne diseases after a disaster	A1. Promoting people to take preventive measures after disasters
		A2. To provide people with safe drinking water
		A3. Monitoring by community groups
	S4. Tasks to improve literacy rate	A1. Promoting "each one teach one"
		A2. To organize literacy campaign
		A3. Promoting informal education within the community
	S5. Task to improve the awareness and knowledge about the threat and impact of disaster	A1. Organizing training program for awareness and knowledge building
		A2. Collaborating with external stakeholders (Delhi Disaster Management Authority, NGOs, etc.)
		A3. Motivating people to participate in the awareness building program
	S6. Task to improve people's participation in community activities	A1. Creating awareness and interest for people's participation
		A2. Forming different groups (women, youth groups, environment groups, etc.)
		A3. Providing incentives and awards for best participation
	S7. Task to improve the acceptance level of community leader	A1. Promoting democratic election
		A2. Promoting transparency
		A3. Promoting monitoring

Resilience dimensions	Tasks	Community-prioritized actions
Social	S8. Tasks to improve consensus building within the community	A1. Having regular interaction within the community
		A2. Promoting common goal or interest in the community
		A3. Participating in the decision-making process
	S9. Tasks to improve community participation in the district decision-making process	A1. Promoting people to build consensus
		A2. Joint ownerships
		A3. Protests and vigilance
	S10. Task to improve the interlinkage between different ethnic groups	A1. Promoting people to participate in all festivals
		A2. Promoting people to get on well together in their local area
		A3. Promoting people to participate in community activities
	S11. Tasks to improve the preparedness of households for disasters in terms of logistics, materials, and management	A1. Promoting people to participate in disaster education program at local level
		A2. Promoting people to develop disaster management plan
		A3. Community drills
	S12. Task for providing shelter and support for affected people after disaster	A1. Promoting people to create community evacuation or relief center
		A2. Collaboration with disaster management authority for support
		A3. Community drills
	S13. Task to improve participation of people in relief work after disasters	A1. Promoting people to participate voluntarily
		A2. Promoting people to form relief group in case of disasters
		A3. Community drills
Economic	E1. Task to reduce the unemployment among the youth	A1. Providing skills and training program to youth
		A2. Community them self-posting jobs for employment opportunities
		A3. Youth cooperative
	E2. Task to increase women employment	A1. Training to women
		A2. Microcredit
		A3. Women themselves posting jobs for employment opportunities
	E3. Task to reduce child labor	A1. Encouraging parents to send their children to school
		A2. Enhancing community awareness
		A3. Child homes

Resilience dimensions	Tasks	Community-prioritized actions
Institutional	I1. Tasks to improve community participation in development plans	A1. Participating in need assessment
		A2. Participating in decision-making
		A3. Participating in implementation and evaluation
Natural	N1. Task to improve the quality of urban biodiversity	A1. Promoting people to create wildlife gardens
		A2. Promoting people to use gardens in wildlife friendly ways
		A3. Partnership schemes with external stakeholders like the Ministry of Environment and Forests, NGOs, etc.
	N2. Tasks to improve urban air quality	A1. Promoting people to get their vehicles pollution check done regularly
		A2. Sensitizing people about the implication of bursting crackers during certain festivals
		A3. Partnership schemes with external stakeholders like the Ministry of Environment and Forests, NGOs, etc.
	N3. Task to improve water quality in rivers	A1. Participating in the cleanup activities
		A2. Monitoring sanitation condition at household level
		A3. Partnership schemes with external stakeholders like the Ministry of Environment and Forests, NGOs, etc.
	N4. Task to improve loss of urban green space	A1. Planting trees
		A2. Involving in park services
		A3. Partnership schemes with external stakeholders like the Ministry of Environment and Forests, NGOs, etc.

References

Ahrens J (2002) Governance and economic development – a comparative institutional approach. Edward Elgar Publications, Cheltenham

Ahrens J, Rudolph PM (2006) The importance of governance in risk reduction and disaster management. J Conting Crisis Manag 14(4):207–220. doi:10.1111/j.1468-5973.2006.00497.x

Arnstein SR (1969) A ladder of citizen participation. J Am Inst Plann 35(4). doi:10.1080/01944366908977225

Aysan Y, Lavell A (2014) Disaster risk governance during the HFA implementation period. Background paper. Prepared for the Global Assessment Report on Disaster Risk Reduction 2015. UNDP. Available http://www.wcdrr.org/wcdrr-data/uploads/847/UNDP%20Thematic%20Review%20_%20Disaster%20Risk%20Governance%20during%20the%20HFA%20implementation%20Period%20(Aysan%20and%20Lavell,%202014).pdf. Assessed 15 Apr 2016

Bhatt M, Gupta M, Sharma A (1999) Action planning from theory to practice. Doc Centre Urban Plann 24(3):16–23

Boyer-Villemaire BUJ, Cooper JAG, Bernatchez P (2014) Analysis of power distribution and participation in sustainable natural hazard risk governance: a call for active participation. Environ Hazards-Hum Policy Dimens 13(1):38–57. doi:10.1080/17477891.2013.864592

Brown, C (2012) The 21st century urban disasters. Issue Brief. Available http://www.globalcommunities.org/urbandisasters. Accessed 15 Apr 2016

Brunner R, Steelman TA, Coe-Juell L, Cromley C, Edwards C, Tucker D (2005) Adaptive governance: integrating science, policy, and decision making. Columbia University Press, New York

Census of India (2001) Population projections for India and states 2001–2026. Office of the Registrar General & Census Commissioner, New Delhi

Chakrabarti PD (2007) How rise of middle class activism in Indian cities is changing the face of local governance. Massachusetts Institute of Technology

Cosens BA (2013) Legitimacy, adaptation, and resilience in ecosystem management. Ecol Soc 18(1). doi:10.5751/es-05093-180103

Coskun A (2013) The expansion of accountability framework and the contribution of supreme audit institutions. Input paper prepared for the 2015 global assessment report on disaster risk reduction. UNISDR, Geneva

Dahiya B (2012) Cities in Asia, 2012: demographics, economics, poverty, environment and governance. Cities 29:S44–S61. doi:10.1016/j.cities.2012.06.013

DDMA (2009) Chapter III: hazard and risk assessment. In Delhi disaster management draft plan. Delhi Disaster Management Authority. Available http://www.delhi.gov.in/wps/wcm/connect/doit_dm/draft/plan. Accessed 28 May 2010

DDMA (n.d.) Draft: Delhi Disaster Management Plan 2014–2015. Revenue Department. Government of NCT of Delhi. Available http://www.delhi.gov.in/wps/wcm/connect/8d935d004829d5ec8becebae1ccb7dea/Delhi+Draft+State+Plan+09042015+Part+I.pdf?MOD=AJPERES&lmod=-300925450. Accessed 15 Apr 2016

Djalante R (2012) Adaptive governance and resilience: the role of multi stakeholder platforms in disaster risk reduction. Nat Hazards Earth Syst Sci 12(9):2923–2942. doi:10.5194/nhess-12-2923-2012

Djalante R, Holley C, Thomalla F (2011) Adaptive governance and managing resilience to natural hazards. Int J Disaster Risk Sci 2(4):1–14. doi:10.1007/s13753-011-0015-6

Frederickson HG, Smith KB (2003) The public administration theory primer. Westview Press, Boulder

Gall M, Cutter SL, Nguyen K (2014) Governance in disaster risk management (IRDR AIRDR Publication no. 3). Integrated Research on Disaster Risk. Beijing

Government of NCT of Delhi (2008) Guidelines and procedure for seeking registration of a society for Residents and Welfare Association (RWAs) under the societies registration act, 1860. Office of the Commissioner of Industries, Delhi

Govt. of NCT of Delhi (n.d.) Economic survey of Delhi: 2014–2015. Planning Department. Available http://delhi.gov.in/wps/wcm/connect/DoIT_Planning/planning/economic+survey+of+dehli/economic+survey+of+delhi+2014+-+2015. Accessed 15 Apr 2016

Hamdi N, Goethert R (1997) Action planning for cities: a guide to community practice. Wiley, England

Hemmati M (2001) Multi-stakeholder processes for governance and sustainability: beyond deadlock and conflict. Earthscan Publications Ltd, London

Highfield WE, Brody SD (2013) Evaluating the effectiveness of local mitigation activities in reducing flood losses. Natural Hazards Review 14(4):229–236. doi:10.1061/(ASCE)NH.1527-6996.0000114

Holley C, Gunningham N, Shearing C (2011) The new environmental governance. Routledge, New York

IPCC (2012) Managing the risks of extreme events and disasters to advance climate change adaptation, A Special Report of Working Groups I and II of the Intergovernmental Panel on Climate Change. Cambridge University Press, New York

Joerin J, Shaw R (2011) Mapping climate and disaster resilience in cities. In: Shaw R, Sharma A (eds) Climate and disaster resilience in cities, Community, environment and disaster risk management, vol 6. Basingstoke, Emerald Group, pp 47–61
Joerin J, Shaw R, Takeuchi U, Krishnamurthy R (2012) Action-oriented resilience assessment of communities in Chennai, India. Environmental Hazards. doi:10.1080/17477891.2012.689248
Mehta S (2009) Economic survey of Delhi 2008–2009. Government of NCT of Delhi. Current Advertising Pvt. Ltd, New Delhi, p 328
Ministry of Law and Justice (2005) The disaster management act, 2005. Available http://www.ndma.gov.in/images/ndma-pdf/DM_act2005.pdf. Accessed 15 Apr 2016
Nolte IM, Boenigk S (2011) Public-nonprofit partnership performance in a disaster context: the case of Haiti. Public Adm 89(4):1385–1402. doi:10.1111/j.1467-9299.2011.01950.x
Osborne D, Gaebler T (1993) Reinventing government. Addison-Wesley, Reading
Ostrom E (2010) Beyond markets and states: polycentric governance of complex economic systems. Am Econ Rev 100(3):641–672
Pahl-Wostl C (2009) A conceptual framework for analyzing adaptive capacity and multi-level learning processes in resource governance regimes. Glob Environ Chang 19(3):354–365
Parvin AG, Joerin J, Prashar S, Shaw R (2011) Climate and disaster resilience mapping at micro-level of cities. In: Shaw R, Sharma A (eds) Climate and disaster resilience in cities, Community, environment and disaster risk management, vol 6. Basingstoke, Emerald Group, pp 103–127
Pelling M (2011) Urban governance and disaster risk reduction in the Caribbean: the experiences of Oxfam GB. Environ Urban 23(2):383–400. doi:10.1177/0956247811410012
Picciotto R (1997) Putting institutional economics to work: from participation to governance. In: Clague C (ed) Institutions and economic development: growth and governance in less-developed and post-socialist countries. John Hopkins University Press, Baltimore, pp 343–367
Planning Department (2001) Delhi urban environment and infrastructure improvement project (DUIIP) Delhi – 21. Ministry of Environment and Forests, Govt. of India
Prashar S (2012) Prioritizing community actions of climate related disaster resilience: a participatory approach in Delhi, India. Kyoto University, Japan
Prashar S, Sharma A, Shaw R (2011) From action planning to community-based adaptation. In: Shaw R, Sharma A (eds) Climate and disaster resilience in cities, Community, environment and disaster risk management, vol 6. Basingstoke, Emerald Group, pp 163–182
Prashar S, Shaw R, Takeuchi Y (2012) Community action planning in East Delhi: a participatory approach to build urban disaster resilience. Mitigation and adaptation strategies for global change. Springer Science. doi:10.1007/s11027-012-9368-4
Renn O (2008) Risk governance: coping with uncertainty in a complex world. Earthscan, UK
Samaratunge R, Coghill K, Herath HMA (2012) Governance in Sri Lanka: lessons from post-tsunami rebuilding. South Asia-J South Asian Stud 35(2):381–407. doi:10.1080/00856401.2012.662713
Shaw R, Goda K (2004) From disaster to sustainable civil society: the Kobe experience. Disasters 28(1):16–40. doi:10.1111/j.0361-3666.2004.00241.x
Tompkins EL, Lemos MC, Boyd E (2008) A less disastrous disaster: managing response to climate-driven hazards in the Cayman Islands and NE Brazil. Glob Environ Chang- Human Policy Dimens 18(4):736–745. doi:10.1016/j.gloenvcha.2008.07.010
UNDP (1997) Governance for sustainable human development. UNDP policy document, New York
UNDP (2013) Issue brief: disaster risk governance.
UN-ESCAP (2016) Disasters without borders: regional resilience for sustainable development. Asia-Pacific Disaster Report 2015. Available http://www.unescap.org/resources/asia-pacific-disaster-report-2015. Accessed 15 Apr 2016
UNISDR (2005) World conference on disaster reduction, 18–2 January 2005, Kobe, Hyogo, Japan. United Nations, Geneva
UNISDR (2009) Progress review of national platforms for DRR in the Asia and Pacific region. Geneva. Available http://www.unisdr.org/files/12082_ProgressReviewofNationalPlatformsfo.pdf. Accessed 15 Apr 2016

UNISDR (2011) Global assessment report on disaster risk reduction: revealing risk, redefining development. Geneva: UNISDR. Available http://www.preventionweb.net/english/hyogo/gar/2011/en/home/download.html. Accessed 15 Apr 2016

UNISDR (2015a) Making development sustainable: the future of disaster risk management. Global assessment report on disaster risk reduction. United Nations Office for Disaster Risk Reduction (UNISDR), Geneva. Available http://www.preventionweb.net/english/hyogo/gar/2015/en/home/download.html. Accessed 15 Apr 2016

UNISDR (2015b) Working Session 1 Governance and development planning at national/local levels (HFA priority 1). Issue Brief. UN World Conference on Disaster Risk Reduction 2015, Sendai Japan. Available http://www.wcdrr.org/wcdrr-data/uploads/847/OFFICIAL%20ISSUE%20BRIEF%20-%20WS1%20Governance%20and%20Development%20Planning.pdf. Accessed 15 Apr 2016

Walker G, Whittle R, Medd W, Watson N (2010) Risk governance and natural hazards, CapHaz-Net WP2 Report. Lancaster Environment Centre, University, Lancaster. (available at: http://caphaz-net.org/outcomes-results/CapHazNet_WP2_Risk-Governance.pdf). Accessed 15 Apr 2016

Wilkinson E (2012) Why 'small is beautiful' in municipal disaster risk reduction: evidence from the Yucatan peninsula, Mexico. Environ Hazards-Human Policy Dimens 11(2):155–171. doi:10.1080/17477891.2011.609878

Zanotti L (2010) Cacophonies of aid, failed state building and NGOs in Haiti: setting the stage for disaster, envisioning the future. Third World Q 31(5):755–771. doi:10.1080/01436597.2010.503567

Chapter 15
Impact of Natural Disasters on Schools in India: Current Scenario in Terms of Policies and Practices

Saswati Paik

Abstract In India, almost 85% of its geographical area is vulnerable to one or multiple natural disasters. Twenty-two states in India have multi-hazard zones. When a disaster strikes, schools as integral part of the community play an important role by sheltering, building and enhancing the resilience and cohesiveness amongst community. Disasters usually result in trauma amongst children in addition to the financial and personal loss to the community. The academic sessions run as per schedule; however, children affected by natural disasters are not able to cope up with the academics leading to exclusion from the school system. Unfortunately, in India, so far the schools located in natural disaster-prone areas are not mapped properly. As children suffer, the challenges faced by them are never addressed by current education system and its policies. This paper has focused on the current status of schools in India located in the natural disaster-prone areas with special focus on the policy discourses to address the issues associated with such schools. It will also make an attempt to highlight few best practices where the effects and losses caused by natural disasters are mitigated by means of 'vulnerability reduction'.

Keywords Vulnerability • Disaster mitigation • Disaster risk reduction • Schools • Community resilience

15.1 Introduction

Hazards are common phenomena across the world and give rise to disasters. As per the definition of United Nations International Strategy for Disaster Risk Reduction (UNISDR), hazards are 'a dangerous phenomenon, substance, human activity or condition that may cause loss of life, injury or other health impacts, property damage, loss of livelihoods and services, social and economic disruption, or

S. Paik (✉)
Faculty, School of Education, Azim Premji University, Bengaluru, India
e-mail: saswati@apu.edu.in

I. Pal, R. Shaw (eds.), *Disaster Risk Governance in India and Cross Cutting Issues*, Disaster Risk Reduction, DOI 10.1007/978-981-10-3310-0_15

environmental damage' (UNISDR 2009). UNISDR defines disaster as 'a serious disruption of the functioning of a community or a society involving widespread human, material, economic or environmental losses and impacts, which exceeds the ability of the affected community or society to cope using its own resources' (UNISDR 2009). Disasters are therefore caused by exposure to any kind of hazard, exposure to the conditions of vulnerability which are present and insufficient capacity to cope with the potential negative consequences. Disasters can be natural or man-made; disasters may occur for a few seconds to a few days depending on the type of it, but all disasters disrupt the community in various ways.

15.2 Disaster Risk Reduction: Major International Initiatives

Across the world, Disaster Risk Reduction (DRR) is considered to be crucial and important. It is also widely recognised that there is a huge significance of knowledge, perception, comprehension and actions for reducing disaster risks in any country.

The 1990s was declared as International Decade of Natural Disaster Reduction (IDNDR). Since then, numerous efforts have been made at various levels to ensure public education on disaster risk reduction. It has been recognised that 'public awareness' and 'education for disaster risk reduction' are both equally important for disaster risk reduction. Later the Hyogo Framework for Action (HFA) 2005–2015 also emphasised on 'knowledge and education'. This framework highlighted 'formal and non-formal education and awareness raising as important components for disaster risk reduction' (Shaw et al. 2011). Both the UN World Disaster Reduction Campaign (2000) and UN/ISDR Campaign on 'Disaster Risk Reduction Begins at School' made an attempt to highlight the importance of integrating disaster risk reduction into formal education and also community participation as a sustainable attempt of disaster risk reduction.

The Sendai Framework for Disaster Risk Reduction 2015–2030 was adopted at the Third UN World Conference in Sendai, Japan, in March 2015. The framework mentioned about four priority areas which are as follows:

- Priority 1: Understanding disaster risk
- Priority 2: Strengthening disaster risk governance to manage disaster risk
- Priority 3: Investing in disaster risk reduction for resilience
- Priority 4: Enhancing disaster preparedness for effective response and to 'Build Back Better' in recovery, rehabilitation and reconstruction (UNISDR 2015).

15.3 Natural Disasters in India

In India, almost 85% of its geographical area is vulnerable to one or multiple natural disasters. Twenty-two states in India have multi-hazard zones. The Ministry of Home Affairs, Government of India, in its report published in 2011 mentioned that 'India due to its geo-climatic and socio-economic condition is prone to various disasters' (Ministry of Home Affairs, Government of India, 2011). As mentioned in the document on National Policy on Disaster Management 2009, 58.6% of the landmass in India is prone to earthquake of moderate to very high intensity, around 12% of land is prone to floods and river erosion, 68% of cultivable areas are drought prone, around 5700 km of coastline (out of 7516 km in total) is prone to cyclones and tsunamis and the hilly areas are prone to avalanches and landslides. These figures indicate that considerable parts of India are prone to one or multiple natural disasters; it is more so because most of these areas are densely populated either due to high fertility of land or due to growing urbanisation. Apart from these, there are other hazards of chemical, biological, radiological and nuclear origin, and there are multiple hazards caused by human conflicts, accidents, etc. which also become disastrous at various points of time; however, those are not being discussed in this chapter.

15.3.1 Regional Trends and Characters of Natural Disasters

The states and union territories in the northeastern part and northern India are frequently affected by earthquakes, landslides, floods and flash floods. On the other hand, the central, southern and eastern parts of India are prone to droughts. Floods occur almost everywhere except for a very few states and UTs (Table 15.1). Data generated by the international agencies like Centre for Research on Epidemiology of Disasters (CRED) show that every year India's multiple regions are affected by different types of disasters causing loss of human life, damage to property and affecting livelihood. It has been observed that if some states of north India suffer from heat waves during summer, in the same year, few areas of the same region along with some others get inundated during the rainy seasons, and both the heat waves and floods have caused substantial damage to property and human lives.

15.3.2 Extent of Damages Due to Disasters: Current Scenario and Future Predictions

During the last 30 years' time span, India has been hit by 431 major disasters resulting into enormous loss to life and property (Ministry of Home Affairs, GoI 2011). The report published by the Ministry of Home Affairs has shown year-wise damage

Table 15.1 Types of disasters that frequently hit various states and UTs in India

Regions	Types of disasters States/UTs	Cyclone	Drought	Earthquake	Flood	Flash flood	Landslides	Tsunami
Northeast	Arunachal Pradesh			v	v		v	
	Assam			v	v	v	v	
	Manipur			v			v	
	Meghalaya			v	v	v	v	
	Mizoram			v	v		v	
	Nagaland			v	v		v	
	Sikkim			v	v		v	
	Tripura			v			v	
East India	Andaman and Nicobar Islands				v			v
	Bihar	v	v		v			
	Jharkhand		v		v			
	Odisha	v	v		v			v
	West Bengal	v	v		v			v
North India	Chandigarh			v				
	Delhi		v	v				
	Haryana		v	v				
	Himachal Pradesh			v	v	v	v	
	Jammu and Kashmir			v	v	v	v	
	Punjab				v			
	Uttarakhand			v	v	v	v	
	Uttar Pradesh		v	v	v		v	
Western part	Dadra and Nagar Haveli	v			v			
	Daman and Diu				v			
	Gujarat	v		v				
	Rajasthan		v	v				
Central India	Chhattisgarh		v	v				
	Madhya Pradesh		v		v	v		
	Maharashtra	v	v	v		v		

(continued)

Table 15.1 (continued)

Regions	Types of disasters States/UTs	Cyclone	Drought	Earthquake	Flood	Flash flood	Landslides	Tsunami
South India	Andhra Pradesh + Telangana	v	v		v			v
	Goa	v						
	Lakshadweep	v			v			
	Karnataka		v		v			
	Kerala	v	v		v			
	Puducherry	v			v			
	Tamil Nadu	v			v			

'v' indicates the type of disaster frequently hitting the state/UT
This table is prepared on the basis of the data and maps published by the National Disaster Management Authority (NDMA) in India

Table 15.2 Year-wise damage caused due to floods, cyclonic storms, landslides, etc. during the last 10 years in India

Year	Lives lost human (in no)	Cattle lost (in no)	Houses damaged (in no)	Cropped area affected (in lakh hectares)
2001–2002	834	21,269	3,46,700	18.72
2002–2003	898	3729	4,62,700	21.00
2003–2004	1992	25,393	6,82,209	31.98
2004–2005	1995	12,389	16,03,300	32.53
2005–2006	2698	1,10,997	21,20,012	35.52
2006–2007	2402	4,55,619	19,34,680	70.87
2007–2008	3764	1,19,218	35,27,041	85.13
2008–2009	3405	53,833	16,46,905	35.56
2009–2010	1677	1,28,452	13,59,726	47.13
2010–2011	2310	48,778	13,38,619	46.25

Source: Ministry of Home Affairs, GoI 2011

caused due to floods, cyclonic storms, landslides, etc. during the last 10 years in India (Table 15.2).

Considering the trends of urbanisation causing congestion in settlement and also the extent of population growth on the one hand and global climate change on the other, it is quite obvious that the impact of such disasters would be worse in the near future. A report prepared for the World Bank by the Potsdam Institute for Climate Impact Research and Climate Analytics has established that 'there is a 40 percent chance of warming exceeding 4 °C by 2100 and a 10 percent chance of it exceeding 5 °C in the same period' (The World Bank 2013, p.15). The report also has stated that 'With global mean warming approaching 4 °C, an increase in intra-seasonal variability in the Indian summer monsoon precipitation of approximately 10% is projected. Large uncertainty, however, remains about the fundamental behavior of the Indian summer monsoon under global warming' (The World Bank 2013, p. 22). The same report also indicates the possibility of increase in dry spell in the eastern part of India, depletion of groundwater level and more frequent droughts in various parts of India which are already drought prone, especially the western parts. At the same time, there might be increase in river run off causing floods in the Gangetic plains. All these predictions are quite alarming.

15.4 Impact of Natural Disasters on Education and Schooling

Natural hazards in vulnerable areas result into disasters, causing huge loss of life, property and resources. Developing countries are more vulnerable to such losses due to their demographic patterns and trends and limitations in terms of various development parameters. To link between disaster and development, often the word 'disaster mitigation' is used. Mitigation measures 'are geared to minimise the detrimental impact of disaster upon life, property and economy' (Modh 2010, p.215). Various agencies and communities are also expected to ensure that their other development initiatives include few components that enable disaster mitigation.

Disaster mitigation measures are of two types: (1) structural measures and (2) nonstructural measures. As described by Modh, structural mitigation includes the measures that reduce the social and economic impacts of hazard agents and also the construction programmes such as construction of dams, windbreaks, terracing, hazard-resistant buildings, etc. On the other hand, nonstructural mitigation refers to mainly policy measures, measures in practices such as disaster forecasting, crop diversification, zoning, land use planning, etc. Apart from these, nonstructural mitigation also includes education, awareness generation, environmental understanding, community organisation, empowerment strategies, etc.

15.4.1 Factors Associated with Type and Intensity of Impact

In many of the developing countries including India, impacts of natural disasters on school system, schooling process and schoolchildren are not focused adequately either in academic researches or in the policy documents. Therefore, the disaster mitigation process is mainly limited within the structural measures, and nonstructural measures are rarely focused in an integrated manner. As a result, in many countries, disaster preparedness suffers and countries learn their lessons through experience. Here is an example:

After Hurricane Katrina, the destructive Atlantic tropical cyclone of 2005, the first elementary school in New Orleans reopened after 3 months. In 2010, the earthquake in Haiti destroyed 4000 schools; the schools slowly began to open in the capital after 3 months. But in Japan, following the massive 8.9 magnitude earthquake in 2011, where schools were physically destroyed and several lives of teachers and children lost, classes commenced in disaster-proof and multi-hazard-resilient buildings just after a week (Greubel et al. 2012). The preparedness for disasters makes this difference. Such preparedness can reduce the impact of natural crisis on schools and children studying there.

Schools, schooling system and schoolchildren are most vulnerable to many natural disasters that leave a drastic toll on the human lives, property, development parameters and many other invisible aspects such as psychosocial aspects. The impact on schools, schooling and schoolchildren can be direct or indirect.

The direct impacts are as follows:

1. School building may collapse due to a disaster causing deaths and damage to building that may paralyse the schooling system for several days, or even months, depending on the scale of damage.
2. The access road to the school may get washed away due to flood or flash flood or may get damaged due to a landslide. In such cases, schooling process cannot start until normalcy is restored.

Indirect impacts are as follows:

1. School building as the strongest building in the locality may be used as rehabilitation centre for the victims of disasters during a disaster and post-disaster period. Hence, schooling process may temporarily get disrupted.
2. Schoolchildren may be located little far from the school, and many of them may not be able to access their school due to the bad road condition after the disaster. This may impact some students, and they will have to miss the schooling process for a period of time which might be unpredictable in many cases.
3. After disasters, many families face economic crisis which may lead to children's dropout from school. School dropout is also caused by family's migration from one place to another as a consequence of disaster, especially in case of droughts. Such indirect impacts add on to other socio-economic problems such as child labour, child marriage, child trafficking, etc.

15.4.2 Scale of Impact on Schools and Schoolchildren in India: Evidences from Recent Disasters

In India, the Right to Education (RTE) Act considers children of 6–14 years under the umbrella of 'free and compulsory education'. The latest census of India shows that around 19.29% of the total population belong to 6–14 age group. If we consider 5–18 as the schooling age (from 1st grade to 12th grade), it is around 28% of the total population in India; the community associated with that population is also not negligible. According to the State Report Card of 2014–2015, there are in total more than 1.4 million schools in India, and the total enrolment as per 2014–2015 District Information System for Education (DISE) data is 197,666,909 (Ministry of Human Resource Development, Government of India 2015). These figures indicate to what extent children, schools and surrounding community get affected by natural disasters.

In 2013, 600 villages in Uttarakhand state faced severe damage; schools were either shut down or used as rehabilitation camps in many areas affected by the flood (Pragya 2013). Few school buildings also got completely damaged during this disaster. Some villages were completely abandoned after the flash flood due to the extent of damage in those areas. In the same year, 5825 school buildings were damaged in Odisha due to cyclone Phailin (NIDM 2014). In 2014, around 4484 villages of four districts in coastal Andhra Pradesh were affected by the cyclone 'Hudhud'; it also damaged 41,269 houses, and 455 buildings including 317 primary and secondary schools, before finally crossing over the state (The Times of India 2014). Numerous schools remained closed in Srinagar, Kashmir, for over a month after the flood in 2014, considering the safety and security of the children (The Indian Express, 2014). In 2015, around 1.90 lakh people of 553 villages were affected by the devastating flood in Assam (The Hindustan Times 2015), and numerous schools remained closed for more than a month in Chennai City during and after the flood in December 2015. In 2016, more than 18 lakh people in 22 districts of Assam were affected by flood; schools were closed in this state as well as in some parts of other states such as Bihar, Madhya Pradesh, Uttar Pradesh and Uttarakhand due to the flood. According to some experts, unprecedented floods are recorded in 2016 in the Gangetic plains which affected millions of people (BBC 2016). In 2015–2016, drought has affected more than 330 million people in more than 2.5 lakh villages of 266 districts from 11 states in India (Singh and Pilla 2016).

15.5 Current Policies with Reference to Education and Schooling in India

The Sendai Framework emphasised on developing effective global and regional campaigns as instruments for public awareness and education, building on the existing ones such as the 'One million safe schools and hospitals' initiative, the 'Making

Cities Resilient: My city is getting ready' campaign, the United Nations Sasakawa Award for Disaster Risk Reduction and the annual United Nations International Day for Disaster Reduction. It also recognised the importance of:

- Promoting a culture of disaster prevention, resilience and responsible citizenship,
- Generating understanding of disaster risk,
- Supporting mutual learning and sharing experiences,
- Encouraging public and private stakeholders to actively engage in such initiatives and to develop new ones at the local, national, regional and global levels (UNISDR 2015).

In India, the central government enacted the Disaster Management Act, 2005, on 23 December 2005. This act envisaged the creation of the National Disaster Management Authority (NDMA), headed by the prime minister, State Disaster Management Authorities (SDMAs) headed by the chief ministers, and District Disaster Management Authorities (DDMAs) headed by the district collector or district magistrate or deputy commissioner as the case may be, to spearhead and adopt a holistic and integrated approach to disaster management. Under this Act of 2005, the National Institute for Disaster Management (NIDM) has been given the responsibility of carrying research works. NIDM is also expected to develop a mechanism to capture the data on disasters including its impact on socio-economic life of the nation and its GDP growth. There are mainly three sources of collecting data at the global level: (i) EM-DAT collected by the Centre for Research on the Epidemiology of Disasters (CRED), (ii) nat cat maintained by Munich Reinsurance Company and (iii) Sigma maintained by Swiss Reinsurance Company (Ministry of Home Affairs, GoI 2011).

Recent major developments in respect of government initiatives towards managing disasters for and in schools include the following:

- Regular training programmes conducted by NIDM for administrative functionaries,
- Involvement of NIDM in input mechanism for policy formulation for disaster management,
- Launch of the National School Safety Plan (NSSP) by NDMA,
- Introducing the subject on disaster management in Social Science Curriculum of Central Board of Secondary Education (CBSE) (started in 2003 for eighth standard students),
- A proposal of introducing an optional paper on disaster management for undergraduate students in 2012 by the University Grants Commission (UGC).

Many of these initiatives are not directly linked with the disaster mitigation for schools, by the schools and within the schools. These are also far from achieving the goal of 'promoting a culture of disaster prevention, resilience and responsible citizenship' as proposed by the Sendai Framework.

15.5.1 Current Practices to Address the Issues of Schools in India: Good Practices from Uttarakhand State

Uttarakhand is a state in north India. This state was carved out of the hilly terrain of Uttar Pradesh state. About 86% of the 53,483 km^2 area is hilly region and about 65% is covered by forest (Government of Uttarakhand 2015). This state is located in a seismically active zone, which is also prone to floods, flash floods, landslides and cloudbursts (Das 2013).

Uttarkashi District of Uttarakhand was worse affected in 2013 during the flash flood of Rudraprayag. In this district (Uttarkashi), there are six administrative blocks (sub-district units) named as Bhatwari, Purola, Naugaon, Mori, Dunda and Chinyalisaur. One of the worst affected blocks during the flash flood was Bhatwari. A study on nine public and private schools in Bhatwari Block conducted by few students of Azim Premji University, Bengaluru (India), revealed that despite being located in such vulnerable area, there is hardly any map of the village and/or block level which details out the locations of schools in that locality.

The schools in this district are categorised as A, B, C, D E and F, depending on their access to the main road; availability of transport facilities, health facilities, post office, educational facilities, commercial centres, telecommunication and general public amenities; and altitude above sea level. As per this categorisation, the schools belonging to 'A' category are considered to have the most advantageous location. The advantages in terms of all parameters sequentially decrease from B to F categories. All the schools surveyed during this study were under 'A' category. Even in these schools, which are located in most advantageous locations, there is no structured plan for disaster mitigation. Part of the issue is associated with the regular livelihood of local people who are accustomed with the geo-climatic situation.

There are many schools located far from the main road (Figs. 15.1 and 15.2). The villages are remotely located and most of the villages have single teacher schools, and many teachers posted in challenging location hardly get transferred to better locations. As a result, they do not get a chance to attend any professional development training during their service period, and finally it results into getting demotivated in their job. Researchers hardly reach them to understand the challenges and issues associated with such schools highly vulnerable to disasters with high possibility of dropout during or after elementary level schooling.

On the other hand, it has been observed that this state has taken some special measures to ensure continuity of education of children in schools during the natural crisis period especially in the areas suffering from chronic natural issues such as snowfall for a few months. There are a few people in the communities in Uttarakhand who stay at the higher altitude areas which experience snowfall for 2–3 months, come downhill and stay in another locality where their children also continue their education in local schools. This is a regular arrangement for these areas. One more systemic change brought here is that the schools located in higher altitude, suffering from extreme weather in the winter, follow little different academic calendar. Their winter break is longer and summer break is shorter to balance with the weather condition (Paik 2015).

Fig. 15.1 A reconstructed school in Uttarkashi, which was washed away during a flash flood in 2013

Fig. 15.2 The main approach road to a school in Uttarkashi

A few more good practices were also identified during this study which are mentioned below:

- *Active participation of DDMO in development and liaison:* The District Disaster Management Office (DDMO) in Uttarkashi district acts as advisory to many departments on providing technical assistance. It also plays host to fortnightly and monthly meetings between various government departments which come together to discuss and update their disaster management plans. DDMO has already finished reconstruction of houses affected during the 2013 flood and has now started assessment of public buildings which also include schools. DDMO acts as liaison agents to the international funding agencies and looks over the disbursal of compensation funds at the ground level (Vishala 2016).
- *Maintenance of the main road:* The roads are usually levelled and repaired using dozers. To ensure the timeliness of this process, the dozers are typically stationed at every 3–4 km^2 on the main road. Ropeways are installed till the bridge is constructed for easy access to and from villages where access roads got damaged during the flood.
- *Active NGOs:* NGOs in Uttarkashi are very active. They have formed a network and work collaboratively. This has reduced duplication of work and has increased their reach to the community. Some of these NGOs include Azim Premji Foundation (APF), Reliance Foundation and Shri Bhuvneshwari Mahila Ashram (SBMA). Volunteers are provided with training from Azim Premji Foundation (APF) to fill in for teachers during times of need. NGOs also contribute to the following areas in this locality:

  - Distributing pamphlets for communities and schoolchildren regarding disaster management for mass awareness generation across villages and schools,
  - Helping the community by providing alternate livelihood options through expert consultation,
  - Helping community revive traditional methods of farming, co-operative buying and selling of raw materials and produce,
  - Fully/partly funding reconstruction of houses, roads and bridges which are damaged due to natural hazards using low-cost and resilient structures,
  - Encouraging entrepreneurship in traditional craft, cuisine and herbiculture,
  - Conducting health campaigns stressing the need for clean and sanitised environment in villages through street plays and placards,
  - Conducting training in village communities and schools regarding disaster management, preparedness, mitigation and response,
  - Demonstrating the importance of safety kits at home, office and school and its maintenance (Vishala 2016).

- *Contribution of mountaineering institute:* The Nehru Institute of Mountaineering (NIM), which is a renowned mountaineering training institute, charts out paths for new roads. It also trains youth volunteers from villages in search and relief operations so that they can act as first responders during emergency.

15.5.2 Some Major Concerns Related to Policies and Practice

The figures mentioned in Sect. 15.3 indicate the importance of having a structured database on the number of schools located in the areas vulnerable to one or multiple natural hazards and database (regularly updated and monitored) on the children enrolled in each of these schools. Also Sect. 15.5.1 reveals some facts from Uttarkashi district (Uttarakhand) that signifies the use of such database. But unfortunately, there is no such structured database available in this state and probably not in any state or UT in India which suffer from natural hazards regularly. There are also multiple other factors that aggravate the magnitude of the issues associated with the vulnerability of schools and schoolchildren. Some of those factors are discussed here to arrive at a possible constructive way forward.

15.5.2.1 Lack of Local Level Map to Show Location of Schools

The Ministry of Home Affairs has accepted that mechanism for preparing a database for different kinds of disasters happening across the country is yet to be developed (The Ministry of Home Affairs 2011). Most of the database available currently are state- or district-level data. Block- and village-level data are still not available which are very important in India as the geophysical environment varies a lot within same district in a state as evident from many areas, especially in hilly regions. So far, no attempt has been made in a holistic manner in India in the following areas which are most crucial:

1. Useful tool such as regional unit level detailed maps to identify and locate the schools in each state and union territory which are located in natural disaster-prone areas,
2. A regular and standard mechanism to estimate how many children study in those schools,
3. Relevant tool to measure the risks and vulnerability of the community associated with those schools,
4. Relevant training of the administrative functionaries working in such vulnerable localities to interpret, monitor and evaluate the maps and data on regular basis.

15.5.2.2 Lack of Research on Socio-economic Issues and School Dropout Issues Linked to Disasters

It is not known to the researchers how many vulnerable students from disaster-prone areas have to migrate every year, have to drop out from the school permanently, how many of them add to the total stock of child labourers and how many get trapped in child trafficking business during or after the natural disasters (Paik and Kumar 2015). Such database is very much important in Indian context given the scale, extent and frequency of natural disasters in various parts of this country.

15.5.2.3 Lack of Attention Towards Vulnerability Measurement for Each School

India is a country of diversity. Its topography, climate and demographic pattern vary across the states. Probably any other particular country's best practices may not fulfil the requirement of the entire country as the requirements in each state and UT are quite unique. Still there is a pattern of types of disasters frequently affecting different geographic regions as shown in Table 15.3. It is important to measure vulnerability of the schools located in areas which are disaster prone and where livelihood as well as children's education are frequently affected by disasters. It must be remembered that the RTE Act can only propose 'free and compulsory education', but it cannot ensure the same if such crucial aspects are continuously neglected.

15.5.2.4 Lack of Micro-level Plans for Schools

Disaster risk reduction and preparedness are gaining priority at the administrative levels; however, the initiatives at the school and community levels in India are still sporadic. One of the main reasons behind this sporadic nature is that the plans are mainly made at the state level which may not be necessarily relevant for each district and block in the same state. In fact, in most of the cases, micro-level (district, block and village level) plans are more useful rather than macro (national level)- or meso-level (state level) plans. Here is an example that would explain this argument. States with more than or nearly 1 lakh schools include Uttar Pradesh (243,014 schools), Madhya Pradesh (142,512 schools), Rajasthan (106,254 schools), Maharashtra (97,084 schools) and West Bengal (95,572 schools) (Table 15.3). Three of these states are located in central India; one is in north India and one in east India. In none of these states, a single plan at the state level can serve the purpose of the whole state as the topographical character varies a lot within these states. Even the smaller states with less number of schools such as Uttarakhand also need micro-level planning rather than meso- or macro-level planning as geophysical conditions vary a lot due to its location.

15.5.2.5 Ignorance and Negligence Towards Children's Mental Health

Disasters have multiple impacts, many of which are directly experienced by the children. Children may suffer from trauma that further impacts their holistic development in the future. In India, mental health-related issues in general and especially that of children are usually neglected. Such negligence causes severe impact on children on which research works are not adequately done, and the issues therefore remain unnoticed and unaddressed for a long time.

Table 15.3 Total number of elementary schools as per DISE data 2014-2015 in the states and union territories located in different geographic regions

Regions	Types of disasters	States/UTs	Number of schools
Northeast	Earthquake, flood, flash flood, landslide	Arunachal Pradesh	3903
		Assam	65,141
		Manipur	4858
		Meghalaya	13,175
		Mizoram	3067
		Nagaland	2963
		Sikkim	1274
		Tripura	4818
Total number of schools			99,199
East India	Cyclone, drought, flood, tsunami	Andaman and Nicobar Islands	410
		Bihar	79,196
		Jharkhand	46,773
		Odisha	68,305
		West Bengal	95,572
Total number of schools			290,256
North India	Drought, earthquake, flood, flash flood, landslide	Chandigarh	197
		Delhi	5739
		Haryana	21,791
		Himachal Pradesh	17,956
		Jammu and Kashmir	28,543
		Punjab	29,023
		Uttarakhand	23,665
		Uttar Pradesh	243,014
Total number of schools			369,928
Western part	Cyclone, drought, earthquake, flood	Dadra and Nagar Haveli	320
		Daman and Diu	120
		Gujarat	43,838
		Lakshadweep	43
		Rajasthan	106,254
Total number of schools			150,575
Central India	Cyclone, drought, earthquake, flood, flash flood	Chhattisgarh	53,299
		Madhya Pradesh	142,512
		Maharashtra	97,084
Total number of schools			292,895

(continued)

Table 15.3 (continued)

Regions	Types of disasters	States/UTs	Number of schools
South India	Cyclone, drought, flood, tsunami	Andhra Pradesh	61,915
		Goa	1478
		Karnataka	61,628
		Kerala	16,419
		Puducherry	722
		Tamil Nadu	57,153
		Telangana	43,839
Total number of schools			243,154

Source: State Report Cards, District Information System for Education (DISE) 2014–2015

15.6 Recommendations and Suggestions

While the schools are seen in India as an isolated institution, schooling does not receive enough attention during any natural disaster and even after that, and also school's role in disaster mitigation is usually perceived as mere school safety measures portrayed through some curriculum-based knowledge generation, stereotype safety measures and drills; it seems that there is utmost necessity to emphasise on building a visible interlinkage and bondage between the school and local community. Here comes the importance of community resilience that must be connected to the school community and above all to the schoolchildren.

'Resilience assumes the idea of factors that promote healthy communities that are able to sustain and rebound from the effects of a hazardous event' (Ronan and Johnston 2005). To ensure a successful framework for community resilience, schools must be considered to be a part of the community and have their place in the framework. Schools also require mechanisms to address loss of teaching-learning hours and the safety of students and teachers. Just like a community needs to be helped to get back to normalcy soon, so does a school require help to get back to normalcy. The networking and professional help that the community gets in a resilience framework is as important in case of schools (Venkiteswaran 2016).

At the policy implementation level, bottom-up approach must be adopted rather than top-down approach, because local-level knowledge is essential to work on a micro-level plan in such domains. It has been evident from various natural disaster events worldwide that people-centred approach has more potential to prepare community for any disaster and mitigate the losses to a great extent. The failures of warning during Hurricane Katrina in New Orleans and Hurricane Mitch in Central America suggest the move from technical approach to a people-centred approach to warning (Villagra'n de Leo'n 2014). To ensure community resilience to locate the schools in disaster-prone areas and plan for such schools, devolution in the policy and decision-making process is highly advisable. Without involving local community, integrating schools and school system-level functionaries with the community, it is impossible in a country like India to work for and with the schools to save schools as organisation, schooling as a process and schoolchildren as future of our nation from the long-term impact of natural disasters.

Acknowledgement A special acknowledgement to my students of Azim Premji University, Surya Venkateswaran, M.S. Vishala, Bansod Swapnil Suresh and Md Aaqib who visited Uttarkashi in Uttarakhand (India) for their 6 weeks' field internship despite knowing the challenges in this project and collected very useful data from the field. All the photographs used here are taken by Surya and Vishala. Also special thanks to Mr. Ashish Tripathi, Mr. Pramod Painuly and all colleagues of Azim Premji Foundation, Uttarkashi (Uttarakhand State, India), for their continuous engagement during the field internship of the students in Uttarkashi.

References

British Broadcasting Corporation (August 30, 2016) India Ganges floods 'break previous records'. www.bbc.com

Das PK (2013) 'The Himalayan Tsunami'- cloudburst, flash flood & death toll: a geographical postmortem. IOSR J Environ Sci Toxicol Food Technol 7(2):33–45

Government of Uttarakhand (2015) State profile. Retrieved from http://uk.gov.in/pages/display/115-state-profile

Greubel, Lauren, Xanthe Ackerman and Rebecca Winthrop (2012) Prioritizing education in the face of natural disasters. Retrieved from www.brookings.edu/blogs/up-front/posts/2012/10/31-natural-disasters-winthrop

Ministry of Home Affairs, Government of India (2011) Disaster management in India. www.undp.org/content/dam/india/docs/disaster_management_in_india.pdf

Ministry of Human Resource Development, Government of India (2015) Elementary state report cards. http://www.dise.in/src.htm

Modh S (2010) Introduction to disaster management. Mcmillan Publishers India Ltd., Delhi

National Disaster Management Authority, Government of India (n.d.-a) National policy on disaster management 2009. https://www.nidm.gov.in/PDF/policies/ndm_policy2009.pdf

National Disaster Management Authority, Government of India (n.d.-b) Some major disasters in India. https://www.ndma.gov.in/en/disaster-data-statistics.html

National Institute of Disaster Management (NIDM) (2014) India disaster report 2013. Ministry of Home Affair, Government of India, Delhi

National University of Education Planning and Administration (2014) School education in India – U-DISE flash statistics 2013–14. NUEPA, New Delhi

Paik, Saswati (2015) Mapping schools in disaster prone areas: geographers' role in policies and plans for school systems. Unpublished paper presented in NAGI Conference. University of Jammu, Jammu, India

Paik, Saswati and C. V. Raghu Kumar (2015) Natural disasters and impact on schools in India. Disaster response and management. Uttarakhand (India): Centre for Disaster Management, LBSNAA, Mussoorie. 3, 1. 30–39.

Pragya (2013) Devastating flash floods in Uttarakhand and Himachal Pradesh. https://www.pragya.org/doc/flashflood_update.pdf

Ronan KR, Johnston DM (2005) Promoting community resilience in disasters: role for schools, Youth and Families. Springer, New York

Shaw R, Takeuchi Y, Gwee QR, Shiwaku K (2011) Disaster education: an introduction. In: Shaw R, Shiwaku K, Takeuchi Y (eds) Community, environment and disaster risk management, Disaster Education, vol 7. Emerald Group Publishing Limited, Bradford, pp 1–22

Singh, Ashish, Raghu Pilla (2016) Indian drought 2015–16: Lessons to be Learnt. http://www.countercurrents.org/

The Hindustan Times (June 10, 2015) Assam flood situation grim, nearly 1.95 lakh hit. www.hindustantimes.com
The Indian Express (Oct 02, 2014) With schools shut in Srinagar, future of thousands hang in balance. indianexpress.com
The Times of India (Oct 25, 2014) Hudhud killed 46 in Andhra Pradesh, 21 lakh families hit. https://www.timesofindia.indiatimes.com
UNISDR (2009) Terminology on disaster risk reduction. United Nations International Strategy for Disaster Reduction, Geneva. *preventionweb.net/files/7817_UNISDRTerminologyEnglish.pdf*
UNISDR (2015) The Sendai framework for disaster risk reduction 2015–2030. Geneva. https://www.preventionweb.net/files/43291_sendaiframeworkfordrren.pdf
Venkiteswaran, Surya (2016) Relevance and use of community resilience framework for holistic development of disaster inflicted areas with special reference to School. Unpublished Internship report. Azim Premji University, Bengaluru, India
Villagran de León J (2014) People-centred early warning. In: López-Carresi A, Fordham M, Wisner B, Kelman I, Gaillard JC (eds) Disaster management: international lessons in risk reduction, response and recovery. Routledge, New York, pp 64–81
Vishala, M.S. (2016) Policies for schools in natural disaster prone areas: a case study from Uttar Kashi. Unpublished Internship report. Azim Premji University, Bengaluru, India
World Bank (2013) Turn down the heat: climate extremes, regional impacts, and the case for resilience, A report for the World Bank by the Potsdam Institute for Climate Impact Research and Climate Analytics. World Bank, Washington, DC

Part IV
Risk Reduction and Crosscutting Issues

Chapter 16
Enforcement of Building Construction Regulations in Urban Centers of India

Chandan Ghosh, Ranit Chatterjee, and Rajib Shaw

Abstract The urban centers in India are experiencing a rapid transformation, both controlled and uncontrolled, supported by the growing economy in the past two decades. However, this rapid urban growth has led to failure of urban authorities to regulate and implement safe building standards and practices, resulting in risk accumulation in the urban centers. Building regulation is an important tool to address the problems of uncontrolled development and increasing disaster risk. In spite of existence of various codes and guidelines (BIS codes, NDMA guidelines, and PWD/CPWD manuals), nothing has proved accountable enough to scale down the magnitude of the vulnerabilities of building stocks. In India, state governments, local bodies (urban and rural), development authorities, special and new town development agencies, etc. are directed to modify, revise, and revamp the existing building regulations, planning, and safety standards in line with the National Building Code (NBC) 2015 with due consideration for the local variations. But implementation has remained a challenge due to weak institutional and financial capacities of the urban local bodies and non-emergence of specified agencies/expertise in the country. The strengthening and capacity building of various building development and regulating agencies with adequate level of expertise for proactive responses need to be supported by the building professionals and builder's lobby. This chapter highlights the importance of compliance of built structure with national guidelines/standards/codes/manuals developed in the country. Further the chapter highlights various gaps based on four factors: (1) legal mechanism, (2) building code development, (3) implementation, and (4) training and awareness. In order to narrow the gaps, possible

C. Ghosh
GeoHazards Division, National Institute of Disaster Management, Ministry of Home Affairs, New Delhi, India
e-mail: cghosh24@gmail.com

R. Chatterjee (✉)
Graduate School of Global Environmental Studies, Kyoto University, Kyoto, Japan
e-mail: ranit13@gmail.com

R. Shaw
Graduate School of Media and Governance, Keio University, Shonan Fujisawa Campus, Fujisawa, Kanagawa, Japan
e-mail: shaw@sfc.keio.ac.jp

I. Pal, R. Shaw (eds.), *Disaster Risk Governance in India and Cross Cutting Issues*, Disaster Risk Reduction, DOI 10.1007/978-981-10-3310-0_16

solutions like expert pooling among contiguously situated urban local bodies, rapid visual surveys, and financial incentives are discussed in the later part of the chapter.

Keywords Building regulations • Urban local bodies • Urban disaster risk • Building byelaws • Building codes • Building typology • Vulnerability • India

16.1 Introduction

India is rapidly urbanizing, and by 2030, 50% of its population is projected to be living in the urban centers. In majority of the Indian cities, 20–30% of buildings are unauthorized and are constructed in complete violation of the byelaws. India's strategy to urbanize for achieving faster and more inclusive growth has put additional pressure on the existing services and infrastructure; at the same time development of new townships has been created with weak institutional capacities. Mukhopadhyay and Revi (2012) suggests in similar tone that weak planning and building regulations increase the vulnerability of the urban India to future disasters, and there is a need to strengthen urban governance and institutional capacities. The United Nation Development Program (UNDP) in the past has identified 38 Indian cities, which lie in seismic zones III, IV, and V for implementing Urban Earthquake Vulnerability Reduction Project (UEVRP), and has focused on strengthening of techno-legal regime across the cities.

India being vulnerable to natural disasters on account of its unique geo-climatic conditions, disaster management in India occupies an important place in the country's policy framework. The national disaster management framework covers a wide range of topics like institutional mechanism, disaster prevention strategy, early warning systems, disaster mitigation, preparedness, and response in which human resource development is an important aspect. In the pre-disaster phase, preparedness is a major encompassing activity to effectively anticipate, respond to, and recover from the impacts of likely hazard events or conditions (UNISDR). In the context of built environment, this will include physical protective measures, mainly engineered, and legislative measures controlling land use, urban planning, and building byelaws supported by formal institutions and budget.

Since the 1988 Bihar-Darbhanga earthquake, eight major earthquakes have resulted in over 30,000 deaths in India. The regions far away from the Himalaya and other inter-plate boundaries, which were once considered to be relatively safe from strong shaking, have also experienced several devastating earthquakes. The huge loss of life and property in the earthquake-prone areas of the country has shown that the built environment is extremely fragile and our ability to respond to these events is extremely inadequate. It has been found that the casualties were caused primarily due to the collapse of buildings that have usually no earthquake-resistant features. This emphasizes the need for strict compliance of town planning byelaws and earthquake-resistant building codes in India.

The goals of prevention are (a) to ensure that all new buildings are designed and constructed with proper engineering intervention taking due care for safety against natural hazards in both urban and rural areas so that no unsafe buildings are added to the existing inventory of unsafe buildings, (b) to ensure upgrading the safety of existing buildings in the public sector by retrofitting techniques and encourage similar actions for buildings in the private sector, and (c) to carry out timely maintenance and repair work of the built structure as and when necessary.

*The built environment is a complex, techno-socio-economic system comprising of two components, namely, physical components (*e.g., *the built environment consists of the neighborhoods, roads, buildings, food sources, and recreational facilities) and socioeconomic system (behavioral trends, byelaws, poverty).* Further resilience has been defined as a function of time and capacity (Bodin and Wiman 2004). Combining both, it can be said that resilience of built environment is the capacity of the physical infrastructure which gets back to its functional capacity with or without human intervention with a stipulated time span (McAllister 2013). In India, disaster management is a state subject. Efforts toward infrastructure development are majorly focused on the key areas of physical and social infrastructure. The effective functioning of government during any emergency lies in collaboration with other departments, stakeholders and agencies, and lifeline facility managers, which is a measure of resiliency against all impending hazards. India, while stepping up in public investment in infrastructure, has been actively engaged in involving private sector to meet the growing demand. India has advanced considerably in developing earthquake-resistant codes of practice and guidelines for constructing RCC and steel-framed buildings, brick or stone masonry buildings, and combination of clay, wood, bamboo, and thatched houses. Yet high level of earthquake risk in India's context is mostly attributed to the unplanned and ill-planned urban infrastructure development. In order to reduce vulnerability, it is important to create proper awareness about earthquake-induced damages and their structural and nonstructural mitigation measures. To make the techno-legal regime in the country enough to ensure safe construction, the Government has already drawn a road map. This includes amends in the existing laws, development control rules, and byelaws. The first step in this road map is the preparation of Model Town and Country Planning Legislation, Zoning Regulations, Development Control, and Building Regulations/Byelaws. Since 1960, India Govt. has been looking forward to establish the necessary techno-legal and techno-financial mechanisms in order to ensure that all stakeholders like owners, builders, architects, engineers, and government departments responsible for regulation and enforcement adopt earthquake-safe construction measures in all design and construction activities. Workshops and further follow-up action including capacity building exercise are the subsequent steps for ensuring adequate and effective techno-legal regime in the country.

Sendai Framework of Action for Disaster Risk Reduction brought the topic of building regulation back on focus of the national and local government by mentioning it as an important criterion for disaster risk governance. Studies on past disasters have shown that compliance to building regulations and codes reduces the economic damage and life loss considerably in the event of a disaster. Considering the fact that 50% of Indians will be living in urban centers by 2030, it is high time that cities' infrastructure and buildings be made safe. With this idea, the study looks into the current building regulations and codes in India and their implementation status. For the analysis, various building code regulations and related studies are referred, and mainly a desk review is done in order to identify the present gaps, and a road map is developed for strengthening the building code implementation.

16.2 Building Regulatory Regime

In the past decade, high-income countries with developed building control mechanisms have experienced 47% of disasters but account for 7% of the total casualties (World Bank and GFDRR report 2015). This emphasizes on the need for strong building regulation and implementation mechanism to reduce the future risk to human lives. Mostly, in low- and middle-income countries, the institutional capacity to support changes in building regulations and implementation is limited or absent. One of the main constrain is lack of information on planning and investment World Bank and GFDRR report (2015). The 1999 Turkey earthquake divulged that about 65% of apartment blocks in Istanbul and other cities had been built in violation of local building codes due to widespread corruption (Moullier 2014). On the other hand, developed countries have brought in change to their building byelaws from time to time specially after major disaster events. Figure 16.1 shows the cases of Japan and the United States and significant change in the building regulations and code as a response to disaster event. Japan developed and adopted seismic coefficient and building height restrictions in 1923, and similarly after the Santa Barbara earthquake in 1925, the United States adopted seismic design provisions into the 1927 Uniform Building Code (Ishiyama 2012; FEMA 1998).

Having said that, implementation of building codes does not always secure life as was seen in the case of East Japan Earthquake and Tsunami in 2011 where safety codes led to creation of false sense of security and supported greater investment in hazard-exposed areas (Hallegatte et al. 2010). Further, in developing and developed countries, one of the major challenges remains in retrofitting of the existing structures, as a considerable amount of investment is required from the owner. The case of Japan conforms to most of the issues.

The government of Japan has enacted various policies supporting retrofitting and certification of engineers and architects (Act for Promotion of Retrofitting, revision to Building Standard Law, Housing Quality Assurance Act). Further, financial support of 2/3 of the risk assessment cost and 1/3 of the retrofitting cost was given, and income tax benefit of 10% and low-interest loans were offered to owners who retrofitted their buildings. In 2005, public poll conducted by the Japanese Cabinet Office on seismic retrofitting showed that about 2/3 of the surveyed population consider their house vulnerable to earthquake in the next 10 years, but less than 10% conducted seismic diagnosis and/or retrofitting of their house. This suggests that there is a need to look beyond making of building codes, getting the community onboard, and understanding their perception of risk, which will govern the building safety implementation.

In order to understand the building regulations and implementation in India, a conceptual framework is developed based on the four key factors, namely, legal administrative mechanism, building code development and maintenance, training

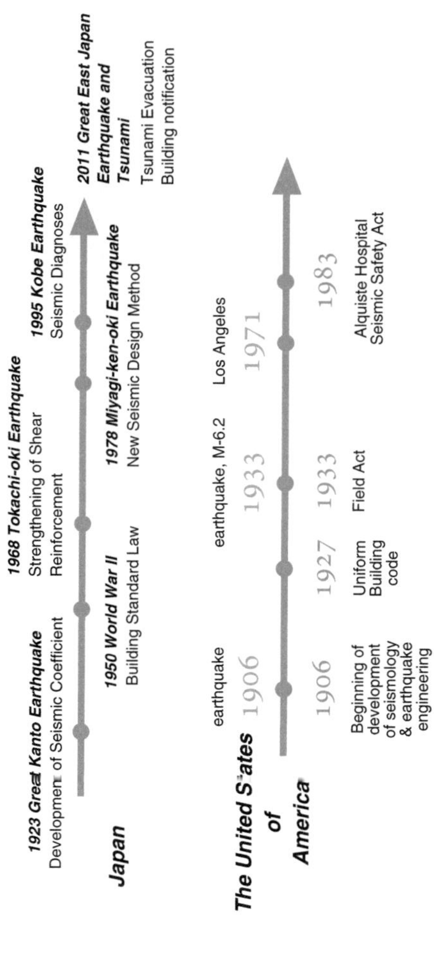

Fig. 16.1 Changes in building codes in Japan and California State, USA (Source: Ishiyama 2012; World Bank and GFDRR report 2015)

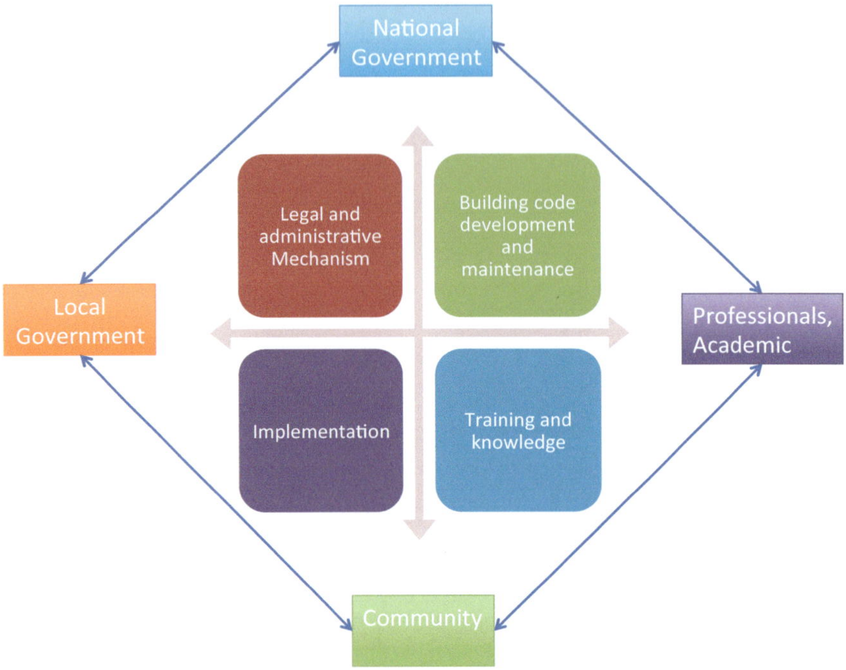

Fig. 16.2 Conceptual framework for understanding building regulations in India (Modified after World Bank and GFDRR report 2015)

and awareness, and implementation. These factors are tested across various stakeholders (national and local government, community, academic institutions, and professional associations like engineer associations, architectural associations, and others) as shown in Fig. 16.2. The national and local governments are main stakeholders in legal and administrative mechanism, while local government and community are two major stakeholders when it comes to implementation. Each of the factors is discussed in the following sections.

16.3 Legal Mechanism: Building and Disaster Guidelines/ Legislations at Various Levels in India

This section gives an overview of the various existing and upcoming guidelines and legislations available at National, State, and local body level. These serve as important entities for institutionalization and enforcement of building construction regulations.

16.3.1 Sendai Framework of Action for Disaster Risk Reduction

The Sendai Framework envisions revision of existing or the development of new building codes and standards and rehabilitation and reconstruction practices at the national or local levels for safer built environment. It also advises to establish mechanisms and incentives for strengthening compliance with the existing safety-enhancing provisions of sectoral laws and regulations, like urban planning and building codes among others.

16.3.2 South Asian Association for Regional Cooperation (SAARC)

SAARC has implemented the Earthquake Risk Reduction and Recovery Preparedness Programme (ERRRP) for five South Asian countries, of which one is India. The main objective of the project was to enhance capacities of government, institutions, and communities by adapting and implementing earthquake preparedness planning and safe construction practice. Further it emphasized on developing/modifying appropriate guidelines for earthquake-resistant building construction and skill training for construction professionals for safe construction.

16.3.3 Building and Disaster Guidelines/Legislations at Various Levels in Asia

In India, the building regulation amends, and implementation has been mainly done by the government departments at national to local level, and awareness generation and training part have been supported by the nonprofit organization and academic institutions. A few of the important regulations are discussed in this section.

16.3.3.1 National Building Code of India (NBC, 2005)

A building code is a set of norms that govern construction of buildings by stipulating minimum standards. Building codes are intended for further safety, welfare, and health of the residents of a building. In India, the guidelines for building activities are set by the National Building Code of India. The standardized norms in the National Building Code of India guide construction of most types of buildings across India. Building codes are expected to lead to safe and orderly development of buildings. The National Building Code of India was created in 1970 later revised in 1983. Three amendments were issued in 1987. It was further revised to create a comprehensive building code in 2005, and after 10 years, it is being revised in 2016. According to the building code of India, workmanship and the materials used in construction should confirm to the Bureau of Indian Standards specification. Buildings should be certified for safety against natural disasters by structural engineers. High-rise buildings and special buildings need two stage permits. Occupied buildings should periodically renew certificates regarding safety from fire, electricity, and structural issues. Development planning for hilly areas and low-income housing projects has special provisions in the code. The building code of India also promotes the usage of new and innovative materials and methods in building technology.

The National Building Code prepared by BIS is advisory in nature and has not been made mandatory so far. For the seismic strengthening of existing buildings, new codes have been prepared in 1993. Prior to seismic strengthening/retrofitting of any existing structure, evaluation of its structural vulnerability in the specified wind/seismic hazard zone needs to be carried out by a registered structural engineer. Further review of structural design by the senior structural designer is also recommended.

16.3.3.2 National Disaster Management Act 2005 and National Disaster Management Policy of 2009

The Disaster Management Act and Policy form important legislations for enforcing disaster management activities in India. The national disaster management policy puts forward a strong step toward safer built environment and proposes updating of building codes in every 5 years as a mandatory requirement and putting them in the public domain. The Bureau of Indian Standards (BIS) will be supporting in developing uniform codes and specifications with the help of professional bodies. Further, self-certification, social audit, and audit by professional agencies are envisioned to bring in strong compliance (Disaster management policy 2009).

16.3.3.3 Vulnerability Atlas of India, 2006

Under the initiative of the Ministry of Urban Development, a Vulnerability Atlas of India was prepared during the period 1994–1997 in which the earthquake, cyclone, and flood hazard maps for every State and union territory of India have been prepared to a scale of 1:2.5 million. In these maps the boundaries of the districts are clearly shown so that the areas of the districts prone to the various intensities of the hazards are clearly visible. Also the vulnerability of the buildings, as per the Census of Housing 1991, has been brought out in tabular form in one sheet for each district. The same was revised in 2006 using 2001 census data, and new version is being made using 2011 census data. This information clearly highlights the risk to buildings of various types in every district when subjected to the different intensities of the three hazards. As an extension, State-wise Vulnerability Atlases have also been prepared including an action plan that the State may adopt for achieving disaster reduction.

16.3.3.4 Building Byelaws in India

In September 2004, a report – "Proposed Amendment in Town and Country Planning Legislations, Regulations for Land Use Zoning, Additional Provisions in Development Control Regulations for Safety & Additional Provisions in Building Regulations/Byelaws for Structural Safety- in Natural Hazard Zones of India" – was prepared along with approval and the same sent to all States/UTs for adoption.

> The Core Group was of the opinion that a mere reference to the BIS code is not sufficient and as such it was considered necessary to make essential elements of the code a part of the building bye-laws. The Core Group was also of the view that byelaws should also cover non-structural aspects, which have implications from the point of view of seismic safety. It was also decided that apart from the building bye-laws State, Town and Country Planning Acts as well as Zoning Regulations would need to be reviewed so as to ensure that these codes were in conformity with the mitigation requirements.

In context of the byelaws, Kumar and Pushpalata (2015) cite the case of existing building regulations of Indian hill towns, which are derived from the National Building Code and Delhi Master Plan(s), but do not appropriately address the geo-environmental and socio-developmental local context, instead refer to Delhi as a model. Further JNNURM optional primer (2011) suggests reforms to expedite approval of JNNURM projects to include mitigation measures for natural disaster in byelaws. Example of Mysore City byelaws has been cited, "As per the Government Order No. Na Aa Ee/94/ACM/2007 dt. 12-11-2007, the Corporation has included the following items in the Building Byelaws."

1. Rainwater harvesting.
2. The facilities for handicapped people
3. Preventive measures to prevent the earthquake seismic forces in buildings
4. Mandatory utilization of solar energy

From the above, the proactive approach of the city administration is appreciable but then raises a concern what to cover under the purview of disaster management and directs one to the limited awareness level of the staff and experts working on such cross-cutting issues.

16.4 Shaping a New Future

In the recent times, new regulations amend to older codes and standards, and new projects have tried to address the issues of weak building regulations. A few of them, which have potentials to change the present scenario, are discussed in this section.

16.4.1 Model Building Byelaws for India, 2016

Originally prepared by TCPO in 2004, the new Model Building Byelaws, recently released by the Ministry of Urban Development for adoption by the entire country, is a small step toward achieving this larger overarching goal. It provides a fundamental framework to help create objective, rational, and updated provisions for building construction as also an online single-window building plan approval process.

The objective of the Byelaw is to help reduce approval time by application and issuance of online no objection certificates. Further it also aims at updating building byelaws to make them comprehensive of contemporary urban trends and makes them more inclusive and user friendly. In context of disasters, in order to reduce urban risks and have graded provisions, a risk-based matrix for different types of buildings has been introduced so that small buildings with low-risk criteria get faster approvals, whereas the high-risk buildings such as malls and multi-storied and big complexes are examined in detail. In addition, provisions for fire safety, structural safety, earthquake, and natural disaster safety have also been incorporated along with special needs of elderly, physically challenged, and children.

16.4.2 Smart City Project

The smart city project in India covering 100 cities in the first phase proposes to apply smart solutions to infrastructure and services in regions for improving their conditions. Two of the target areas of smart city are affordable housing and sustainable and safe cities. This provides a good platform to revisit the existing building code and regulations mainly at the city level and enforce new regime for a safer city in the future.

16.4.3 Rajiv Awas Yojana (RAY) and Jawaharlal Nehru National Urban Renewal Mission (JNNURM)

The RAY project is an extension of JNNURM to redress the shortfalls of the formal system, which leads to slum reduction/removal through planning for affordable housing stock targeting the urban poor and supporting policy changes. As the project focuses on housing schemes and is supported by the city government, it is a good platform for mainstreaming of DRR into developmental planning. The detailed project report (DPR) of RAY projects includes disaster management plan, risk assessment of the building and site, accessibility of relief and response services, and implementation of disaster-resistant techniques (RAY report 2013).

16.5 Discussions

The Latur earthquake in 1993 and Gujarat earthquake in 2001 lead to training of construction workers on safe construction techniques, but with no incentive to enforce implementation, the envisioned effectiveness of such trainings is not realized on field.

16.5.1 Legal Mechanism and Implementation Challenges

This involves institutions that deliver enabling legislation for the establishment and enforcement of land-use and building regulations, including dispute resolution mechanisms, insurance, and certification among others. The legislations generally involve the government offices and political parties as a representative of the citizen.

16.5.1.1 Techno-Legal and Techno-Financial Regime for Safer Buildings in India

The paradigm shift in disaster management proposes a shift from response to risk reduction and resilience-centric development, and techno-legal regime plays an important role in realizing this dream. In India, earthquake-resistant building measures were developed in 1962 as standard design criteria which later led to the development of earthquake-resistant building construction code of practice. But since these are not mandatory in nature and have no legal framework backing them, many unsafe buildings have come up in the last half century (BMTPC 2000). Disaster management is a State subject and so is building regulations as the State is competent to legislate and make laws on such subjects. But there are exceptions where the central government legislates on such subjects and parliament makes laws. These are mostly applicable to union territories and in the States such as Delhi, where land-use planning is under the central government (Ministry of Home affairs and BMTCPC report n. d.).

The techno-legal regime for buildings can be broadly classified into two types: (1) exiting buildings and (2) new constructions. Each of them can be further subclassified under usage/occupancy, material, and hazard risk. A composite of this might be essential to decide the necessary action related to repairing/retrofitting as well as supervision and architectural and structural designing of the buildings. Akhilesh et al. (2011) identify three main factors for low adherence to building byelaws: (1) Training level of most construction is low, (2) the urban local bodies have less capacity to monitor the rapid construction activities, and (3) the owners do not have awareness on building byelaws leading to poor compliance. Though there have been initiatives from the government and nonprofit organization in training and awareness generation, techno-financial issues become a constraint in implementation. Further, most of the cities have no micro hazard zonation done for multi-hazards, and hence the approach is still response centric or reactive. The engineers

Table 16.1 Various constraints for building code implementation

Stakeholder	Constrains
Government	Limited capacity of local government/agency
	Limited skill of building control officers
	Underpaid staff at local bodies
	High rate of corruptions
Community/house owners	Social/economic obstacles
	Myth of high cost if codes are followed
	Lack of awareness in public about codes
	Large ratio of self-built construction
Professionals (construction worker, builders, and site supervisors)	No professional trainings and continuing education
	Lack of skill/understanding in designers
	Petty contractors and artisan
	Not enough motivation among engineers for building code implementation

Source: Modified after Subedi (2008)

and owners both to a higher extent are unaware about the hazard risk of the site where the construction is being done. Geotechnical surveys are generally not carried out for small residential buildings, and hence the foundation design does not adhere to the need. Narafu (2008) in his findings mentioned a huge difference between engineering methods and actual construction conducted by manual laborers on site. Ando (2008) from his study on 2005 Pakistan earthquakes suggests that building with good compliance and workmanship survived the earthquake with minor damages. In addition to this, his study on 2007 Peru earthquake highlights the need for enforcement and application of the code, monitoring of its performance, the advancement of the level of understanding, and the specific preparation of design/supervision by professionals to achieve the essential objectives of the code earthquake safety. He further stresses on the need for capacity building of local governments for effective enforcement as well as guidance of citizens for the building code compliance. Subedi (2008) lists out various constraints in building code implementation across various countries based on a questionnaire survey done in 2006 on building safety regulations and status of implementation and dissemination. The constraints can be categorized under various heads (Table 16.1).

16.5.2 Implementation

It refers to the action on ground zero, which is responsible for effective reduction of hazard risks. Building regulation implementation comprises of activities such as preconstruction plan review, on-site building inspection, and permitting, maintenance, and retrofitting.

While the provisions are indeed updated and developed through a methodology of consultations with professionals across the country, the crux of the issue is that these in themselves are not binding on the state governments. Building regulation in India is a state subject, and therefore, states have to adopt these in their respective statutes. They are also free to make some changes to suit local situations, and this may even be a desirable thing to do. Therefore, the key to implementation of these Model Building Byelaws lies with the state governments. Further, the actual ground implementation has to be done by the urban local bodies, which enjoy substantial autonomy as per the 74th Constitution Amendment Act of 1992. Therefore, there is an urgent need to create awareness across the country about the need for states and local bodies to update their building byelaws. In their own interest of development of their respective states, the state governments need to see the embedded benefits of improving the building permit systems in towns and cities in their respective states as that would only help in attracting more investments, more employment opportunities, and prosperity. Further, the advantage of having a similar, if not same, system of building byelaws in all the states also helps in improving the "ease of doing business." Therefore, awareness campaigns are necessary to educate the states and municipalities about the benefits of adopting the Model Building Byelaws. In this regard, the urban development ministry has decided to hold workshops to discuss these model byelaws with various states for adoption and implementation. Compliance regulation by law is necessary. However, in order to ensure that people abide by the laws, penal provisions alone are not sufficient. One of the ways to ensure compliance is through incentives. While provisions in the Model Building Byelaws may be laudable, the ground implementation has to be radically improved. Strict enforcement and the fear of the law are absent in most urban areas. As a result, once the plan approvals are completed by way of complying with the building byelaws, during actual construction, people tend to invariably deviate and even violate. There are many who continue constructing buildings without permissions and with complete violation. Most Indian cities have around 20–30% of unauthorized colonization in which all the buildings are constructed in complete violation of the laws. Further, in urban villages, engulfed within the urban milieu, building byelaws are seldom enforced. So is the case in the inner/old city area. Therefore, the key to success is to frame byelaws specific to areas. However, the local bodies can suitably develop and include clauses to suit such special needs for specific parts of the city. Even so, the key to effective outcome of such legislation still remains at the level of enforcement and many other possible drawbacks (Table 16.2).

16.5.2.1 Socioeconomics of Building Regulation Implementation

The retrofitting of existing building stocks would require investment from the owner's end for assessment and structural changes. In India though codes and regulations have been developed, affordability still remains as a gap considering a huge slum population in majority of the urban areas in India. As seen in the case of Japan, the government supports the assessment and retrofitting expenses to an extent and offers tax benefits and lower interest rates which acts as an incentive for the owners.

The other important factor which governs the implementation of building regulations is the attitude of the citizen toward the government and the level of trust in the

Table 16.2 Building and disaster guidelines/legislations at various levels in India

Level	Legislations	Drawbacks
National	National Building Codes 2005	Are mere guidelines not mandatory by law
		Need legal framework for compliance
		Do not always translate into building byelaws and no monitoring mechanism to check its integration at local level
	Model Regional and Town Planning and Development Law, 1985 (MRTP)	Haryana, Rajasthan, and union territories of Chandigarh and Lakshadweep do not have Town and Country Planning Acts
		State governments do not integrate the guidelines in the Town and Country Planning Acts into the state level plans though UDFI 1996 suggests so
		Human resources and expertise at state level are lower leading to weak MPC and DPC
	National Disaster Management Act 2005 and Policy 2009	Do not emphasize on the implementation of building byelaws at city level
		The building code updation and revision have not been done as suggested in the policy
	Vulnerability Atlas of India, 2006	The available data set are of 2004 and updation is not regular
		The micro hazard zonation maps at city, village, and cluster level are not available which is useful for urban master planning
		Awareness among community and engineers is low
		The extension of the urban and peri-urban areas in the last 10 years is not accounted
	Reserve bank of India National Disaster Management Guidelines on Ensuring Disaster Resilient Construction of Buildings and Infrastructure	No mechanism to check the implementation of the guidelines at the bank level
		This does not consider the existing buildings and also does not ensure the proper maintenance
	Central Public Works Department Maintenance Manual 2000	The manual does not include the provision of inspecting the structural safety on the basis of seismic and other hazards' risk
	Handbook of Conservation of Heritage Buildings 2013	There is no provision made for retrofitting/repairing of the building specially heritage building as they need special attention, and many properties are under CPWD supervision in the 2013 Handbook of Conservation of Heritage Buildings

(continued)

Table 16.2 (continued)

Level	Legislations	Drawbacks
State	The Environment Protection Act, The Land Acquisition Act, Transfer of Property Act, Regional and Town Planning Act, Housing Society Act, Fire Act, Industry and Warehouse Act, Shops and Establishment Act, State Disaster Management Act, Metropolitan Region Development Authority Act	At the state level, too many legislations across various departments lead to overlaps and often lead to negation of legislations
		The enforcing agencies are the local authorities or respective agencies (fire brigade) whose expertise is limited, and there is no monitoring mechanism at the state level
Local	Development Control Regulations and Building Byelaws	The National Disaster Management Division made recommendations for changes in building byelaws in 2004–2007 but has not been incorporated in the majority of the cities
		Do not consider hazard proneness of the area
		Due to limited expertise in the city level, DCRs are modeled on the earlier available DCR of other cities hence missing out on the geophysical need of the place
		Amending the byelaw involves political process and is time consuming
		Increase level of corruption leading to weak implementation
	Development plans/master plans	Out of 7933 towns in India, only about 25% of the towns have master plans (1483 notified +549 under preparation) as per 2014

Sources: NBC (2005), CPWD Maintenance Manual (2000), DM Act 2005, Vulnerability Atlas (2006), MRTP (1985), Kshirsagar and Srinivas (2014)

government. In India, the level of trust on the local government is relatively high (76.8%) based on a working paper by Serra for OECD (2013) on social capital. A similar result is shared by Blinda (2010) where 64% of the population in India trust the political leadership. Having said that, this does not necessarily mean that citizens' actions reflect a conscious knowledge of government actions and trustworthiness. In this case this also reflects that there is a lacuna in administration and political will to reinforce the building byelaws in India.

16.5.2.2 Adherence of Building Codes in Developmental Projects

Developmental projects like RAY, smart city, and others can change the building safety implementation to a great extent by adhering to the prescribed norms. But most of them relate to the development control regulations of the local authority. Further, weak monitoring of the work and no certification process for the contractor

Table 16.3 Condition of houses built under Indira Awas Yojna in six states of India

States	Building condition		
	Good	**Livable**	**Dilapidated**
Odisha	25%	64%	11%
Gujarat	68%	30%	2%
Tamil Nadu	62%	36%	2%
Uttar Pradesh	29%	61%	9%
Uttarakhand	25%	64%	11%
Himachal Pradesh	70%	28%	2%

Source: Unnati report (2012)

and labors lead to increased risk. A study of Indira Awas Yojna by Unnati report (2012) in six states of India shows no compliance to building codes due to inadequacy of funds with the homeowner, limited understanding of the homeowner regarding issues of risk and vulnerability, as well as limited availability of trained manpower for correct execution (Table 16.3).

16.5.2.3 Building Code Development and Maintenance

In order to benchmark minimum requirements for safe construction of new buildings and repair/retrofit of existing buildings, participatory approach involving all relevant stakeholders is necessary. It would ensure regulations that represent the values and resources of the relevant community. In India, most of the code development is done at the central level by involving government institutions and academic institutions without much of community involvement.

The building code development and updates have been regular at the national level, but at the state and city levels, the integration of various overlapping codes and updation of codes and regulations hardly consider the national guidelines. Further the political will is necessary to approve the amends in the codes, and the real estate and politics nexus play an important factor in acceptance of stricter codes limiting construction works in high-risk areas.

16.5.2.4 Training and Awareness

The training and awareness generations are important factors for strengthening building regulations, implementation, and acceptance. The academic, nonprofit organizations and professional associations are key to this. Unless the community is aware of various policies and schemes of the government, has basic knowledge of the building safety, and is aware of the eminent risk, it is impossible to implement building regulations; however, well-developed codes might there be in place.

In addition, to tackle the issue of limited human resources and expertise in the local authorities and other related agencies, training of school teachers and house owners in rapid visual survey of buildings can be effective to maintain the buildings. As earthquake-resistant design has recently become a part of technical courses, training of registered architects and civil engineers through national, regional, or local associations is necessary.

16.6 Way Forward

This section proposes different actions to better the building regulation enforcement based on the present gaps identified in the earlier part of the chapter.

16.6.1 Strengthening Legal Mechanism for Buildings Regulations

Strengthening the legal mechanism is one of the priorities for mandating the construction of safe buildings and enabling the construction process to be effective in managing urban risk and prevent generation of future risks. Further this will also ensure additional financial support for the local governments.

16.6.1.1 Peri-urban Areas Need Different Set of Building Regulations from the Core City Area

Peri-urban areas are transition areas linking urban and rural but are predominantly urban in nature and have a different governance structure and capacities than the core city area. This is supported by study done by Nilsson et al. (2013) on Chinese PLUREL case study showing conflicts and the pressures in respect to peri-urban areas that can be strategically managed in different development and regulatory contexts. As the cities expand rapidly, peri-urban areas also expand at the same pace, and most of their building regulations follow that of the bigger cities. As the building typologies differ from the core area and so are the socioeconomic and institutional conditions, a customized building regulation is necessary for these areas. This can be done through capacity building of the staff of urban local bodies and monitoring by district planning committee (DPC). Academic researches can prove beneficial for identifying the local needs and also capacities of the local government.

16.6.1.2 Legal Frameworks to Support Building Regulations and Strong Monitoring for Compliance

As discussed earlier in the chapter, majority of the regulations are in the form of advisory to the state and local government which are often overlooked. A legal framework backing the guidelines will set a minimum level of compliance to them.

Further, strengthening of the state and district level planning authorities and empowering them will bring in compliance at the city level.

16.6.1.3 Revising Policy for Empanelment of Architects and Engineers

Knowledge updation is an important part of learning process, and in order to make the architects and engineers stay updated on the recent development in the field of construction and make them accountable for their actions, a gradation system can be introduced at the city level. Earlier system of empaneling of engineers and architects based on company's turnover and experience of similar works can be revised with addition of qualification and training for empanelment. As seen in the case of Japan, Kenchikushi (Architect) Law was amended to create three classifications (first-class Kenchikushi, second-class Kenchikushi, and Mokuzo Kenchikushi), which restrict the use, scale, and structure of buildings for which the Kenchikushi is qualified to offer a professional service.

16.6.2 Building Code Development and Maintenance

In building code development, updation is an important task which needs cutting-edge research to be associated with practices, thus bringing in academicians, practitioners, and legislators on the same table. The following section highlights a few areas which can be looked into in this respect.

16.6.2.1 Risk Management Approach Based on Hazard Microzonation of Urban Centers

Given the interdependencies and connectedness among critical infrastructures, a disruption of any one service could have a cascading effect across essential services or systems. A "risk management" approach to critical infrastructure refers to the continuous, proactive, and systematic process to understand, manage, and communicate risks, threats, vulnerabilities, and interdependencies across the critical infrastructure community, including the owners and operators of critical infrastructure. Having a strong situational awareness of the risks and interdependencies that confront critical infrastructure is the first step toward a comprehensive risk management process.

16.6.3 Implementation

In building byelaws, implementation is the biggest challenge mainly faced by the local government and hence leads to creation of new risks in the urban areas. The community and local government have created a closed group based on both legal and mutual trust. Close monitoring of the developmental works in the urban areas specially areas which are vulnerable to various disaster risk can reduce the risk considerably.

16.6.3.1 Financial Incentive for Retrofitting and Risk Assessment

Specially for existing structures, sharing/subsidizing retrofitting and risk assessment charges in seismic zones IV and V and cyclone-prone zones between government and community will reduce the risk considerably. The subsidization can be done through channelizing the corporate social responsibility funds of the private sector and also by using the development fund of local politicians on a case basis. Urban local bodies in various states of India have the freedom to provide incentives to individuals for rainwater harvesting and rooftop solar power generation by various mechanisms, including discounts and rebate in property tax. These measures will go a long way in ensuring that greater compliance is achieved.

16.6.3.2 Trust Building and Involving Communities in Decision-Making Process

Trust building of the professional associations and community in order to implement building regulations is important. As already seen, India has no dearth of regulations but falls short when it comes to implementations. A common problem in the city government is rampant corruption and red-tapeism. Social auditing, citizen surveys, and third-party audit are possible solutions to reduce corruption and bring the citizens on board.

16.6.3.3 Stricter Structural Checks for Loan Approvals

The Reserve Bank of India (RBI) in 2011 came up with a notification to all banks to link loan approvals to the structural design of the proposed buildings and structures. This was based on the observation that majority of the structures are not completed before submitting the application for a bank loan and no processes are in place at the banks to ensure that disaster resilience has indeed been incorporated in the assets during the design process at least before the construction begins. Along with this there has been no mechanism setup to monitor and evaluate the effectiveness and compliance of building regulations in the structure. Academic institutions and research institutions in collaboration with financial institutions can support such initiatives.

16.6.3.4 Integration of Safe Construction in Development Projects

Although the development projects undertaken by the government (RAY, Smart city) have disaster management as a part of their mandate, but on ground since most of the work for construction is entrusted to the city government and contractor, there is a need for a third-party check or citizen check done to ensure adherence of the building norms to the built structures.

16.6.4 Training and Awareness of Community

Considering that construction technology is used by many but understood in its both technical and social conjunction by a few, training and awareness become an important tool to create knowledge of safe constructions across societies.

16.6.4.1 Rapid Visual Survey of Buildings by Occupants

Considering that city government is understaffed and expertise level is low, it is necessary that occupants themselves are trained in rapid visual surveys to check if a particular building requires further evaluation for assessment of its seismic vulnerability. Doing this will ease the burden on the government officials and make the owner/occupants accountable to the building conditions. RVS can be effectively done for housing societies, schools, universities, hospitals, and other public and commercial buildings. The nonprofit organizations, academic institutions, and private sectors can come together to set up training activities in each city.

16.6.4.2 Capacity Building of Human Resources in the Building Department in State and Local Authorities

Capacity building of the existing staff through training and certification is required to bring them at the similar level of understanding. For the new staff recruitment, a minimum educational standard needs to be set. As for strengthening MPC and DPC, the training and awareness campaigns by professional associations from time to time and also through Administrative Training Institutes, private sectors, and other quasi government agencies would benefit in the long run. This will also facilitate expert pooling among contiguously situated urban local bodies and the related regulating agencies in order to overcome the human resource and budgetary constraints of smaller level local bodies dealing with building regulation work.

16.6.4.3 Technical Advisory for House Owners from the City Corporations

Technical advisory for perspective house owners from the city corporations to educate them about the safe construction techniques is the need of the hour. This service can be started by involving the local architectural and engineering associations. Further, technology park demonstrating safe construction techniques can be supported by private sector and builder lobby for educating citizens about basic structural elements. Involving the media for dissemination of safe construction practices can have wider reach.

The most important step to ensure safety in building constructions is to conform to the building codes and make it safe from possible future hazard (earthquake, cyclone, landslide, fire, and others). There is a need to prioritize for creating expertise at the local level to authorize the building codes available since 1962 in India. Non-implementation or poor enforcement of building byelaws or development control rules by responsible authorities is often discussed, but building codes in the country are the prescriptions according to the experience gathered over long periods in civil engineering and allied discipline. Following them in practice along with now debatable Engineer's Bill will not take ground until self-regulations start taking root in the mind of people toward safe building constructions.

16.7 Conclusion

Sendai Framework of Action for Disaster Risk Reduction puts disaster risk governance as a priority for the coming years at national and local level across sectors. The strengthening and enforcement of building regulations is a much-desired parameter under this especially in countries where the building sector is rapidly expanding in tandem with a fast-growing economy. As discussed in this chapter, a coordinate approach across various disciplines backed by interdisciplinary researches connecting the innovations in engineering and community perceptions and capacities is required to fill the gaps existing on ground. A country like India, which is rapidly urbanizing, needs the building regulation enforcement as a measure to sustain the growth at the same time set a good example for the other countries to tow the line in the future.

Acknowledgment The second author thankfully acknowledges the Japanese Government (Monbukagakusho) Scholarship and GSS research grant for conducting research in Graduate School of Global Environmental Studies in Kyoto University.

References

Akhilesh S, Anshu S, Rajib S (2011) Understanding urban resilience. In: Rajib S, Sharma A (eds) Climate and disaster resilience in cities. Community, environment and disaster risk management, vol 6. Emerald Publishers, London, pp 17–45

Ando S (2008) Experience of damages in recent earthquakes, from Code to Practice Challenges for Building Code Implementation and the further direction of Housing Earthquake Safety, Records and outcomes of International Symposium 2008 on Earthquake Safe Housing 28–29 November 2008, Tokyo, Japan, pp 51–60

Blinda PK (2010) In: Shabbir Cheemand G, Popovski V (eds) Building trust in government: innovations in governance reform in Asia. Tokyo, UNU Press

Building Materials & Technology Promotion Council (BMTPC) (2000) Building a new techno-legal regime for safer India. Ministry of Urban Employment & Poverty Alleviation, Government of India
Bodin P, Wiman B (2004) Resilience and other stability concepts in ecology: notes on their origin, validity, and usefulness. ESS Bull 2:33–43
Federal Emergency Management Agency (FEMA) (1998) History and principles of seismic design. Appendix A: 33–42.
Hallegatte S, Ranger N, Bhattacharya S, Bachu M, Priya S, Dhore K, Rafique F, Mathur P, Naville N, Henriet F, Patwardhan A, Narayanan K, Ghosh S, Karmakar S, Patnaik U, Abhayankar A, Pohit S, Corfee-Morlot J, Herweijer C (2010) Flood risks, climate change impacts and adaptation benefits in Mumbai: an initial assessment of socio-economic consequences of present and climate change-induced flood risks and of possible adaptation options, OECD Environment Working Papers No. 27. OECD Publishing, Paris
Ishiyama Y (2012) Earthquake damage and seismic code for buildings in Japan. Accessed online at http://ares.tu.chiba-u.jp/peru/pdf/meeting/120817/M6_Ishiyama.pdf on 8th June 2016
JNNURM Optional primer report (2011) Revision of building bylaws to streamline the approval process-optional reform. Accessed online at http://jnnurm.nic.in/wp-content/uploads/2011/01/Optional_Primer_primers.building.pdf on7th June 2016
Kshirsagar J, Srinivas R (2014) Town planning and development laws: evolution and current amendments, presentation made on 25.9.2014. Accessed online at http://spa.ac.in/writereaddata/tcpo.pdf on 7th June 2016
Kumar A, Pushplata (2015) Building regulations for hill towns of India. HBRC J 11:275–284. Accessed online at http://ac.elscdn.com/S1687404814000492/1-s2.0-S1687404814000492-main.pdf?_tid=20bd0b94-2de5-11e6-8bce-00000aab0f02&acdnat=1465437460_f53994b12c872719d7b520651115768c. Accessed on 7 June 2016
McAllister T (2013) Developing guidelines and standards for disaster resilience of the built environment: a research needs assessment, NIST Technical Note 1795. Accessed online at http://nvlpubs.nist.gov/nistpubs/TechnicalNotes/NIST.TN.1795.pdf on 18th September 2016
Ministry of Home Affairs report (2011) National progress reports on the implementation of the Hyogo Framework for Action 2011–13. Accessed online at http://www.preventionweb.net/files/28592_ind_NationalHFAprogress_2011-13.pdf on 7th June 2016
Moullier T (2014) Improving building code implementation and compliance for more resilient buildings in developing countries: considerations for policy makers. Accesed online at https://www.gfdrr.org/sites/gfdrr/files/Built%20Environment%20Project%20Description.pdf. Accessed on 17 June 2017
Mukhopadhyay P, Revi A (2012) Climate change and urbanization in India. In: Dubash NK (ed) A handbook of climate change and India: development, politics and governance. Oxford University Press, New Delhi, pp 303–316
Narafu T (2008) Bridging gap between engineering and construction. From Code to Practice Challenges for Building Code Implementation and the Further Direction of Housing Earthquake Safety, Records and outcomes of International Symposium 2008 on Earthquake Safe Housing 28–29 November 2008, Tokyo, Japan, pp 31–39
National Policy on Disaster Management (2009) Ministry of Home Affairs, Government of India. Accessed online at http://www.ndma.gov.in/images/guidelines/national-dm-policy2009.pdf on 7th June 2016
Nilsson K, Pauleit S, Bell S, Aalbers C, Nielsen TS (eds) (2013) Peri-urban futures; scenarios and models for land use change in Europe. Springer, New York
OECD (2013) "Trust in government, policy effectiveness and the governance agenda", in Government at a Glance 2013, OECD Publishing. http://dx.doi.org/10.1787/gov_glance-2013-6-en
Rajiv Awas Yojana report (RAY) (2013) Guidelines for preparation of detailed project reports 2013–2022. Accessed online at http://www.apmepma.gov.in/files/RAYGuidelinesDPR.pdf on7th June 2016

Safety, records and outcomes of international symposium 2008 on earthquake safe housing 28–29 November 2008, Tokyo, Japan, p 67–74

Smart City report (2015) Smart cities mission statement & guidelines, Government of India Ministry of Urban Development, June, 2015. Accessed online at http://smartcities.gov.in/writereaddata/smartcityguidelines.pdf on 7th June 2016

Subedi JK (2008) Framework for building code implementation; an experience of Housing Earthquake Safety Initiative (HESI). From Code to Practice Challenges for Building Code Implementation and the Further Direction of Housing Earthquake, UNCRD. Accessed online at http://www.preventionweb.net/files/10591_HESITokyoPapers.pdf on 17th June 2017

Unnati Report (2012) Disaster resilience of Indira Awaas Yojana Houses- pilot study. Accessed online at http://www.unnati.org/pdfs/indira-awaasyojana-houses.pdf

World Bank and GFDRR report (2015) Building regulation for resilience managing risks for safer cities. Accessed online at https://www.gfdrr.org/sites/default/files/publication/BRR%20report.pdf on 18 September 2016

Chapter 17
Catastrophic Insurance in South Asia: Scope in India

Subir Sen

Abstract According to the World Bank (2013), financial protection allows for accelerated resource mobilization in an emergency or pre-emergency situation, through contingency funds and credit, and a set of risk transfer and insurance instruments, which include disaster micro-insurance, agriculture insurance, private property insurance and public asset insurance. The insurance industry in South Asia is developing and is characterised by low insurance penetration and low insurance density compared to global standards. This region every year is affected by natural disasters like tropical cyclones, floods, earthquakes, etc. Post-disaster, government support and external aid mostly support the restoration and rehabilitation process. Swiss Re (2012) highlighted that innovative insurance solutions are the most cost-effective way to deal with low-probability, high-severity weather events, providing a mechanism to finance a disaster before it strikes. Prominent examples include programmes in the United States of America, Japan, France, Australia, the United Kingdom and New Zealand. However, experiences of three existing initiatives in comparatively less developed insurance markets, Turkey, the Caribbean countries and China, are important. Like economies in South Asia, these economies are highly vulnerable to climate-related natural catastrophes, in particular earthquake, hurricane/cyclones and floods respectively. Therefore, the framework adopted in these economies to mitigate disasters may aid design and formulation of catastrophic insurance for South Asia with India playing a major role.

Keywords Natural disasters • Risk reduction • Mitigation • Insurance • South Asia • India

S. Sen (✉)
Department of Humanities and Social Sciences, Indian Institute of Technology (IIT), Roorkee, India
e-mail: subirfhs@iitr.ac.in; subir.econ@gmail.com

I. Pal, R. Shaw (eds.), *Disaster Risk Governance in India and Cross Cutting Issues*, Disaster Risk Reduction, DOI 10.1007/978-981-10-3310-0_17

17.1 Introduction

The main objective of this chapter is to highlight the scope and urgency for a market-based catastrophic insurance programme complementing the disaster risk reduction (DRR) initiatives in South Asia with a special focus on scope and role for India. Market-based insurance mechanisms are accepted in both climate change adaptation (CCA) and DRR literatures as the most efficient and effective ways to transfer risks of extreme events. In the absence of insurance, often the burden of the disaster in economic and financial terms is borne by the government. There is growing evidence on how natural disasters impairs growth and that disasters' impact on GDP is 20 times higher in developing countries than in industrialised nations, and such adverse relationship poses a major threat to the sustainable development goals especially of eradicating poverty (World Bank and United Nations 2010).There are further evidences showing that developing economies allocate limited funds for DRR. For example, for every USD 1 spent on DRR, more than USD 160,000 were spent on response (Kellett and Caravani 2013). The disaster risk management framework calls for resilience building through risk identification, risk reduction, preparedness and financial protection and planning for disaster recovery. Insurance is one of the tools for financial protection in this context.

According to the Notre Dame Global Adaptation Index (2014), India and its South Asian neighbours (Bangladesh, Nepal, Pakistan and Sri Lanka) are listed among the most vulnerable economies highly exposed to natural disasters. Therefore, in the absence of proactive measures from the governments in the region, the public expenditures following natural disasters may rise to an unmanageable level. The Nepal earthquake 2015 caused about USD 2 billion losses in Nepal and a fraction of it was insured. Additional USD 800 million damages were reported in adjoining India. But again less than 1% was insured. On the contrary, New Zealand in 2011 suffered a major earthquake. The city of Christchurch was rebuilt very quickly because almost 80% of its assets were insured. There are two important reasons why this chapter explores the possibilities for India and its role in DRR activities in South Asia. First, a large proportion of global population resides in the region and the predicted exposure to natural events is high. Second, the region has almost negligible social protection to cope with natural disasters and precariously low DRR per capita. In many countries, the government acts as reinsurer for catastrophic insurance programme like the Australian Reinsurance Pool Corporation, Japan Earthquake Reinsurance Co. Ltd., Nederlandse Herverzekeringsmaatschappij voor Terrorismeschaden (NHT), Thailand Natural Catastrophe Insurance Fund, Pool Re and Flood Re in the United Kingdom, among many others. Another role that the government may play is of being a direct insurer, for example, Iceland Catastrophic Insurance, Residential Storm and Flood Coverage in Korea, Earthquake Commission in New Zealand, Sasria Limited in South Africa, TCIP in Turkey and California Earthquake Authority in the United States, to mention a few. As India is looking forward to the goals set in the National Disaster Management Plan 2016, it is timely to consider the importance of risk transfer through a feasible market-based catastrophic insurance.

In the next section we present a brief discussion on the economic consequences of natural disasters in India and South Asia. A brief observation on the current stage of development in the insurance industry is presented following which we discuss the three identified catastrophic insurance policies in Turkey, Caribbean islands and China. In the last section, we present some of the broader issues related to feasibility of a catastrophic insurance in South Asia and the role that India may play.

17.2 Catastrophic Events in South Asia and Economic Consequences

The Intergovernmental Panel on Climate Change (IPCC) (2007, 2012) forecasted that the intensity and frequency of natural disasters would increase in the coming years. There is an increase in the number of natural catastrophes in recent years affecting lives and livelihoods of people especially in developing and less developed economies (IRIN 2005). Every year, some part of South Asia is affected by natural disaster, such as cyclones, floods, droughts or even heatwave and cold wave. The worst catastrophic events in the last decade affecting in particular the economies in South Asia are presented in Table 17.1.

If we analyse the distribution of different disaster events in South Asia, in the last decade (2005–2015), (Fig. 17.1) we observe that the most frequently occurring disasters were floods. Moreover, floods not only occurred during the rainy season or the monsoon months of the year but floods were also triggered by unseasonal heavy rainfall. The Indian state of Assam, northern Bihar, eastern Orissa, West Bengal and the low-lying areas of Bangladesh are few areas which are inundated almost every year during the monsoon season. The report of the Working Group on Flood Management and Region Specific Issues for the Twelfth Plan (2011) reported that during the years 1953–2010, the floods resulted in an annual damage of more than INR 18 billion (at 2010 price level) besides the loss of precious human lives and cattle in India. The Central Water Commission (2015) revised this figure for the period from 1953 to 2013 to more than double, that is, INR 38 billion based on 2013 prices. It is also important to note that taking note of the severity, the flood management expenditure in XI Plan (2007–2012) was increased significantly by more than 159% as compared to the X Plan (2002–2007). The unseasonal rains are also harmful for the economy and in particular the agricultural sector. Prolonged and heavy unseasonal rains and hail storm, damaged wheat and other Rabi crops in many parts of India when they were ready for harvest (Chand et al. 2015).

According to Dewan (2015), the common types of floods in Bangladesh include flash floods from the overflowing hilly rivers, rain floods due to poor drainage, monsoon floods in the flood plains of major rivers and coastal floods following storm surge. About one-fifth to one-third of Bangladesh is flooded in varying degrees each year from June through September (IPCC 2012) (cited from Paul and Mahmood 2016). Although the exact average yearly flood damage estimates were not available

Table 17.1 South Asia: 25 worst natural disasters during 2006–2015

Victims	Insured loss[a]	Year	Event	Country
8960	160	2015	Earthquake Mw 7.8, avalanche on Mt. Everest, aftershocks	Nepal, India, China, Bangladesh
5748	500	2013	Floods caused by heavy rains	India
4234	–	2007	Cyclone Sidr, floods	Bangladesh, India
2248	–	2015	Heatwave	India
1980	–	2010	Floods caused by heavy rains	Pakistan
1500	–	2007	Floods caused by heavy rains	India, Bangladesh
1270	–	2015	Heatwave	Pakistan
950	–	2008	Floods caused by monsoon rains	India
678	–	2007	Floods caused by heavy rains	Bangladesh
665	–	2014	Severe monsoon floods	India, Pakistan
605	–	2006	Floods, mudslides, heavy rains	Nepal
600	–	2006	Heavy storm causing floods	India, Bangladesh
531	–	2013	Heatwave	India
520	–	2009	Floods caused by monsoon rains	India
456	–	2011	Floods caused by heavy rains	Pakistan
450	–	2006	Floods caused by heavy rains	India
399	–	2013	Earthquake Mw 7.7, aftershocks	Pakistan
399	–	2015	Earthquake Mw 7.5	Afghanistan, Pakistan, India
388	–	2013	Cold wave	India, Bangladesh, Nepal
361	–	2012	Floods after monsoon rains	Pakistan
350	407	2006	Floods caused by monsoon rains	India
340	–	2007	Cyclone Yemyin, heavy rains	Pakistan
300	–	2008	Earthquake Mw 6.4	Pakistan
300	51	2009	Floods caused by heavy rains	India
289	755	2015	Severe flash floods in Chennai	India

[a]In million USD
Source: Compiled from data released by Swiss Re (various issues)

for Bangladesh and Nepal, Mirza et al. (2001) report that the damage estimation methods in Bangladesh, India and Nepal were only based on direct losses.

The intensity of precipitation is likely to be affected by climate change with climate models predicting a statistically significant relationship between the saturation water vapour pressure in the atmosphere and surface temperature (O'Gorman 2015). Although the relationship may vary across regions and affected by other conditions, it is largely believed that changes in short-term precipitation events lead to hydrological impacts such as flash floods, erosion, landslides and debris flows (Ban et al. 2015). This implies that heavy rains have led to landslides and mudslides not only in the mountainous regions but also in hilly regions of southern India. In the coming years, incidence of flood may increase not only in India but also in Bangladesh,

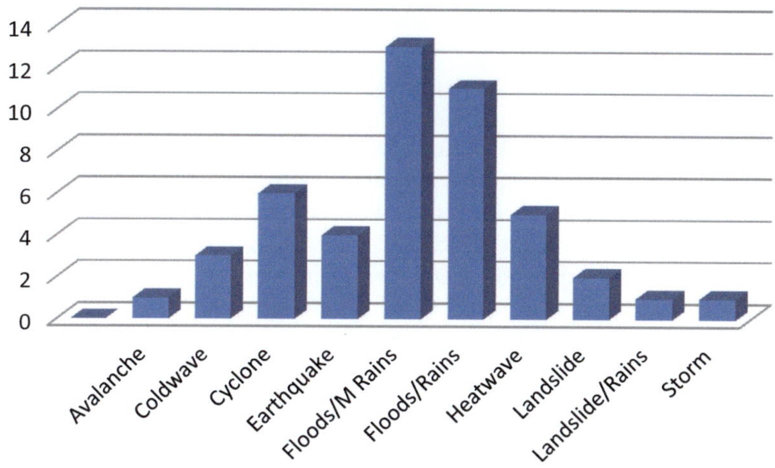

Fig. 17.1 Number of disasters events (2005–2015) (Data Source: EMDAT 2016)

Pakistan, Nepal and Sri Lanka. These hydrological events have impacted private properties and public infrastructures.

The second most frequently occurring disaster in South Asia are cyclones with India, Bangladesh and Sri Lanka being the worst affected. Together with floods, they were responsible for 90% of the weather-related damages worldwide (Kunreuther and Michel-Kerjan 2013). In recent years, the cyclones had varying wind speeds but recorded very limited mortality due to enhanced warning systems and disaster preparedness. The number of people exposed to tropical cyclones have been increasing, and by 2050, there will be 246 million city dwellers in cyclone-prone areas in South Asia (World Bank 2010). The eastern coast of India along the Bay of Bengal (the states of West Bengal, Orissa, Andhra Pradesh and Tamil Nadu) and also western parts of Maharashtra and Gujarat, areas adjoining the Arabian Sea were affected by cyclones and storms. The world's highest surges have been consistently observed in the Bay of Bengal in the northern Indian Ocean and also recorded credible storm tide levels and wind speed (Needham et al. 2015). These cyclones caused huge economic losses in Bangladesh, India and Sri Lanka. For example, cyclone Hudhud in 2014 is recorded as one of the ten costliest tropical cyclones[1] which claimed only 84 lives but inflicted an overall economic damage of USD 7000 million (Munich Re 2016). Studies have pointed out that there may be a sharp increase in the socio-economic impact of tropical cyclones in the form of increasing property damages (Raghavan and Rajesh 2003). Following hurricane Katrina in 2005, the scientific community turned their attention on understanding trends in hurricanes and tropical cyclones, to analyse their destructive potential. According to Wing et al. (2015), few studies, though inconclusive, indicate that tropical cyclone intensities have changed (Emanuel 2005; Elsner et al. 2008; Kossin et al. 2013).

[1]The tropical cyclones are selected based on the overall economic losses.

The cover page for the Munich Re publication, Topics Geo (2016), was titled *The earth's hotting up*. The world is experiencing the warmest years and the direct link with disasters is the increased incidence of heatwaves in the region. A major threat is that of forest fires in the region which causes huge economic losses. The abnormal above average daily temperatures have claimed many lives and adversely affected livelihood. In particular, the impact of heatwaves on the agricultural sector has triggered fear of droughts that may threaten food security.

In the above paragraphs, the trends and forecasts of natural disasters have been discussed. An important question is whether the impact of these disasters is uniform across different sections of the population. The impacts are often severe for those who are least prepared or adapted for the situation. According to Gonsalves and Mohan (2011), the vulnerable section of the community in Bangladesh, India and Sri Lanka resides in low-lying areas, and the three South Asian economies have large coastal communities. Large section of the population is highly dependent on coastal resources for their livelihood such as coastal fisheries. For example, the fishery sector provides employment to about 250,000 people in Sri Lanka. Disasters affecting lives and livelihood of these communities will have severe impact on the economy at large. Similar is the case with India having the longest coastline of the three countries. The impact of natural disasters in Bangladesh has been widely studied. For example, Paul and Routray (2011) studied the impacts of cyclones and sea/ocean surges on infrastructure, environment as well as on people's livelihoods and showed that indigenous coping strategies can significantly minimize the cyclone vulnerability. Ahmed et al. (2013) studied the impact of natural disaster especially on shrimp cultivation, and the study revealed that climate change has severe effects on the ecosystem of post-larvae fishing sites thereby directly affecting the livelihood of the fishing communities.

It is important to note that South Asian economies are at different levels of growth but with regard to adaptation and mitigation strategies, they are almost at similar levels. In Fig. 17.2, we compare the number of events and the number of lives affected in South Asia, Bangladesh, India, Nepal, Pakistan and Sri Lanka. The first graph shows that in South Asia although the number of natural disasters is on a rise, the number of lives affected has actually decreased. However, whether or not such decrease is due to enhanced disaster management, preparedness and rehabilitation is a question for further enquiry. The frequency and severity of disasters do clearly vary across economies. Therefore, the impact of the 2004 tsunami in India and Sri Lanka was similar, and the post-disaster rehabilitation experiences were also common in many ways. Lack of life and property insurance and lack or limited number of post-disaster shelters are some examples.

Studies on the relationship between impact of natural disasters on economy and economic growth are largely inconclusive. Rasmussen (2004) finds empirical evidence that natural disasters lead to a median reduction of 2.2% in the same year real GDP growth rate. On the similar lines, Hochrainer (2009) found that natural disasters have a negative, around 0.5%, impact on the GDP growth in 1 year time and almost −4% impact after 5 years. But Caselli and Malhotra (2004) fail to establish a negative relationship, and Cavallo and Noy (2011) highlight that many studies

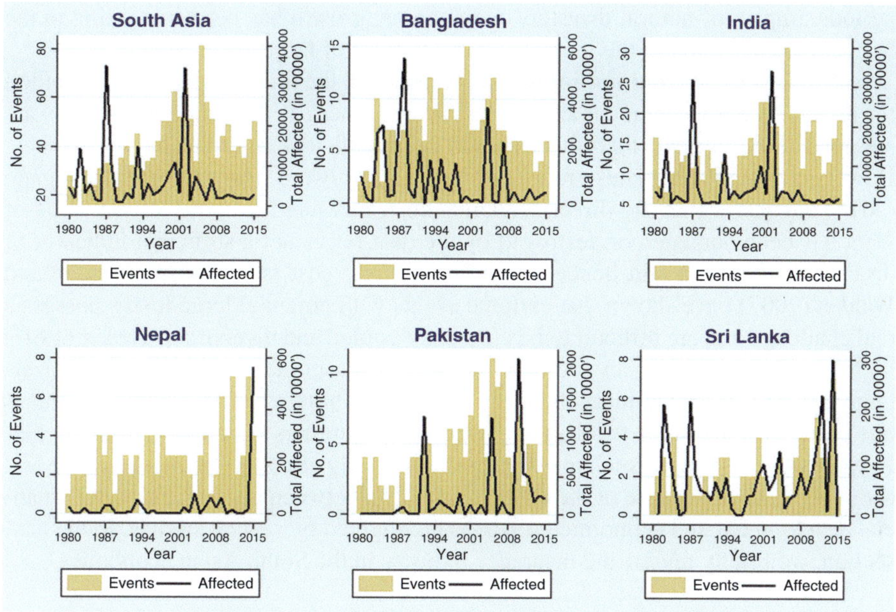

Fig. 17.2 Natural catastrophes and total affected (Data Source: EMDAT 2016)

have largely remained inconclusive. But the general perception and result suggest that the period following a natural disaster is marked by a period of instability, leading to severe macroeconomic crisis (Noy 2009) since they have a negative relationship with long-run growth (Raddatz 2007) and per capita GDP (Skidmore and Toya 2002). Loayz et al. (2012) studied the relationship between economic growth and natural disasters to conclude that the relationship is not always negative, and the effects differ substantially across disaster and economic sector. Lazzaroni and van Bergeijk (2014) looked at the macroeconomic impact of natural disasters and found that disasters have negative impact in terms of direct costs but are statistically insignificant in terms of indirect costs. They further highlight that future studies should take into account the resilience factors such as investment, education and openness for direct cost studies, population and institutions for indirect cost in order to provide more plausible explanation for a negative or linear relationship between natural disaster and economic growth.

Felbermayr and Gröschl (2014) using a database of disaster events and their intensities from primary geophysical and meteorological information found a negative relationship between natural disasters and economic output. This study, in particular, raised the issue of endogeneity in econometric analysis in earlier studies using damage data. Klomp and Valckx (2014) analysed 750 natural disaster estimates and conclude that a seemingly negative effect on growth exists especially when developing countries are observed. Although such effects vary across time periods, impact of natural disasters on economic growth has been increasing in the last four decades due to the increase in the number of large-scale events.

What does these relationships or effects mean for the financial sector intervention for disaster risk management and especially insurance? If the damage estimates are assumed to be correct, then there should be an efficient risk transfer mechanism. However, damage estimates are often not straightforward. For example, Hellegatte (2015) advocates that the direct economic cost, measured in terms of the value of what has been damaged or destroyed by the disaster, is not a sufficient indicator of disaster seriousness, and hence estimating indirect cost is crucial. Ibragimov and Walden (2007) have shown that extreme events with potential large losses possess a real challenge and are difficult to be efficiently pooled and diversified. Tucker (1997) highlighted that there is an additional cost to the insurance industry due to uncertainty associated with climate change. Many studies have advocated that insurance is one of the best measures to adapt against climate extremes (Kunreuther and Michel-Kerjan 2009). However, such mechanisms may finally affect the financial system and economy at large because of the close relationship between insurance activity, financial development and economic growth as highlighted by several studies. In the next section, we briefly review the insurance industry in the South Asian economies.

17.3 The Insurance Industry in South Asia

In 2015, the Emerging Asia[2] share in the world insurance market was 11.51% (Swiss Re 2016b). If we further consider only South Asia, the share falls below 2%[3] in 2015 indicating that the insurance industry is relatively underdeveloped. For example, there is hardly any information on insurance outreach in Bhutan. The Royal Monetary Authority reports that finance and insurance sector contributed 5.4% to the country's nominal GDP in 2014 with two insurance companies providing life and non-life insurance coverage to little over 100,000 policyholders in a country of less than a million population.[4]

Insurance industry in Nepal is also underdeveloped. Nepal reported lowest insurance penetration[5] of 1.4% in 2011–2012 (Beema Samiti 2012), compared to other South Asian economies. The total premium collected by life and non-life insurers amounted to NRs[6] 244 million in 1990 and increased to NRs 30.28 billion in 2013–2014.

[2] Excludes advanced economies in South and East Asia, namely, Hong Kong, Singapore, South Korea and Taiwan

[3] Share of India, Pakistan, Bangladesh and Sri Lanka in the world insurance industry based on total premium volume from all lines (in USD) for 2015 was 1.68 %.

[4] Population of Bhutan is 0.77 million as of 2015 (World Bank: World Development Indicators, http://data.worldbank.org/country/bhutan).

[5] Total insurance premium as a percentage of gross domestic product (GDP)

[6] Nepali rupees

Table 17.2 Insurance density and penetration (Source: Swiss Re various issues)

Density			Penetration	
	2005	2015	2005	2015
Bangladesh	2.9	9.1	0.77%	0.67%
India	25.6	54.7	3.58%	3.44%
Pakistan	4.7	11.5	0.71%	0.82%
Sri Lanka	19	43.1	1.52%	1.15%
Asia	196.9	311.7	6.82%	5.34%
World	520.0	621.2	7.47%	6.23%

Insurance density: Total premium s per capita in USD
Insurance penetration: Total premium s as % of GDP

Following liberalization of the insurance industries in Bangladesh (1990), Pakistan (1992), Nepal (1993), India (1999) and Sri Lanka (2001), the size and number of insurance companies have increased, and the industry has become more stable. In Table 17.2, the insurance density and penetration figures for selected emerging South Asian economies are compared with Asia and world figures. The data show that the insurance indicators are lower for the selected economies when compared with Asia and world figures. The indicators reported by Swiss Re (2016b) suggest that during the last decade (2005–2015), contribution of insurance to the GDP has fallen marginally. Although "insurance density" is criticized for not being the best measure to assess insurance outreach and spread because of the absence of suitable alternative, premium per capita continues to be the parameter to assess insurance consumption level.

The per capita consumption or expenditure on insurance continues to remain at lower levels in South Asia. The premium for life insurance is higher than that of non-life insurance. For example, in Bangladesh, India and Pakistan, the share of life insurance premium in total business for 2015 are 73.5%, 79% and 66.5%. For Sri Lanka, the share of life insurance in total business in 2015 was 43%. Moreover, the non-life insurance penetration is considerably lower than that of life insurance penetration. Therefore, the insurance industry has scope for increasing non-life insurance service delivery and raises general awareness for insuring property and assets in these economies. Noteworthy, the non-life insurance policies do not cover damages due to natural disasters until and otherwise mentioned in the policy statement. In India, the premium rates for property (residential or commercial dwellings) insurance policy with additional coverage for earthquake are higher compared to a traditional fire policy which does not cover indemnification for disaster-related losses. These exclusions are often due to demand-side problems attributed to adverse selection, undervaluation of risks, ignorance, etc. But given the threats of natural disasters on the financial sustainability of the insurance companies, the supply side also seems reluctant in selling such insurance policies.

According to Swiss Re (2015, 2016a), total economic losses from natural disasters have averaged around USD 180 billion annually in the last decade with 70% of that uninsured. Figure 17.3 above shows that the uninsured losses as a percent of

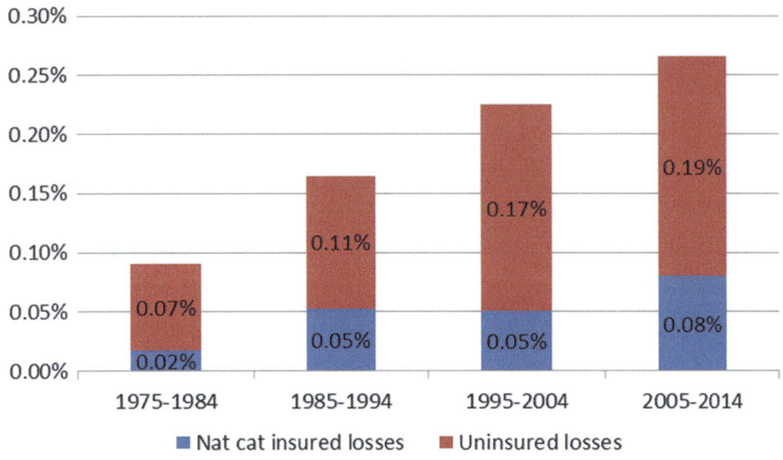

Fig. 17.3 Global insured and uninsured natural catastrophe losses as a percent of GDP, 1975–2014 (Adopted: Swiss Re 2015) [nat cat refers to natural catastrophe]

GDP during 1975–2014 has been widening. The major reason being underinsurance of property risks is largely because of individual perception of risk, individual knowledge, affordability, reliance on government post-disaster relief, limited trust in insurers, supply-side problems and delays in the claim settlement process. In recent times, large disasters had very little impact on insurers. For example, the 2015 Nepal earthquake resulted in direct economic losses in the range of USD 3 billion to USD 3.5 billion (CEDIM 2015) but the actual insured loss was USD 160 million. Similarly, Aon Benfield's Impact Forecasting report on global catastrophe indicate that rainfall in southern India and Sri Lanka during November 2015 caused an estimated USD 3 billion loss, Chennai being the worst affected urban centre. But the reported insurance claims were USD 300 million (Aon Benfield 2015). The Uttarakhand disaster which caused over 6000 reported deaths and economic losses of more than 50 billion had lesser number of insured claims in comparison to claims after monsoon flooding in Mumbai (largest in 2005) because the city has high density of non-life insurance coverage.

The financial sector performance in South Asia has improved but access to and use of insurance services remain limited or concentrated only in selected pockets. A World Bank study on financial inclusion on Nepal reported that only 26% of households have a bank account, and 45% of these households prefer to save at home, while 53% prefer to borrow from the informal sector (Ferrari et al. 2007). The study highlighted that microfinance institutions may extend insurance services to enhance density and has not discussed the problems or remedy for underinsurance in the economy. The major hurdles towards growth of the insurance industry are lack of product innovation and availability of limited products. Although insurance is a luxury good in this part of the world where eradication of poverty is a major challenge, studies on developing societies and climate change suggest that urbanization and climate adaptation strategies may foster demand for insurance. For example, the growth in the non-life insurance is fuelled by the increase in the sale of vehicles. Life insurance products are viewed similar to other competing savings instruments and at the same time, premiums paid entitle partial tax exemption, and preference for life products has increased over time. However, policy initiatives and regulatory intervention may promote non-life insurers to boost sale of "property-liability" insurance covering risks related to weather variability and extreme events. This would help building a weather-resilient economy that would help save money in the long term.

17.4 Catastrophic Insurance: Experience of Turkey, Caribbean Islands and China

According to Swiss Re (2010), innovative insurance solutions often present the most cost-effective way to deal with low-probability, high-severity weather events, providing a mechanism to finance a disaster before it strikes. The World Bank and the Global Facility for Disaster Reduction and Recovery (GFDRR) have developed a framework with an aim to guide governments in designing a practical and comprehensive disaster risk management approach. Disaster risk financing and insurance are identified as key components of the framework. Financial protection and insurance are expected to help governments' manage disasters without compromising economic growth, fiscal stability and well-being (World Bank 2014).

The insurance industry can work with governments to design and deploy new and innovative risk transfer solutions that can play their part in helping the most vulnerable societies to cope with natural disaster impacts, especially in the developing world. In this context, the catastrophe insurance/reinsurance was conceived in economies where the insurance industry had limited capacity and was unwilling to underwrite catastrophic risks (Stone 1973, studied the capacity constraint of the insurance companies in the event of catastrophic risks). According to Collier (2008), insurance coverage for catastrophe is a recent development and notes that:

> Despite some early efforts to develop actuarial frameworks for such events, attention to the problem of natural hazards risk assessment was limited well into the post-World War II period. Insurers either did not provide coverage for natural hazard risk or folded it into general property insurance

According to Zeckhauser (1995), "catastrophes provide a principal justification for insurance. One pays premiums to secure financial protection against low-probability high-consequence events-what we normally label catastrophes". Jaffee and Russell (1997) argue whether or not catastrophic risks are insurable or not and conclude that given various limiting factors, the capital generated through insurer earnings and additional capital support from the government may lead to a feasible and sustainable catastrophic insurance market. Catastrophic insurance market developed in Australia, Canada, the Caribbean, Europe (Austria, Belgium, Denmark, France, Germany, Iceland, Netherland, Norway, Romania, Spain, Switzerland and Turkey),

Japan, Mexico, New Zealand, Taiwan, the United Kingdom and the United States, in the initial years.[7] Catastrophe risk insurance facilities have also been developed for the southeastern and central European countries to pool disaster risks.[8] These initiatives were aimed at helping economies in the region to cope with climate risks. The Caribbean Catastrophe Risk Insurance Facility provides 16 Caribbean governments with short-term liquidity in the event of hurricanes and earthquakes, which pose a significant threat to communities in the region. Earthquake insurance in building resilience is popular in Japan and Turkey as well. What is important to note here is the fact that catastrophic events impact buildings and houses and therefore have long-term consequences on the community post-disaster (Lashley 2012).

The important point that emerges is that insurance is one of the mechanisms towards enhancing resilience. For example, in the context of property insurance in disaster-prone areas, it is equally important to emphasize the need to construct disaster- or climate-resilient houses. The engineering and architectural expertise would be further insured by a catastrophic insurance policy or a very specific climate-resilient housing insurance policy. Therefore, a proper planning by the government towards building resilience is essential. The Haitian catastrophe has highlighted the potential of parametric insurance to help countries plan for and prefinance natural disasters as part of a comprehensive disaster risk management strategy.

Risk transfer and risk prevention are mutually reinforcing, and insurance may be helpful in hedging unforeseen losses if and only if the insurance prices are kept in control through prevention. In Samoa, defences against storm surge waves provided dual benefits of reducing exposure to storm perils and ensuring that insurance to continue at a price providing incentives to further invest in prevention activities. Therefore, in proposing a climate-resilient housing insurance, we have to build the resilient infrastructure to ensure that the insurance is viable in the long term. Attracting private-sector financing in the housing sector would be possible in the presence of such insurance initiatives, which would further drive innovation and create sustainable habitat solutions. Extending insurance cover would therefore have particularly far-reaching benefits in the most vulnerable regions of the developing world, where resources are scarce and the potential impact of climate change fierce.

For insurers and reinsurers to deliver commercially viable solutions, public sector authorities need to put in place an appropriate enabling environment (Reichenmiller et al. 2010). The government is responsible for developing and implementing policies, design regulatory and legal frameworks within which insurers operate and compete. These frameworks are critical to how effectively and efficiently the insurance industry is able to manage its risk exposures. Regulatory frameworks, incentives and public-private collaboration are also critical to provid-

[7] A detail discussion of catastrophic insurance policies and pools is presented in Consorcio de Compensacion de Seguros (2008). Available at http://www.wfcatprogrammes.com/c/document_library/get_file?folderId=13442&name=DLFE-553.pdf

[8] As a part of the disaster risk reduction/climate change adaptation projects supported by the World Bank in Albania, Croatia and Moldova

ing vulnerable communities' access to risk management services and risk transfer products offered by insurers.[9] Therefore, a mutually dependent relationship exists between governments, society and the insurance industry. The role of World Bank in facilitating climate risk insurance has also led to successful synergy between the government, society and insurance industry. For example, Romania and Turkey have established national catastrophe risk insurance pools to protect homeowners against natural disasters; Mongolia created a livestock insurance pool to protect herders against harsh winters; and the Indian government is moving towards market-based crop insurance. The proposed climate-resilient housing insurance can therefore be based on the guiding principles of these policies.

Several studies have selected, studied and elaborated the programmes in the United States, Japan, France, Australia, the United Kingdom, Taiwan and New Zealand. However, we would be drawing from the experiences of three existing initiatives in comparatively less developed insurance markets: Turkey, the Caribbean countries and China. These economies are highly vulnerable to climate-related natural catastrophes, in particular earthquake, landslides/erosion, hurricane/cyclones and floods. Most of these programmes were initiated by external funding agencies, prominent being the World Bank and subsequently managed by local governments. The three programmes in these countries are discussed next.

About 70% population and 75% industrial facilities in Turkey are exposed to high magnitude earthquake. In order to control the fiscal burden on the government to meet post-disaster refinancing, the Turkish Emergency Flood and Earthquake Recovery Programme was undertaken, and under this initiative, the Turkish Catastrophe Insurance Pool (TCIP) was developed (Fig. 17.4). Although the target was to insure infrastructure and lives against multiple hazards, the programme was launched with coverage for only earthquake (earthquake physical damages, fire, explosions and landslides following earthquake) on 27 September 2000. This immediately led to the abolition of the government's obligation to extend credit and construction of dwellings as required under the Disaster Law (implemented since 27 March 2001). The TCIP is a legal public entity providing "compulsory" property earthquake insurance for owners of "private" dwellings built legally on registered land. This compulsory insurance is for all dwellings paying property tax in the municipal boundaries[10] and encourages physical risk mitigation and safer construction practices and aimed at reducing citizen's dependence on government fund for post-disaster restoration.

The Caribbean Catastrophe Risk Insurance Facility (CCRIF) is a risk pooling facility owned, operated and registered in the Caribbean for Caribbean governments. This facility was developed under the technical leadership of the World Bank and with a grant from the Government of Japan, among several other donor agen-

[9] For details refer:
http://www.wcdrr.org/wcdrr-data/uploads/888/OFFICIAL%20ISSUE%20BRIEF_Working%20Session%20-%20Disaster%20Risk%20Transfer%20and%20Insurance.docx.

[10] Industrial and commercial risks as well as residential buildings in small villages can be managed on a voluntary basis.

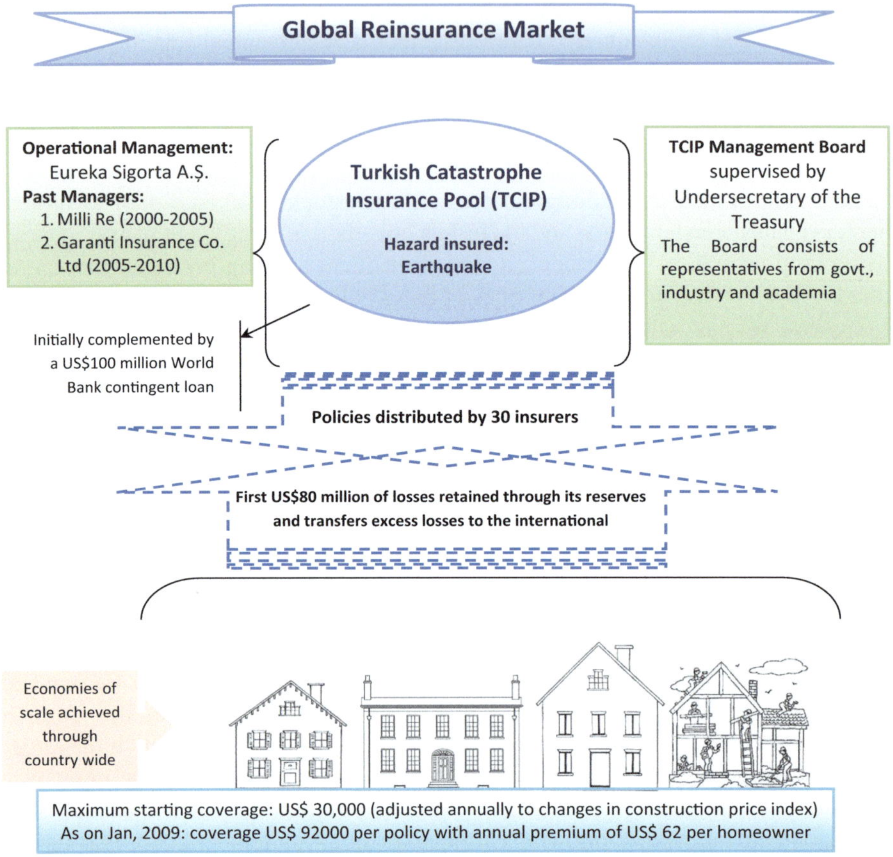

Fig. 17.4 Framework for the Turkish Catastrophe Insurance Pool (TCIP)

cies. The Caribbean island economies are severely exposed to the vagaries of hurricanes, flash floods and sea erosion. Sixteen Caribbean island countries (Anguilla, Antigua and Barbuda, Bahamas, Barbados, Belize, Bermuda, Cayman Islands, Dominica, Grenada, Haiti, Jamaica, St. Kitts and Nevis, St. Lucia, St. Vincent and the Grenadines, Trinidad and Tobago and Turks and Caicos Islands) joined hand to initiate the first of its kind and only existing *regional catastrophe insurance fund* since 2007. The facility provides coverage for tropical cyclones and earthquakes and from June 2013 provides additional coverage for climate risk and in particular excess rainfall.

The CCRIF's insurance policy is parametric in nature with payouts calculated a priori using a catastrophe risk model (multi-risk peril estimation system (MPRES)). Moreover, payouts are proportional to the estimated impact of an event on each country's budget. Through risk pooling, CCRIF provides coverage to countries at a significantly lower cost than individual governments would incur if they had to maintain their own reserves or if they were to independently purchase insurance in

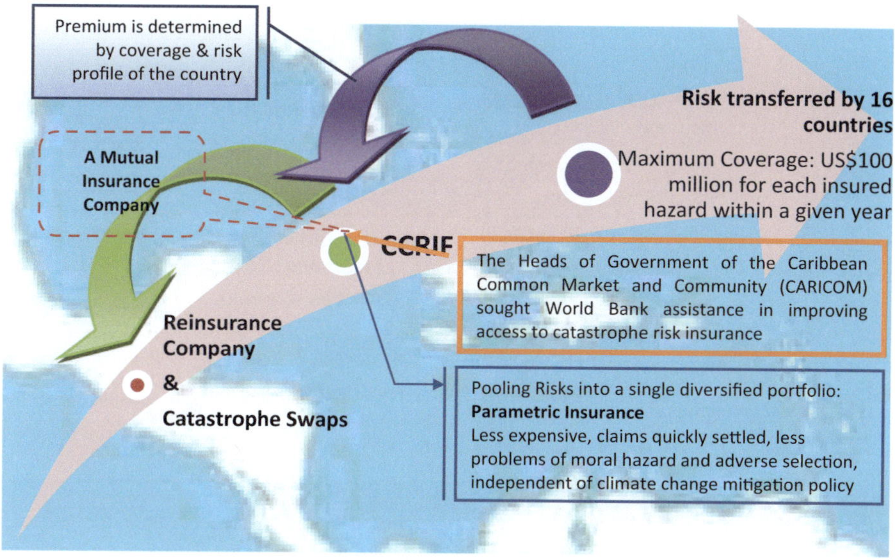

Fig. 17.5 Framework for the Caribbean Catastrophe Risk Insurance Facility (CCRIF)

the open market (Fig. 17.5). The premium is determined by the amount of coverage (maximum limit fixed at USD 100 million) a country decides to take, the attachment and exhaustion points of that policy opted by the country and the risk profile of the country. Risk mitigation practices are promoted in the island economies and a precondition before joining the facility.

Finally, our third example is the Chinese framework for catastrophic disaster risk reduction. In September 2013, the China Insurance Regulatory Commission (CIRC) approved Shenzhen as one of the pilot areas for setting up of a *catastrophe insurance pool*.[11] The area is famous for one of the most successful models of special economic zones but is under threat of natural disasters. The experiences of the PRC in post-disaster rehabilitation following major earthquakes and floods, especially in the Yunnan province, prompted a pilot project to be implemented for wider experimentation. On December 2013, the Shenzhen Catastrophe Insurance Pool was approved by the Shenzhen Municipal Government. The insurance would cover the whole of the population within the Shenzhen administrative district and cover for the following identified perils: earthquake, hurricane, tsunami, mudslide, sudden landslide, water logging, etc. as well as against nuclear accidents triggered by natural disasters (Fig. 17.6).

The three selected catastrophic insurance discussed in the above paragraphs, in short, show how such may be helpful for countries, preparing for increased climate

---

[11] There are two pilot programmes proposed: Yunnan selected for an earthquake insurance, guaranteed mainly for residential housing, and Shenzhen to test insurance against typhoon, flood and other great disasters through the comprehensive insurance for the residents to provide personal and property security.

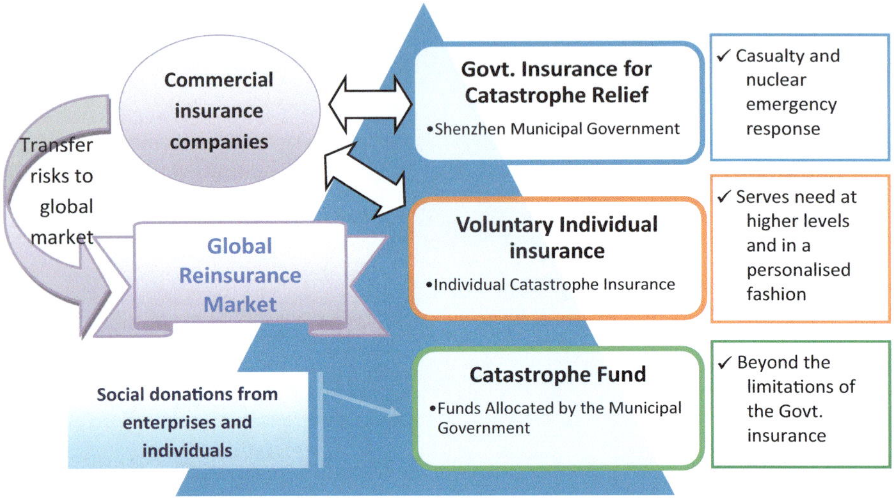

Fig. 17.6 Framework for the Shenzhen Catastrophe Insurance Pool

variability and natural disasters associated with climate change. The catastrophe insurance also highlights the role to be played by the local governments in the three countries. These disaster insurance schemes have provided significant liquid funds to the government for immediate relief and rehabilitation in the aftermath of natural disasters.

17.5 Learnings for South Asian Economies

Section 17.2 of this chapter presented the economic losses borne by the selected South Asian economies focussing especially on natural events during the past decade. Further, in Sect. 17.3, an overview of the state of the insurance industry in the selected economies was presented. The objective of this chapter was to assess the scope for catastrophic insurance in the selected South Asian economies. The reason behind such an exploration is the absence of a catastrophic insurance mechanism in these economies, as discussed under Sect. 17.4, neither at the country level nor at the regional level. It is noteworthy that almost all documents pertaining to direct assessment of losses due to future disaster events or even studies focussing on the cost of climate change adaptation identify the South Asian region as one of the vulnerable regions lacking proper facilities with respect to disaster risk financing and insurance.

In the foregoing discussion, we also referred to climate change adaptation (CCA), although the discussion so far was with regard to disaster risk reduction (DRR). There is sufficient literature to highlight the subtle differences between the two (refer to Linnerooth-Bayer and Hochrainer-Stigler 2015; Mechler and Bouwer

2015; Linnerooth-Bayer and Mechler 2007 for more details). However, it is only a matter of perspective whether or not the two can be kept in isolation, without the possibility of any overlaps. The standpoint this chapter adopts is that the boundary between the two is seemingly blurred in economies, such as those in South Asia where the need is to inculcate an atmosphere of disaster risk financing and of insurance, no matter what it targets, viz. disaster risk reduction or adaptation to climate change. Shaw et al. (2016) present a case for implementing an integrated approach that may complement DRR and CCA.

In a recent estimate, Ahmed and Suphachalasai (2014) show that the costs related to climate change would average to 1.8% of GDP in the business as a usual scenario. Moreover, if there is no action taken to adapt to climate change in the long run, the costs would escalate to about 8.8% of GDP. These forecasted costs are further going to increase public expenditure for disaster risk reduction and post-disaster relief and rehabilitation. Therefore, for fiscal stability, it is time for the government in the South Asian region to plan, design and implement either a local-level catastrophic insurance or a regional-level catastrophic pool in line with CCRIF discussed before. In a discussion paper prepared jointly by the Indian insurance regulator IRDAI (Insurance Regulatory and Development Authority of India) and the National Disaster Management Authority (NDMA), it was noted that the current level of funding for the State Disaster Relief Force (SDRF) and the National Disaster Relief Force (NDRF) was inadequate to ensure long-term risk reduction. Insurance has been identified as an alternative source of funding. The discussion paper furthermore draws lessons from international best practices, in particular the designing, implementation and monitoring of insurance schemes, and proposes that states should encourage insurance consumption behaviour, insurance for individual lives, public and private property and critical infrastructures such as school buildings, hospitals, roads, bridges, etc. (IRDA and NDMA 2013). Since the state governments in India are allocating budgetary funds for the function of SDRF along with the support from the central government, they may as well consider creation of an insurance pool. The state or the central government may either choose to perform the role of a reinsurer or subsidize the premiums for the catastrophe insurance avaiable in the market.

The Thirteenth Finance Commission in India observed that for low-frequency, high-impact disasters, financing through insurance mechanism is certainly a feasible option. A range of risk transfer mechanisms such as catastrophe bonds, catastrophe pools, index-based insurance, index-based insurance and micro-insurance schemes are available. Few experiments are being carried out in the form of pilot insurance programmes especially to tackle impact of catastrophic events in the agricultural sector. However, the urgent need is for a wider insurance programme supported by public policy. A step in this direction may also be aligned with the global risk reduction policy and commitments under the Sendai Framework for Disaster Risk Reduction (SFDRR) 2015–2030. Taking into account the role played by IRDAI post-liberalization of the Indian insurance industry, the regulator may take up additional responsibilities of promoting a market-based catastrophic insurance policy closely with the state and central governments.

The economic progress in India and the urgency for enhancing disaster risk resilience may justify a case for a catastrophic insurance. However, the smaller economies in South Asia may face difficulties in designing and implementing such insurance mechanisms. There are several hurdles other than the fact that such an initiative is highly capital intensive and that the insurance markets are underdeveloped. The development of affordable stand-alone catastrophe insurance products including parametric insurance solutions insuring natural disaster risks requires precise information on past disasters. Unavailability or scarcity of historical data on natural disasters and the constantly changing pattern or probabilities of disasters are further complicating the actuarial methods. The actuarial skills and technological know-how is also limited in these economies. On the supply side, the unavailability or reinsurance facilities for such disasters pose additional challenges. Finally, there are demand-side uncertainties with these insurance since the general perception is the premium for such products would be high. The following points may be considered for designing a catastrophic insurance programme in the South Asian economies.

Firstly, there is a need to identify either the reinsurer or the commercial insurer or both for implementation of the programme. Secondly, the catastrophic insurance framework needs to be holistic and involve different stakeholders so as to efficiently pool risk. This is because individuals may self-identify themselves as being risk-neutral and a potential for adverse selection exists. Thirdly, the objectives, management composition, coverage limit and clauses (like deductibles, exhaustion point, etc.), perils and hazards, among others need to be clearly defined in the catastrophic insurance programme. Fourthly, the information relating to probability of loss from a disaster needs to be precise. Finally, premiums need to be affordable and there should be incentives in place to inculcate disaster mitigation practices.

A question may be asked, whether or not India can play a lead role in the South Asian region. Given the existence of SAARC, such a step towards disaster risk reduction may be a win-win outcome for all the SAARC member economies. The benefits of risk pooling and diversification could be fully exploited in such a case. Although CCRIF may act as a guiding example, there will be challenges. The financial markets and the insurance industries in the South Asian economies are not at the same stage of development. The economies are also different in terms of governance structure. However, the economies do have common beliefs and aversion from private insurance and heavy dependence on state-sponsored social protection. If the efforts of country-specific regulators and governments be aligned, the objectives of disaster risk mitigation through insurance may be both possible and sustainable.

References

Ahmed M, Suphachalasai S (2014) Assessing the costs of climate change and adaptation in South Asia. Asian Development Bank, Mandaluyong City

Ahmed N, Occhipinti-Ambrogi A, Muir JF (2013) The impact of climate change on prawn post larvae fishing in coastal Bangladesh: socioeconomic and ecological perspectives. Mar Policy 39:224–233

Aon Bonfield (2015) Annual global climate and catastrophe report: impact forecasting, Aon Benfield Analytics
Ban N, Schmidli J, Schar C (2015) Heavy precipitation in a changing climate: does short-term summer precipitation increase faster? Geophys Res Lett 42:1165–1172
Beema Samiti (2012). Existing scenario of Nepalese insurance industry. Nepal Insurance Board
Caselli F, Malhotra P (2004) Natural disasters and growth: from thought experiment to natural experiment. International Monetary Fund, Washington, DC
Cavallo EA, Noy I (2011) Natural disasters and the economy – a survey. Int Rev Environ Res Econ 5:63–102
Central Water Commission (2015) Water and related statistics. Water Resources Information System Directorate Information System Organisation Water Planning and Projects Wing Central Water Commission, New Delhi
Chand R, Saxena R, Rana S (2015) Estimates and analysis of farm income in India, 1983–84 to 2011–12. Econ Polit Wkly L 50:139–145
Collier SJ (2008) Enacting catastrophe: preparedness, insurance, budgetary rationalization. Econ Soc 37:224–250
Daniell, JEA; Wenzel F, Khazai B, Brink S, Kunz-Plapp T, Anhorn J, Floerchinger V, Schaefer A, Hoechner A, Schmidt H-M, Girard T, Mühr B, Dittrich A, Vervaeck A, Kunz M, Trieselmann W, Strader A, Schorlemmer D, Beutin T, Fanselow F, Gourraud C, Bendimerad F (2015) Nepal earthquake. Report No.3, Technical Report. Center for Disaster Management and Risk Reduction Technology (CEDIM), Potsdam - Karlsruhe. https://www.cedim.de/download/CEDIM_ImpactSummary_EarthquakeNepal2015_Report3.pdf
Dewan TH (2015) Societal impacts and vulnerability to floods in Bangladesh and Nepal. Weather Clim Extremes 7:36–42
Elsner JB, Kossin JP, Jagger TH (2008) The increasing intensity of the strongest tropical cyclones. Nature 455:92–95
Emanuel K (2005) Increasing destructiveness of tropical cyclones over the past 30 years. Nature 436:686–688
Felbermayr G, Gröschl J (2014) Naturally negative: the growth effects of natural disasters. J Dev Econ 111:92–106
Ferrari A, Shrestha SR, Jaffrin G (2007) Access to financial services in Nepal. The World Bank, South Asia Region
Gonsalves J, Mohan P (eds) (2011) Strengthening resilience in post-disaster situations: stories, experience and lessons from South Asia. Academic Foundation, New Delhi
Guha-Sapir D, Below R, Hoyois P (2016) EM-DAT: international disaster database – www.emdat.be. Université Catholique de Louvain, Brussels
Hallegatte S (2015) The indirect cost of natural disasters and an economic definition of macroeconomic resilience. World Bank Policy Research Working Paper 7357, The World Bank
Hochrainer S (2009) Assessing the macroeconomic impacts of natural disasters: are there any? World Bank Policy Research Working Paper Series 4968, The World Bank
Ibragimov R, Walden J (2007) The limits of diversification when losses may be large. J Bank Finance 31:2551–2569
IPCC (2007) Climate change: the physical science basis. Intergovernmental Panel on Climate Change, Geneva
IPCC (2012) In: Field CB, Barros V, Stocker TF, Qin D, Dokken DJ, Ebi KL, Mastrandrea MD, Mach KJ, Plattner G-K, Allen SK, Tignor M, Midgley PM (eds) Managing the risks of extreme events and disasters to advance climate change adaptation. A special report of Working groups I and II of the Intergovernmental Panel on Climate Change. Cambridge University Press, Cambridge, UK
IRDAI and NDMA (2013) Disaster relief and risk transfer through insurance. Discussion Paper, July. Hyderabad
IRIN (2005) Disaster reduction and the human cost of disaster. Available at: http://www.irinnews.org/IndepthMain.aspx?IndepthId=14&ReportId=62446. Accessed on 7th Nov 2011

Jaffee DM, Russell T (1997) Catastrophe insurance, capital markets, and uninsurable risks. J Risk Insur 64:205–230
Kellett J, Caravani A (2013) Financing disaster risk reduction: a 20 years story of international aid. ODI, London
Klomp J, Valckx K (2014) Natural disasters and economic growth: a meta-analysis. Glob Environ Chang 26:183–195
Kossin JP, Olander TL, Knapp KR (2013) Trend analysis with a new global record of tropical cyclone intensity. J Clim 26:9960–9976
Kunreuther HC, Michel-Kerjan EO (2009) At war with the weather managing large-scale risks in a new era of catastrophes. The MIT Press, London
Kunreuther HC, Michel-Kerjan EO (2013) Costs (economic) of natural hazards and disasters. In: Bobrowsky PT (ed) Encyclopedia of natural hazards. Springer, New York
Lashley J (2012) Weather-related insurance and risk management. A demand study in the Caribbean. Deutsche Gesellschaft für Internationale Zusammenarbeit (GIZ) GmbH, Munich Climate Insurance Initiative (MCII) e.V. Eschborn, Bonn
Lazzaroni S, van Bergeijk PAG (2014) Natural disasters' impact, factors of resilience and development: a meta-analysis of the macroeconomic literature. Ecol Econ 107:333–346
Linnerooth-Bayer J, Hochrainer-Stigler S (2015) Financial instruments for disaster risk management and climate change adaptation. Climate Change 133:85–100
Linnerooth-Bayer J, Mechler R (2007) Disaster safety nets for developing countries: extending public-private partnerships. Environ Hazards 7:54–61
Loayza NV, Olaberria E, Rigolini J, Christiaensen L (2012) Natural disasters and growth: going beyond the averages. World dev 40:1317–1336
Mechler R, Bouwer LM (2015) Understanding trends and projections of disaster losses and climate change: is vulnerability the missing link. Climate Change 133:23–35
Mirza MMQ, Warrick RA, Ericksen NJ, Kenny GJ (2001) Are floods getting worse in the Ganges, Brahmaputra and Meghna basin? Environ Hazards 3:37–48
Munich Re (2016) Natural catastrophes 2015: analyses, assessments, positions. Topics Geo Münchener Rückversicherungs-Gesellschaft, München
Needham HF, Keim BD, Sathiaraj D (2015) A review of tropical cyclone-generated storm surges: global data sources, observations, and impacts. Rev Geophys 53:545–591
Noy I (2009) The macroeconomic consequences of disasters. J Dev Econ 88:221–231
O'Gorman PA (2015) Precipitation extremes under climate change. Curr Clim Chang Rep 1:49–59
Paul BK, Mahmood S (2016) Selected physical parameters as determinants of flood fatalities in Bangladesh, 1972–2013. Nat Hazards 83:1703–1715
Paul SK, Routray JK (2011) Household response to cyclone and induced surge in coastal Bangladesh: coping strategies and explanatory variables. Nat Hazards 57:477–499
Raddatz C (2007) Are external shocks responsible for the instability of output in low-income countries? J Dev Econ 84:155–187
Raghavan S, Rajesh S (2003) Trends in tropical cyclone impact: a study in Andhra Pradesh, India. Bull Am Meteorol Soc 84:635–644
Rasmussen TN (2004) Macroeconomic implications of natural disasters in the Caribbean. IMF Working Paper 04/224, International Monetary Fund
Reichenmiller P, Spiegel A, Bresch D, Schnarwiler R (2010) Weathering climate change: insurance solutions for more resilient communities. Swiss Re, Zurich
Shaw R, Pravakar SVRK, Chiba Y (2016) SDGs, DRR and CCA: potential for strengthening inter-linkages, policy brief 34. Institute for Global Environment Strategies, Kanagawa
Skidmore M, Toya H (2002) Do natural disasters promote long-run growth? Econ Inq 40:664–687
Stone JM (1973) A theory of capacity and the insurance of catastrophe risks (part I). J Risk Insur 40:231–243
Swiss Re (2010) Weathering climate change: insurance solutions for more resilient communities. Swiss Re Ltd., Zurich. (http://europa.eu/epc/pdf/workshop/2-3_pub_climate_adaption_en.pdf)
Swiss Re (2015) Underinsurance of property risks: closing the gap. Sigma 5. Swiss Re Ltd., Zurich

Swiss Re (2016a) Natural catastrophes and man-made disasters in 2015: Asia suffers substantial losses. Sigma 1. Swiss Re Ltd., Zurich

Swiss Re (2016b) World insurance in 2015: steady growth amid regional disparities. Sigma 3. Swiss Re Ltd., Zurich

Tucker M (1997) Climate change and the insurance industry: the cost of increased risk and the impetus for action. Ecol Econ 22:85–96

UNISDR (2015) 2015 global assessment report on disaster risk reduction (GAR15). UNISDR, Geneva. http://www.preventionweb.net/english/hyogo/gar/2015/

Wing AA, Emanuel K, Solomon S (2015) On the factors affecting trends and variability in tropical cyclone potential intensity. Geophys Res Lett 42:8669–8677

World Bank and United Nations (2010) Natural hazards, unnatural disasters: the economics of effective prevention. World Bank Group, Washington, DC. https://www.gfdrr.org/sites/gfdrr/files/publication/NHUD-Report_Full.pdf

World Bank Group (2014) Financial protection against natural disasters: an operational framework for disaster risk financing and insurance. World Bank Group, Washington, DC. https://openknowledge.worldbank.org/handle/10986/21725

Zeckhauser R (1995) Insurance and catastrophes. Geneva Pap Risk Insur Theory 20:157–175

Chapter 18
Landslide Risk Assessment in Darjeeling Hills Using Multi-criteria Decision Support System: A Bayesian Network Approach

Debashis Mitra, Chandrani Bhandery, Anirban Mukhopadhyay, Abhra Chanda, and Sugata Hazra

Abstract Landslide is one of the most recurrent and hazardous events of hilly slopes all over the world and particularly in the hilly regions of Himalayas. Darjeeling district of northern West Bengal, being the most important hill station in terms of tourism and trance-boundary strategic location, experiences landslide very often which causes intermittent loss of tourism revenues and is a problem for national security. In order to assess the landslide risk and accordingly prepare a landslide-risk zonation for the Darjeeling district, factors like slope, drainage density, rainfall soil depth, land use/land cover and geology have been considered. The factors responsible for landslide and their interdependency have been critically evaluated. In the present study, Bayesian network model has been implemented which is a probabilistic statistical graphical model that represents a set of variables and their conditional dependencies. Bayesian network was applied to assess the influences of the factors, and accordingly weightage and ranking of the contributing factors for landslide have been calculated. Finally, using multi-criteria decision support system (MCDSS) in GIS environment, landslide-risk zonation of the Darjeeling district has been prepared. Validation has been done taking into account 25 historical landslide locations, and more than 92% accuracy has been achieved. Rangli Rangliot is the most landslide-susceptible block of Darjeeling district. Kalimpong I, Kalimpong II, Mirik, Jorebunglow Sukhiapokhri and Bijanbari also come under the ambit of the highly susceptible areas.

Keywords Landslide • Darjeeling • Bayesian network • MCDSS

D. Mitra
Indian Institute of Remote Sensing, Dehradun, Uttarakhand, India

C. Bhandery • A. Mukhopadhyay (✉) • A. Chanda • S. Hazra
School of Oceanographic Studies, Jadavpur University, Kolkata, West Bengal, India
e-mail: anirban_iirs@yahoo.com

I. Pal, R. Shaw (eds.), *Disaster Risk Governance in India and Cross Cutting Issues*, Disaster Risk Reduction, DOI 10.1007/978-981-10-3310-0_18

18.1 Introduction

Landslides are one of the most damaging and threatening natural hazards that mostly take place in the hilly regions (Kanungo et al. 2006). The study of landslides has drawn global attention mainly due to increasing awareness of its socio-economic impacts and also increasing pressure of urbanization on the mountain environment (Aleotti and Chowdhury 1999). Landslides in mountainous terrains are often caused by the interaction of diverse environmental factors in association with heavy rainfall. Topography along with geomorphic and geologic features and vegetation cover plays a crucial role in the occurrence of landslides (Cubito et al. 2005; Moreiras 2005). Landslides exclusively accounted for 4.89% of the natural disasters that occurred worldwide during the years 1990–2005 (www.em-dat.net). Ercanoglu and Gokceoglu (2004) predicted that this trend would continue in the future due to increased unplanned urbanization and development, continued deforestation and increased regional precipitation in landslide-prone areas due to changing climatic patterns. In reality, with unplanned urbanization and increasing rainfall in the hilly region, occurrence of landslides has increased worldwide.

Almost 25% of geographical area of India comes under mountainous terrain (Dubey et al. 2005); being the youngest mountain chain, it is geologically/tectonically very unstable and seismically active. Owing to high population density and rapid increase in the developmental activities, the loss of human life and significant economic loss of the order of US$500 million per year is experienced by India due to landslides (Dubey et al. 2005). Especially the fragile and geologically young Himalayan mountains are characterized by a high-relief, high-intensity monsoon rainfall as well as earth tremors, which in turn accounts for natural disasters such as landslides and floods. As mentioned earlier, the losses caused by these natural disasters are incessantly increasing day by day as a result of accelerating rapid population growth and infrastructural development activities in these areas (Kayastha et al. 2013). Moreover, large-scale deforestation, unplanned land use systems and the construction of physical infrastructures, such as irrigation canals, roads and dams in hazardous mountainous regions, have contributed to increasing problems of landslides, debris flows, soil erosion and floods (Upreti and Dhital 1996).

Due to the above-mentioned concern of landslides, the identification of landslide-prone areas is a prerequisite for safer strategic planning of future developmental activities in the particular region (Kanungo et al. 2006). Hence, landslide susceptibility zonation (LSZ) of an area becomes important whereby the area may be divided into near-homogeneous domains and ranked according to degrees of potential hazard due to mass movements (Varnes 1984). The identification of landslide-susceptible regions is one of the most important stages in landslide hazard mitigation since the use of these maps can significantly reduce losses such as fatality, injury and property damage (Ercanoglu and Temiz 2011). These zonation maps highlight the spatial distribution of potentially unstable zones (Ayalew et al. 2004). Knowing the relationships between various causative environmental factors and landslide occurrence is important for the effective quantitative assessment of landslide hazards and/or susceptibility (Choi et al. 2012). If the geo-environmental conditions can be properly taken into account, landslide susceptibility maps are capable of providing a scenario depicting the relative spatial likelihood of slope failures of a particular region (van Westen et al. 2008). Moreover, they provide decision makers with critical information support for urban development and land use planning (Erener et al. 2016). Reliable landslide risk assessment depends on the quality and scale of the available data and the selection of appropriate methodology for analysis and modelling (Choi et al. 2012). The acquisition of landslide-related data is extremely difficult because datasets from poorly accessible mountainous areas are often limited and labour-intensive. Acquiring data using field survey methods is both time-intensive and expensive, while remote sensing techniques can be sometimes more cost-effective (Choi et al. 2012). However, in addition to remotely sensed data, a proper ground truthing is always essential to derive meaningful inferences.

Landslide susceptibility maps are prepared by a wide variety of methods and techniques which can be mainly categorized as qualitative and quantitative methods (Keefer and Larsen 2007). Qualitative methods are based on heuristic approaches, which depend mainly on the experience of experts (Ruff and Czurda 2008; Dwi Wahono 2010), whereas quantitative methods rely on obtaining probability of sliding from quantitative techniques like regression and neural networks (Ercanoglu and Temiz 2011; Sezer et al. 2011; Osna et al. 2014). Statistical methods are becoming popular day by day, especially with the use of GIS since GIS is a powerful tool to integrate geo-environmental data (vanWesten et al. 2006). Logistic regression (LR) model is one of the most widely applied methods among them (Kavzoglu et al. 2014). Multi-criteria evaluation (MCE) and multi-criteria decision analysis (MCDA) have been successfully used by different authors in landslide susceptibility mapping (Ozdemir and Altural 2013). Another direction in landslide susceptibility analysis and mapping is the use of association rule mining (ARM) method (Nefeslioglu et al. 2010).

In the present study, another very strong and upcoming approach, namely, the Bayesian network model, has been implemented to derive the landslide susceptibility zonation. A Bayesian network (Henrion 1989, Goldman and Cherniak 1993) is a powerful tool which can be used for the fusion of information coming from dissimilar sources having varying degrees of reliability. In these networks, each node represents a variable, the label of which is either given or has to be assessed, and the relationships between the nodes are expressed in terms of conditional probabilities. Each variable can be labelled with a label from a corresponding finite set with a certain degree of confidence. The input data may concern any of the nodes, and once they are fed into the right node, a mechanism is provided for the propagation of their information to all other nodes of the system. The use of Bayesian networks in geography, however, has been very limited with a few notable exceptions (Ducksbury 1993; Hass 1991).

18.2 Study Area

Darjeeling is the northernmost district of the state of West Bengal situated in the eastern India in the foothills of Himalaya (Fig. 18.1). It is one of the most renowned hill stations of India due to its scenic beauty. It lies adjacent to Sikkim in the north; in the south, it is bordered by Kishanganj (a district of the state of Bihar), in the east by the Jalpaiguri district and in the east by Nepal. The Darjeeling district has a length from north to south of 29 km and a breadth from east to west of 26 km. It

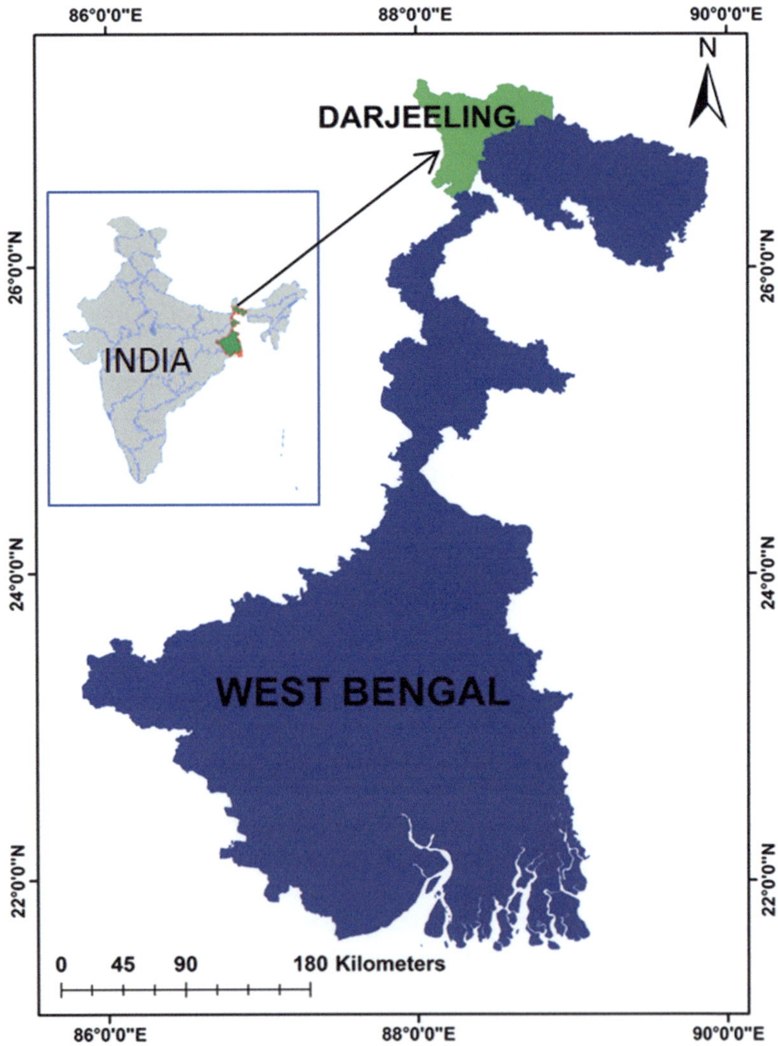

Fig. 18.1 Location map of Darjeeling district, West Bengal

covers a total area of ~3000 km^2, and it is characterized by a steep rise from the alluvial plains of West Bengal and attains a maximum elevation of about 2600 m (Kanungo et al. 2006). The Darjeeling Himalayas is situated within the Lesser and sub-Himalayan belts, and the tectonic units in the area occur in inverted order of stratigraphic superposition due to thrusting. Several rock groups have been named here locally like the Daling group, the Darjeeling group and the Paro subgroup consisting of slate, phyllite, schist, quartzite, greywacke and epidiorite and foliated gneisses (Acharya 1989).

18.2.1 Site Characteristics

The present study has been conducted exclusively in the Darjeeling hills (26° 56′ N – 27° 08′ N latitude; 88° 10′ E – 88° 25′ E longitude) encompassing an area of about 254 km^2. Though it is one of the least populous districts of West Bengal, areas like Darjeeling, Sonada and Sukhiapokhri are fairly inhabited. This region is characterized by highly dissected ridges and valleys. Tiger Hill marks the highest elevation of 2584 m. The area is dominated by slopes ranging between 15 and 35°, while steep slopes of >35° occupy smaller area. In general, the gentle slopes of 0–15° were found on the ridges and at places in the region of lower relief also (Kanungo et al. 2006). Binnie et al. (2007) also observed that slopes are gentler on the ridge tops but become increasingly steeper downward to the streams.

The annual rainfall in the area is of the order of 2000–5000 mm (Soja and Starkel 2007). The rainfall pattern is highly seasonal with a maximum rainfall during the monsoon season from June to October. The main land use practice in the study area is tea plantation. The agriculture land is mostly developed around the habitat areas. In general, the area is dominated by thick forest particularly in the eastern part (Kanungo et al. 2006).

A detailed review of reported landslides is tabulated in Table 18.1. The dates of the respective landslides, identified causes, affected areas and tentative loss and damages accounted are listed in this table. This survey is very much crucial from the perspective of the present study, since taking into account the Bayesian approach and the overlaying thematic maps prepared for the different causative factors, the final landslide susceptibility and risk assessment have been done for the present study area. The regions of landslide occurrences are illustrated in Fig. 18.2.

Table 18.1 Details of the previously reported landslides that took place in the Darjeeling hills

Date	Cause	Affected area	Loss/damages
22–25 September 1899	Incessant rainfall 1065.5 mm	Darjeeling town, Kurseong, Kalimpong, Ghum, Tindharia, Bloomfield and Tukvar tea gardens	72 lives were lost in Darjeeling town. Houses along the eastern slope of the hills were severely affected
15 January 1934	Bihar-Bengal earthquake	Darjeeling town, Sonada, Sukhiapokrii, Ghum and Kurseong	Mild destruction was reported in Darjeeling hills
11–12 June 1950	Rainfall 820 mm	Darjeeling town, Happy Valley, Kalimpong, Mahanadi, Paglajhora, Tindharia, Kurseong, Takdah, Glenburn, etc.	172 lives lost, thousands of people were rendered homeless, slides breached many arterial roads and Siliguri-Kalimpong Railway line was lost forever
3–5 October 1968	Rainfall 1121.4 mm	Darjeeling town (Toongsoong, Kotwali, Rajbari, Kalijhora, Bucher busty), Mainpuri, Lebong, Ambootia, Kalimpong, Tista bazaar, Rambi, Aligarh, Pesoke, etc.	Hill Cart Road and NH 31 were completely wiped out; 667 lives were lost, and great destruction in tea gardens
3–4 September 1980	Rainfall 299.1 mm	Rimbik, Lodhama, Bijanbari, Darjeeling town, Sukhiapokhri, Ghum, Mane Bhanjang, Sonada, Lebong, Tukvar, Tindharia, Happy Valley, Ambootia, etc.	Over 215 lives were lost at Rimbik and Lodhama; property amounting to Rs 100 million was destroyed
16 September 1991	Rainfall 462.5 mm	Darjeeling town (North point Toongsoong, Singamari), Ging, Tukvar, Bannockburn, Bloomfield, Paglajhora, Chunabhatti, etc.	Two people were killed; rail connection between Siliguri and Darjeeling was snapped for 5 months
13 July 1993	Rainfall 211.3 mm	Mangpoo, Takdah, Pesoke, Rongtong, Tindharia, Pankhabari, Darjeeling town, etc.	15 lives were lost at Mangpoo
8–9 July 2003	Rainfall 112 mm	Hill Cart Road (NH-55), Kurseong, Darjeeling town area between Pulbazar, Gokhabazar and Pankhabari	Seven lives were lost near Paglajhora, and rail connection was lost for 3 months

(continued)

Table 18.1 (continued)

Date	Cause	Affected area	Loss/damages
7 July 2004	Rainfall >300 mm	NH-55, Kurseong, Gayabari, Badamtam, Tukvar, Rishihat, Ging, Bannockburn, Pulbazar, Hill Cart and other roads of Darjeeling hills	11 people were killed in Gayabari roads and settlement were largely affected
30 August 2005	Rainfall >350 mm	Dali, Kalimpong and Darjeeling subdivisions	Two people were killed, and many houses were destroyed
24 September 2006	Rainfall >500 mm	Toongsoong area and many other parts of three hill divisions	Nine people were killed, the injured and houses collapsed
16 July 2007	Heavy rains on the Hill Cart Road	Darjeeling, Hill Cart Road, St Mary Hills, Kalimpong	50% of the Hill Cart Road, which serves as NH-55, was damaged. National Highway-31A connecting Siliguri with Gangtok was blocked
7 September 2007	Incessant Rains	Darjeeling district, Kalimpong subdivision	Six lives were lost in three villages in Kalimpong subdivision
20 June 2008	Heavy rains	Three subdivisions of Darjeeling, Kurseong and Kalimpong and road links of NH-55 and NH-31A Sukhiapokhri, Matigara, Khori bari, Phansidewa blocks, etc.	NH-31 was closed several times. Tista the major river was also in space and nearly to danger mark at Domohani (Population affected, 16,674; death, 03; houses damaged, 3173)
26 May 2009	Cyclone Aila, which has triggered heavy rains in the Darjeeling hills	555 villages in eight blocks and four municipalities, Bijanbari, Sukhiapokhri, Takdah, Kurseong, Mirik, Kalimpong I and II, Gorubathan blocks and Darjeeling	Darjeeling remained cut off from Siliguri in the plains as both Hill Cart Road and Pankhabari Road from the north Bengal town remained blocked by landslides at Ghum and at 100 other places Population affected 1,15,758; death-41; houses affected-26,595

(continued)

Table 18.1 (continued)

Date	Cause	Affected area	Loss/damages
16 June 2010	Overnight rains	Kurseong subdivision of Darjeeling district, Rongtong and Tindharia, Siliguri and New Jalpaiguri	A 21-m length of the National Highway 55 was destroyed, besides partially damaging another 22-m stretch. The landslide has resulted in disruption of supply lines between Siliguri and Darjeeling districts of West Bengal and the suspension of the toy train services between Kurseong and New Jalpaiguri due to the damage of the narrow gauge tracks between Rongtong and Tindharia
16 June 2010	Heavy rainfall	Tripi road of Upper Gumfa area in Kalimpong Darjeeling district,	Death of three people and serious injury of two others
3 July 2010	Heavy rainfall	National Highway 55	Damage of the narrow gauge train tracks. This has led to the suspension of the services of Darjeeling Himalayan Railway (DHR)
5 August 2010	Heavy rainfall	Gayabari, Darjeeling District,	The National Highway 55 was disrupted, and the narrow gauge rail track was severely damaged. The landslides have upset all road links from Siliguri to various other places in Darjeeling district such as the hill resort of Darjeeling, Kalimpong, Kurseong and Sukna
26–27 March 2011	Heavy rainfall	Sevoke area of Kurseong, Darjeeling District, West Bengal	Two people were killed, and an equal number were injured
17–19 June 2011	180 mm rainfall	Kurseong, Darjeeling	Vehicles damaged, road breached at Dhobi Khola on NH-55, four lives lost
23 June 2011	576 mm rainfall	Goethals, Kurseong, St Mary Hills	Four lives lost, houses damaged
24 August 2011	Incessant rains 129 mm	Darjeeling-Bhutia busty	No casualties reported
26 September 2011	Earthquake (6.8R)	Darjeeling-Lebong Cart Road	30-m stretch of Lebong

(continued)

Table 18.1 (continued)

Date	Cause	Affected area	Loss/damages
17 April 2012	Construction activity	Heart of Darjeeling town	No such casualties reported
14 June 2012	Heavy rainfall	Tindharia NH-55, Siliguri	100-year-old railway workshop perched, 150-meter highway washed away
14 July 2012	100 mm rainfall over 12-h period	Kurseong, NH-31A between Gangtok and Siliguri and Darjeeling	NH-31A between Siliguri and Darjeeling was blocked, and Darjeeling was cut off from Siliguri for some time
19 July 2012	100 mm rainfall	Tindharia (Kurseong)	Portion of ground adjacent to Himalayan Kurseong, railway workshop at Tindharia collapsed endangering the famous railway
15 September 2012	Heavy overnight rain in the hills	Takdah and Lopchu gardens, Bannockburn, Phoobshering, Ging and Pussimbing	Extensive damage in six tea gardens, while the National Highway 31A was blocked; Takdah has been washed away in four places, and five culverts have been damaged
11–17 September 2012	410 mm rainfall	Darjeeling and North Sikkim	Fatalities occurred only in North Sikkim
18 April 2013	Mountain Stream (Botey jhora), being overloaded by surface runoff	Kalimpong	No such casualties occurred
August 2014	Heavy rainfall	Kalimpong	35 families were homeless
July 2015	Heavy rainfall washed away loose landmass	Mirik and Kalimpong	30 persons died, loss of properties

Source: Starkel and Basu 2000; Basu and De 2003, Darjeeling Times.com, The Hindu, The Economic Times

18.3 Materials and Methods

18.3.1 Selection of Causative Factors

Six major causative factors or criteria potentially linked with landslide phenomenon of this study have been considered for the present study. They are, namely, slope character, drainage density, soil depth, rainfall pattern, land use/land cover classes and geological characteristics.

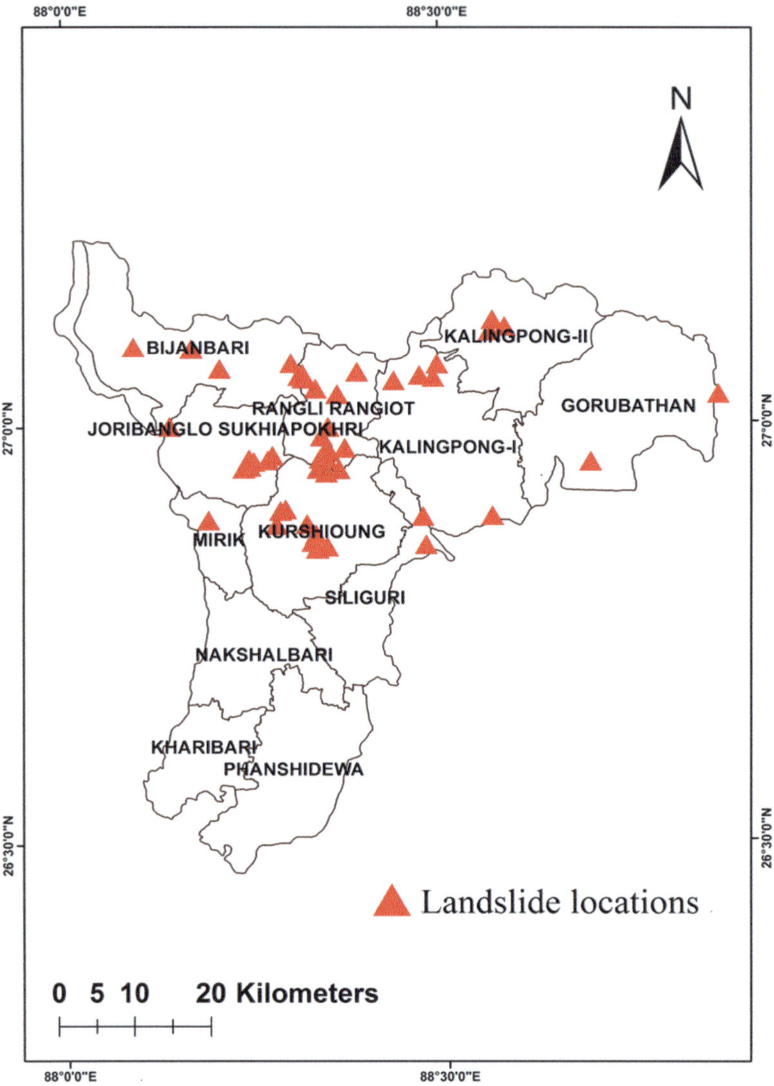

Fig. 18.2 Map showing the locations of occurrences of landslides

18.3.2 Overall Methodology Adopted in a Nutshell

Thematic data layers and maps corresponding to the above-mentioned factors are individually prepared in a GIS environment. These layers were finally overlaid upon one another along with the thematic layer showing the occurrences of the landslides in order to correlate and preliminarily identify the order of the causative factors responsible for the landslides. The quantification of the contributing percentage of the factors responsible and their interdependency is very significant. In the present study, Bayesian network model has been applied which is a statistical probabilistic graphical model that characterizes a set of variables and their conditional dependencies to obtain the final causative percentage of the factors. After compiling all the factors with their relative weightages and rankings using multi-criteria decision support system (MCDSS) in GIS environment, landslide-risk zonation of the Darjeeling district has been prepared, and validation has been done taking into account 25 previously recorded historical landslide locations within the realm of the present study area.

18.3.3 Thematic Map Preparation and Data Used

River and stream network maps have been prepared with the use of DEM data (ASTER 30 m resolution) and LANDSAT TM imageries. Slope character has been also characterized from ASTER data. Drainage density maps were also prepared from the overlay of the above three thematic maps in a GIS environment. Rainfall data has been acquired from TRMM of the last 13 years. Geological thematic data layer has been prepared by digitizing the geological maps according to GIS and Sarkar and Kanungo (2004). The soil map along with the sand, silt, clay (i.e. textural composition), pH, CEC and organic carbon and depth data were extracted from the soil series of West Bengal, National Bureau of Soil Survey and Land Use Planning. Land use/land cover classification is done by means of classifying the LANDSAT imageries.

18.3.4 Preliminary Ranking of the Causative Factors

The slope of the terrain is one of the most critical factors from the perspective of landslides. Since where the steepness of the slope is more, the potential of sliding under normal circumstances due to the action of gravity is also high. Hence, slope has been given the first rank in the present study. Drainage density based on the river and stream flow of the terrain has been given the second rank since it delineates the natural course of materialistic flow in hilly landscape and regulates the landslides to a large extent. The third most crucial aspect is the rainfall. Especially heavy rainfall has been associated with many recorded landslide events as it temporarily destabilizes the immediate topsoil stratum. Soil has been given the fourth rank since their depth plays a vital role in withholding the intact nature of the soil beneath by means of their root system. Land use/land cover is given the fifth rank followed by the geological character being the sixth, respectively. Earthquakes often trigger landslides that are eminent otherwise. Therefore, they are not considered separately in the present study.

18.4 Characterization of the Causative Factors

18.4.1 Slope

Slope is an important factor which governs the stability of a terrain. With the increasing slope possibility of the slope, failure gets enhanced. However, various types of soil depth and strength are two other factors which affect the failure of land

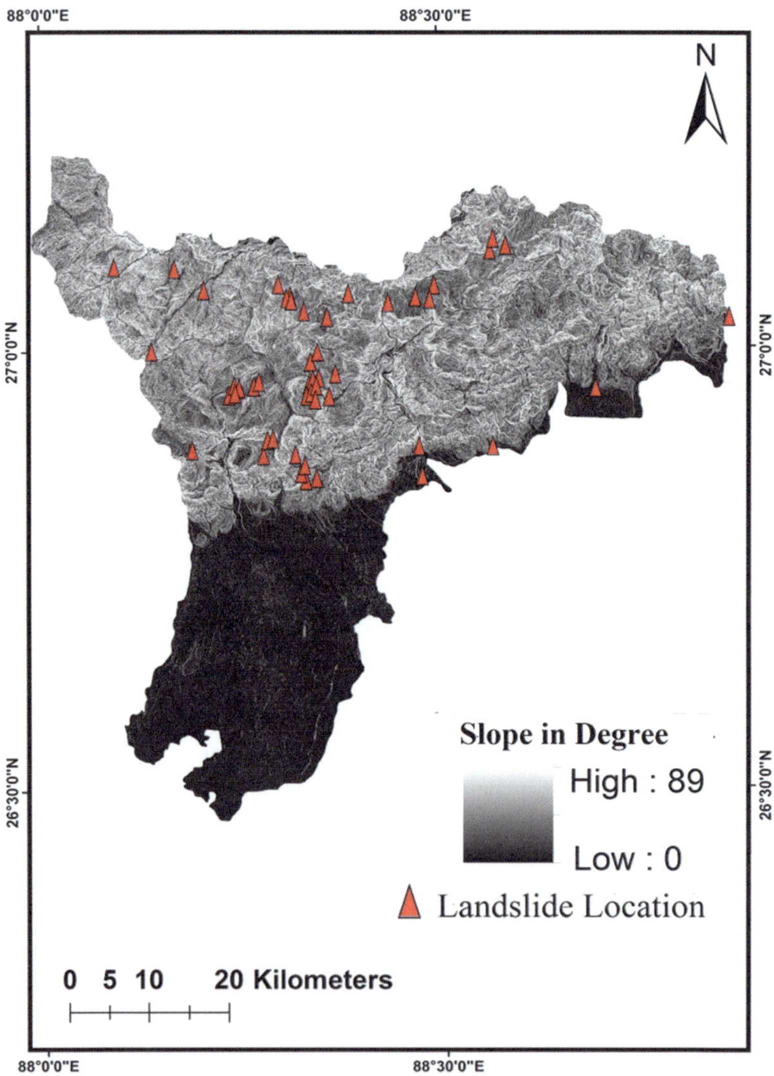

Fig. 18.3 The map depicting the slope character of the study area prepared from ASTER GDEM

mass. In any hilly region, slope characterizes the relief as well as the stability of various landforms. Slope morphology can determine the susceptibility of a landslide in several ways. The steepness of slope indicates the direction and the amount of surface runoff or subsurface drainage reaching a site (Dai et al. 2001). Concentration of subsurface drainage with in a concave slope, resulting in higher pore water pressure in the axial areas compared to the flanks, is one of the possible mechanisms. The slope character of the present study area is portrayed in Fig. 18.3. On the whole, the higher degree slopes are found in the northern part of the study area where more landslides are recorded, while in the southern parts, the slopes are fairly on the gentler side. From the perspective of landslide vulnerability study in this area, it has been taken that the higher the slope, the higher the susceptibility, while gentle slopes are less susceptible.

18.4.2 Drainage Density

In order to identify the landslide-prone regions, drainage density of this study area has been considered. Major rivers of Darjeeling district are Tista, Rangit, Mahanadi, Lishi, Balason, Mechi, Chel, Ramman, Murti, Jaldhaka and so forth. The hill areas of Darjeeling district are located within the Lesser and sub-Himalayan belts of the Eastern Himalayas (Kanungo et al. 2008). The southern foothill belt is demarcated by a highly dissipated platform of terrace deposits extending along the east-west axis. The inner belt is defined by a ridgeline stretching from the Darjeeling hill to the west and Kalimpong hill to the east, over the southerly flowing Tista valley in between. Prominent rivulets contributing to the Rammam-Rangit basin dissipate the northern slope of Darjeeling hills. The Kalimpong hill is rather rugged in topography and is dissipated by radically descending gullies and streams that contribute to the Tista and Jaldhaka river system. On the whole, it can be seen that the northern belt is having a much dense narrow width river network in the valleys and ridges, whereas in the southern portion, comparatively few but wider rivers are found flowing from north-northeast to south (Fig. 18.4).

Drainage density provides the better quantitative expression to the dissection and analysis, which in turn gives us an insight about the landform, lithology, structures and relief history of a particular region. Drainage density can be used as an indirect indicator to explain the morphogenesis of a landform (Pareta et al. 2012). Drainage network has been prepared using LANDSAT optical image and ASTER DEM. The drainage density in and around the Balason, Tista, Mahanadi, Mechi, Neiorar, Lodhoma, Gish, Chel and Rangit rivers are found to be moderate to high, and these regions coincided with sites of previously recorded landslide events. Rangli Rangliot is found to fall under very-high-density area followed by the Kalimpong and Bijanbari regions. The drainage density varies from 0 to 78 m/km^2 in this area. The high weightage has been assigned to the areas with high drainage density: >50 m/km^2 has been given the highest weightage; 25–50 m/km^2 has been given moderate weightage, and 0–25 m/km^2 has been given the least weightage.

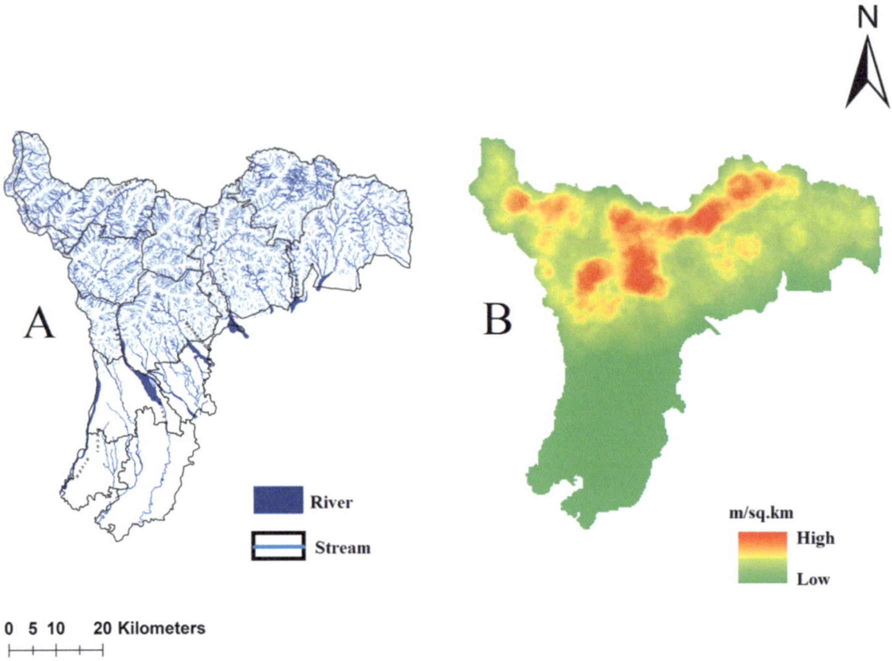

Fig. 18.4 The river and stream network **a** and drainage density map **b** of Darjeeling prepared from ASTER GDEM and LANDSAT satellite images

18.4.2.1 Rainfall

Rainfall is one of the important factors which control the landslide in Darjeeling district (Sing et al. 2011). Enormous rainfall is the cause of landslide but it is difficult to define precisely. Rainfall indirectly affects the pore water conditions in the slope materials, which in turn regulates the slope instability (Caine 1980). The amount of rainfall plays a very important role in causing instability of slopes. A very high intensity of rainfall within a short span of time is not uncommon in Darjeeling hill areas. It is found in the old records that this natural phenomenon has occurred about 42 times during the period from 2000 to 2012 (Chatterjee 1983). The isohyets, i.e. the map prepared on the basis of average annual rainfall during last 13 years in Darjeeling district (Fig. 18.5), shows that the value increases from south-east to north-west. A maximum concentration of landslides falls where the annual rainfall is more than 300 cm.

Heavy rainfall decreases the cohesiveness of the soil leading to soil erosion and landslide (Sarkar et al. 2013; Mondal and Maiti 2013). The yearly rainfall distribution map has been prepared from TRMM rainfall data over the last 13 years, and a rainfall zonation map has been prepared. The area has been divided into three zones: very high rainfall greater than 400 cm per year, high rainfalls 300–400 cm and moderate rainfall less than 300 cm per year. Based on the rainfall pattern and historic

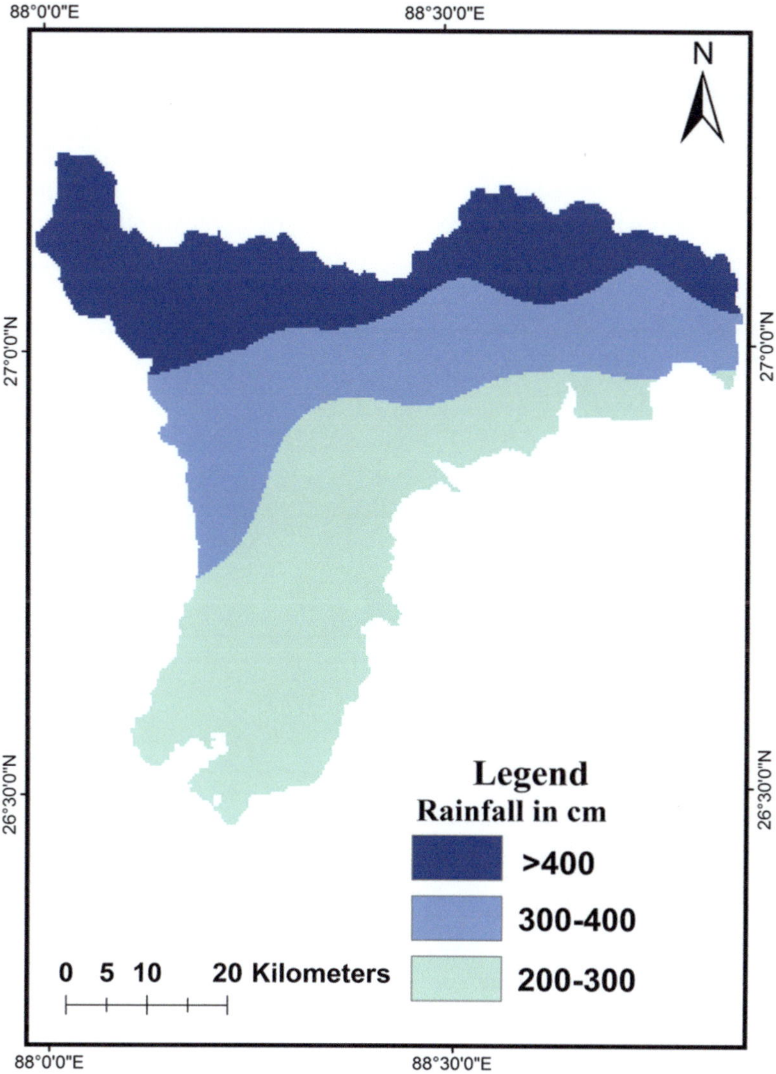

Fig. 18.5 The rainfall distribution map of the study area from TRMM data NASA and IMD

landslide occurrence, maximum weightage has been given to the area corresponding to the very high annual rainfall based on the hypothesis – the higher the rainfall, the higher the susceptibility.

18.4.3 Soil

The soils of Darjeeling hill (Fig. 18.6) area have been developed depending upon the underlying geology. But, in general, the soils have been developed by both fluvial action and lithological disintegration. The soils that have developed in the Kalimpong area are predominantly reddish in colour. Occasional dark soils are found due to extensive existence of phyllites and schist. Soils in the highlands stretching from the

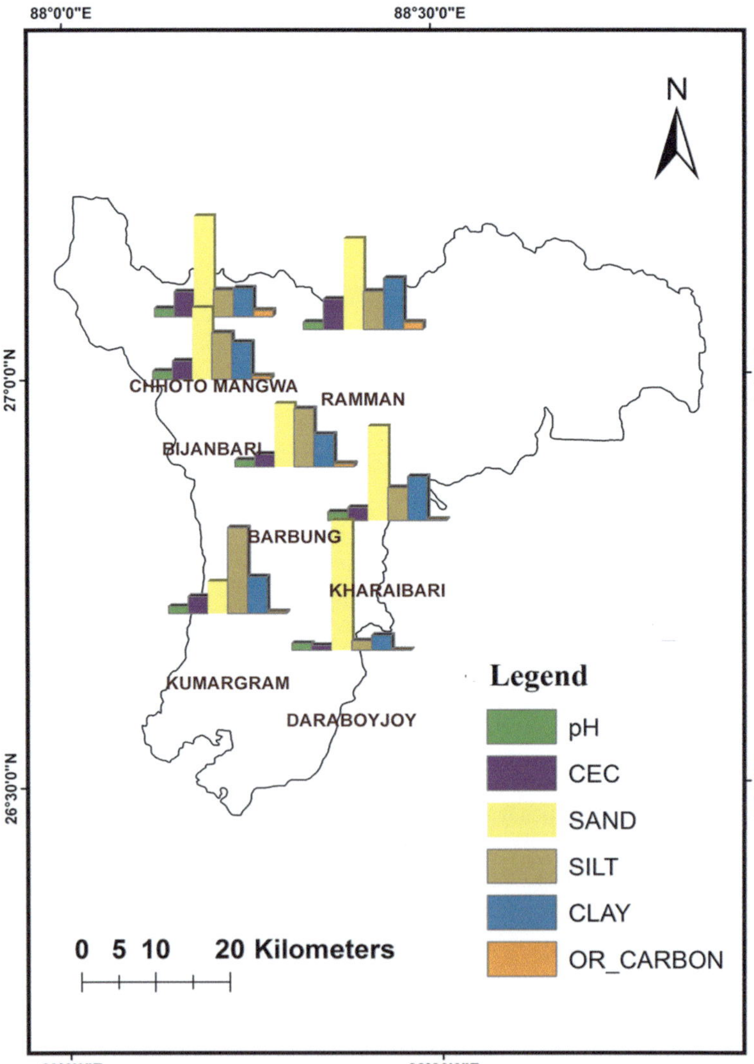

Fig. 18.6 Soil characteristics of the study area (after NBSS&LUP; NATMO; FAO)

west to the east of the district along most of the interfluvial areas are mainly mixed sandy loam and loamy, while those on the southern slopes of Mirik and Kurseong are mainly clayey loam and reddish in colour. Sandy soils are mainly found in the east of the river Tista.

All the soils are acidic in nature and the lowering of pH is found to increase with depth in most cases. This indicates the lack of base materials in the surface stratum. The weathering of lateritic type is the substantial mechanism in the transformation of the substratum. The variable thickness of the regolith and soils depends on the rate of weathering and gradient of the longitudinal slope profiles and intensity/gravity of mass movements. The basic soil types are yellow soils, red-brown soils and brown forest soils. Red and yellow soils have been developed from gneiss, while brown soils from schist and shale. Coarse pale yellow to red-brown soils are found on the Siwalik, while clayey dark soils are developed on Daling series.

The character of the bedrock is reflected only in the grain size composition of the soil. On the Darjeeling gneiss, very coarse-grained (50–80%) particles are found. In Daling series, the percentage of sandy and coarse particles in the soils is high. On the Siwalik, silt-clay fraction is higher. The chemical content of the soil over Darjeeling gneiss is characterized by a high proportion of potassium derived from feldspar and muscovite mica. This soil is poor in lime, magnesium, iron oxides, phosphorous and nitrogen. Therefore, lime is used in the tea plantation areas.

The soil depth (Fig. 18.7) is a very important parameter from the viewpoint of landslide susceptibility analysis. It has been noticed that where the soil is shallow, their relief variation is very high (Sarkar et al. 2013), which catalyses the landslide probability. In this study, the soil depth has been classified in three classes from 0–50 m to 50–100 m and higher than 100 m (Fig. 18.7a), and the weightage has been assigned according to the depth, i.e. the more the depth of soil, the lesser is the landslide susceptibility.

18.4.4 Land Use/Land Cover

The thematic map of the LULCC is portrayed in Fig. 18.8. In terms of percentage of area covered, tea plantations account for a wide majority followed by forest and agricultural lands. The tea gardens are situated almost from the north to south stretch, whereas forests are mostly situated in the eastern and western parts. Agricultural lands are scarcely distributed in between. Settlements rank fourth after the above-mentioned classes, and they are mostly found dense in the southern end of the area; however, in the north also, some scattered but dense settlements are observed. Rivers followed by Cinchona plantation rank at the last.

The weightage has been assigned according to the susceptibility of each land cover classes (Sarkar et al. 2013; Mondal and Maiti 2013; Kanungo et al. 2008). The barren lands have been given maximum weightage followed by agricultural land, tea garden, cinchona plantation, settlement, forest and river.

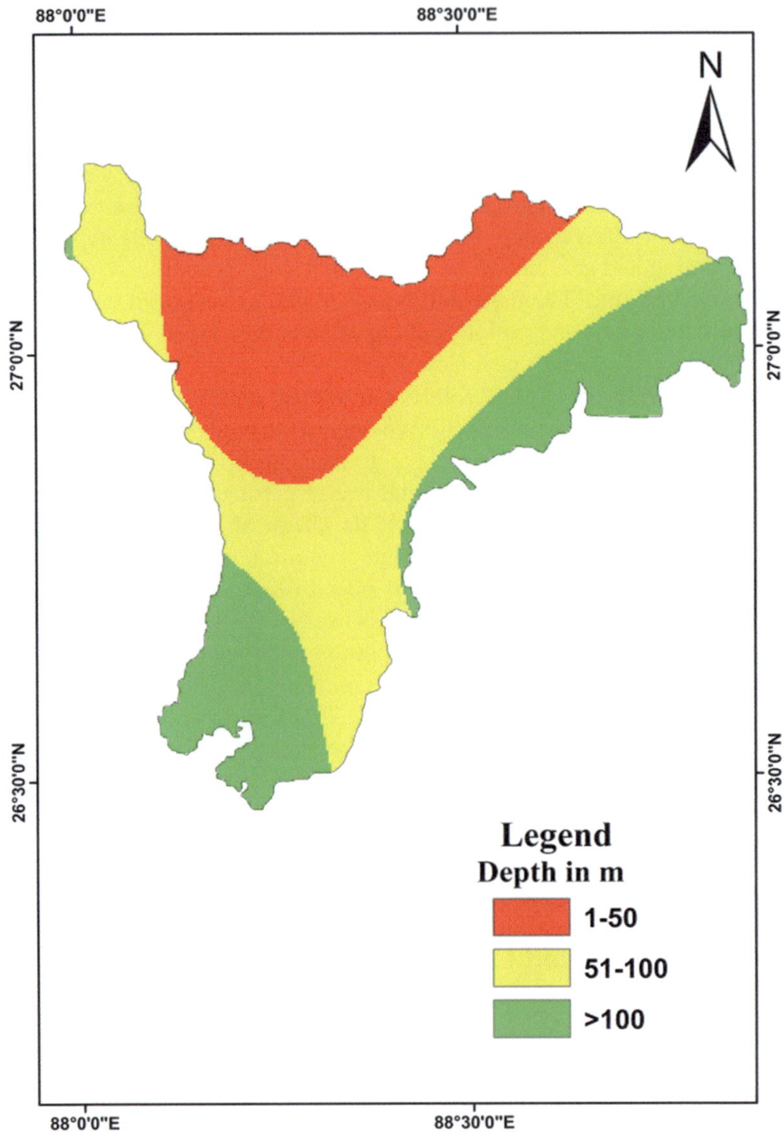

Fig. 18.7 The soil depth of the study area (After NBSS&LUP; FAO)

18.4.5 Geology

From geological perspective, Darjeeling is unique in its environmental eco-perception. This region is mainly an erosion-prone area. Here eroded landforms are mainly found to exist. These eroded landforms are developed by southerly flowing streams, which have exposed fall cross section of different tectonic units. The form

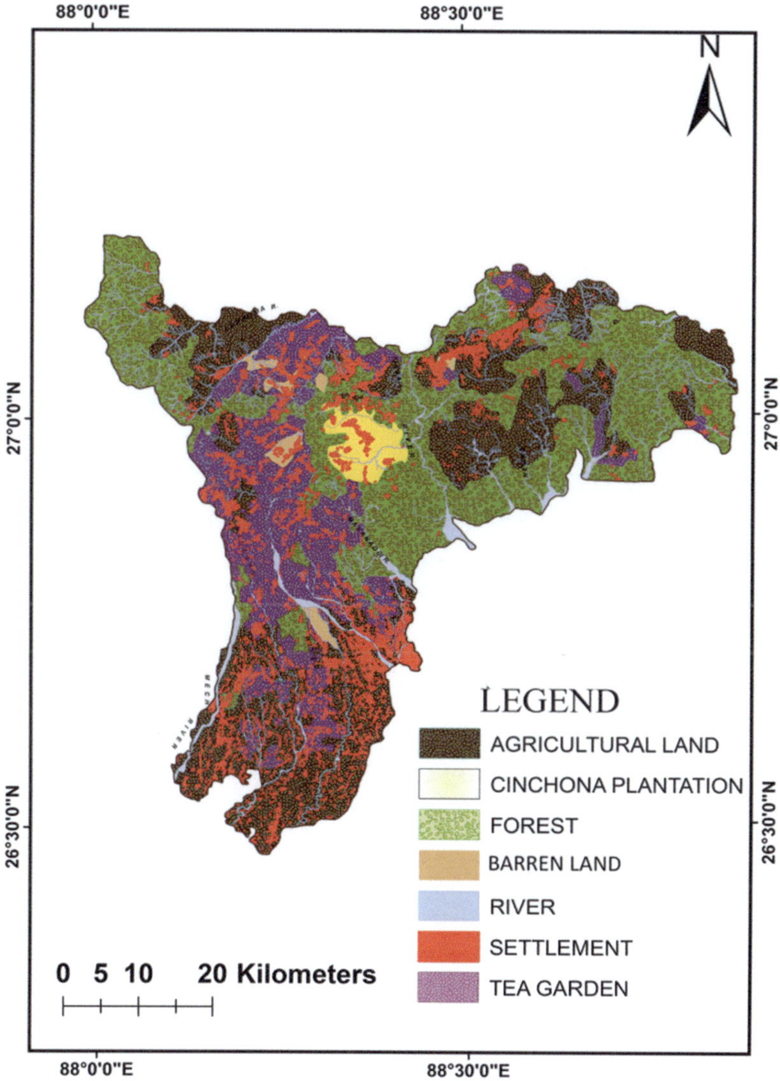

Fig. 18.8 Map showing the land use/land cover classes of the study area (Prepared from LANDSAT, OLI image of the year 2015 and consulting Desai (2006) and Kanungo (2008))

units are, however, approximately the same throughout the hill area, having more or less uniform lithology, structure, climate, soil and vegetation corners. According to Mallet (1875), the tectonic units are found in the reverse order of stratigraphic superimposition and are represented by Siwalik group of rocks. The contact between different groups of rocks is represented by thrusts, dipping at high angles towards north. A brief description of various formations of the Darjeeling Himalaya is given below. The geological scenario of the study area is portrayed in Fig. 18.9.

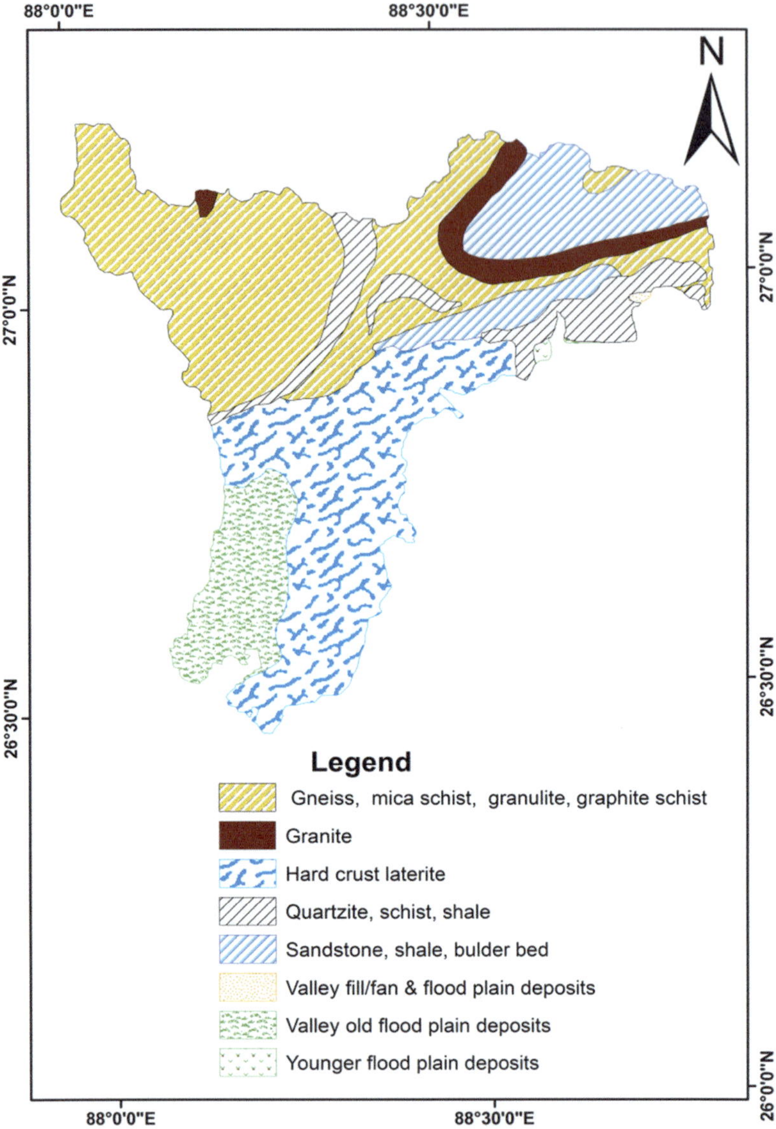

Fig. 18.9 The geological map of the study area after GSI and SI

The probability of the landslide occurrence in this area is very high for the lithological composition of gneiss, mica schist, granulite and graphite schist (Mondal and Maiti 2013). Hence, for the weightage calculation, the lithological units containing gneiss, mica schist, granulite and graphite schist have been given the maximum importance than all other classes.

18.5 Preparing Conditional Probability by Means of Bayesian Network

Bayesian networks (BN) are widely used for knowledge representation and reasoning under uncertainty in intelligent systems (Pearl and Russell 1998; Russell and Norvig 1995). In a general form, the structure of a BN is a directed acyclic graph (DAG) in which nodes correspond to random variables of interest, and directed arcs represent direct causal or influential relation between nodes. The uncertainty of the interdependence of the variables is represented locally by the conditional probability table (CPT) Pr (xi|πi) associated with each node xi, where πi is the parent set of xi. An independence assumption is also made with BN that xi, given its parents πi, is independent of any other variables except its descendants. The graphical structure of BN allows an unambiguous representation of interdependency between variables. This, together with the independence assumption, leads to one of the most important features of BN: the joint probability distribution of X = (xi, K, xn) can be factored out as a product of the conditional distributions in the network. The Bayesian network analysis has been done using Genie 2.1 software from University of Pittsburgh. All of the six parameters and their preassigned weightages have been incorporated in the Genie software (Fig. 18.10).

As the drainage density and depth of soil are dependent on slope, their interdependency was calculated using deterministic functions, and combined effect of slope/drainage density and slope/soil depth has been incorporated. Other factors with their relative ranking and weightage (Sects. 18.3 and 18.4) have been incorporated in the network. Finally, the landslide occurrence probability has been calculated (based on the ranking of the several factors considered during the previous landslide occurrences and their extent taken into account in this study (Fig. 18.2) and saved as Interchange file (.dsc) or Hugin file (.net) which is the input of the multicriteria analysis in GIS environment.

18.6 Final Preparation of the Landslide Susceptibility Map and Conclusion

Taking into account all the thematic maps prepared based on the possible causative factors that play a crucial role behind the occurrence of landslides and implementing the Bayesian network approach, the landslide susceptibility zonation of the study area has been accomplished, and the map is portrayed in Fig. 18.11. Clearly, there is a distinct north-south variation in terms of susceptibility observed from the analysis. The entire southern part of the study area falls under low susceptibility zone (green colour) which covers regions like Nakshalbari, Kharibari, Phansidewa and Siliguri. While in the northern part of the study area, most of the areas are moderately to highly susceptible (brown to white colour). Rangli Rangliot is the most landslide-prone block of Darjeeling district. Kalimpong I, Kalimpong II, Mirik,

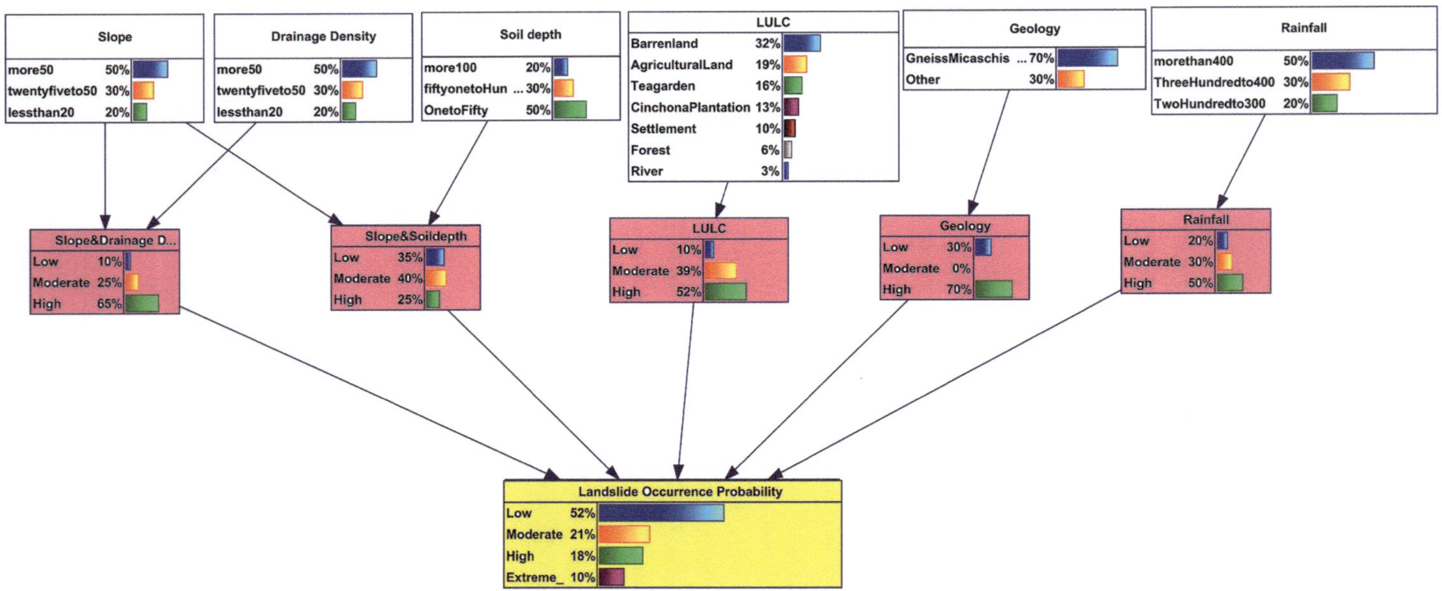

Fig. 18.10 Bayesian network framework for landslide occurrence probability assessment

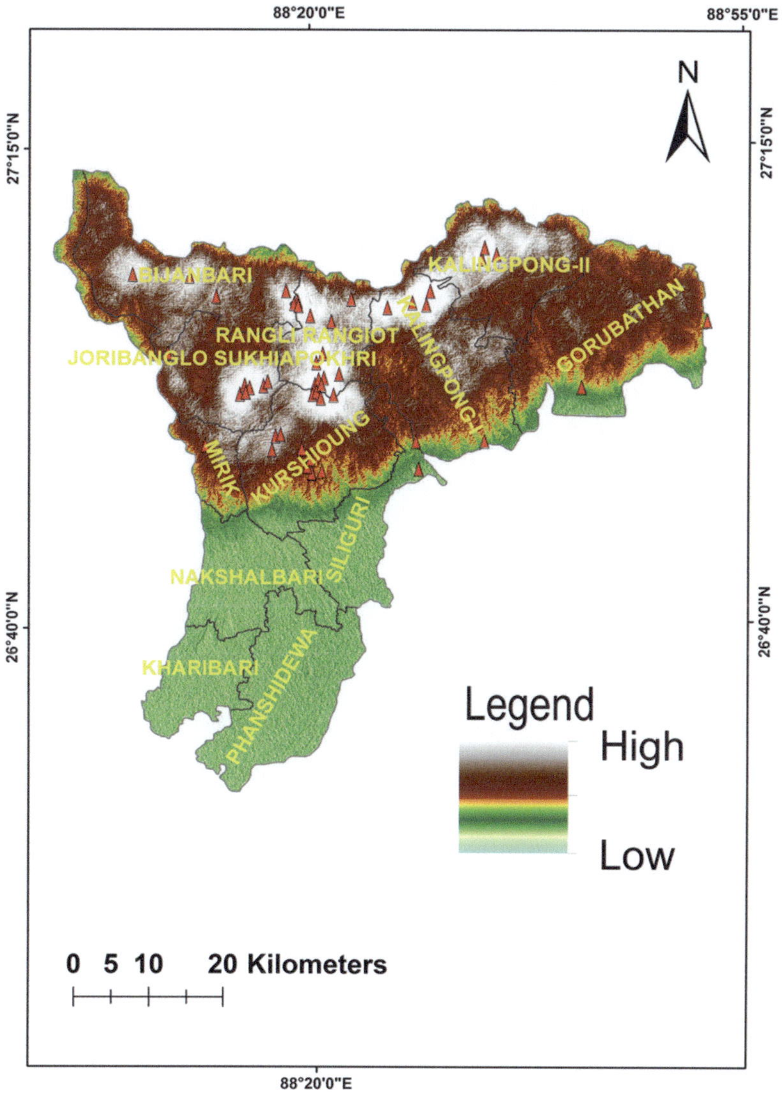

Fig. 18.11 The landslide susceptibility zonation map of the study area

Jorebunglow Sukhiapokhri and Bijanbari also come under the arena of the highly susceptible areas. It is worth mentioning that the regions where the susceptibility is found highest, previous landslides coincide with those places. This in turn proves that the usage of Bayesian network along with multi-criteria decision support system on the causative factors has led to a promising output. The regions of high landslide susceptibility were also found to coincide with an optimum drainage density, high slope, ridges full of streams and partially deforested areas. This implies that the above-mentioned criteria are playing a decisive role behind the occurrence of landslides.

These types of studies are extremely crucial and need to be carried out in most of the landslide-prone areas. GIS-based analysis should be immensely incorporated in such studies as outcomes from these studies would help the policy makers to prioritize the regions that need more attention followed by the other regions which are comparatively less vulnerable. Phenomena like landslides occur naturally; however, their intensity and frequency of occurrence is steadily increasing all over the globe leading to severe loss of life and property. Hence, land acquisition and allotment to common people as well as industrial sectors should be done after carrying out such pilot studies taking into account the above-discussed parameters.

Acknowledgements Authors are grateful to GSI, SI, NBSS&LUP, GLOVIS, GLCF, NASA, FAO, NATMO, IMD, Govt. of West Bengal for providing data and satellite images and the University of Pittsburgh for Genie 2.1 software. Anirban Mukhopadhyay is grateful to BC3, Bilbao, Spain, for providing training on Bayesian network.

References

Acharya SK (1989) The Daling Group, its nomenclature, tectonostratigraphy and structural grain: with notes on their possible equivalents. Geol Surv Indones 22(Spec Publ):5–13

Aleotti P, Chowdhury R (1999) Landslide hazard assessment: summary, review and new perspectives. Bull Eng Geol Environ 58:21–44

Ayalew L, Yamagishi H, Ugawa N (2004) Landslide susceptibility mapping using GIS based weighted linear combination, the case in Tsugawa area of Agano River, Niigata prefecture, Japan. Landslides 1(1):73–81

Basu SR, DE SK (2003) Causes and consequences of landslides in the Darjiling-Sikkim Himalayas, India. Geogr pol 76(2):37–52

Binnie SA, Phillips WM, Summerfield MA, Fifield LK (2007) Tectonic uplift, threshold hill slopes and denudation rates in developing mountain range. Geology 35:743–746

Caine N (1980) The rainfall intensity: duration control of shallow landslides and debris flows. Geogr Ann Ser A Phys Geogr:23–27

Chatterjee B (1983) A geological approach to the landslide hazard zonation in Darjeeling Himalaya, West Bengal, India. Geol Surv India Unpub Rep (F.S. 1981–83)

Choi J, Oh H-J, Lee H-J, Lee C, Lee S (2012) Combining landslide susceptibility maps obtained from frequency ratio, logistic regression, and artificial neural network models using ASTER images and GIS. Eng Geol 124:12–23

Cubito A, Ferrara V, Pappalardo G (2005) Landslide hazard in the Nebrodi Mountains (Northeastern Sicily). Geomorphology 66:359–372

Dai FC, Lee CF, Li J, Xu ZW (2001) Assessment of landslide susceptibility on the natural terrain of Lantau Island, Hong Kong. Environ Geol 40:381–391

Dubey CS, Chaudhry M, Sharma BK, Pandey AC, Singh B (2005) Visualization of 3-D digital elevation model for landslide assessment and prediction in mountainous terrain: a case study of Chandmari landslide, Sikkim, eastern Himalayas. Geosci J 9(4):363–373

Ducksbury PG 1993 Parallel texture region segmentation using a Pearl Bayes Network. In: Proceedings of the 4th British Machine Vision Conference, Surrey, UK, Sept pp 187–195

Dwi Wahono BF 2010 Applications of statistical and heuristic methods for landslide susceptibility assessments: a case study. Wadas Lintang Sub District, Wonosobo Regency, Central Java

Province, Indonesia. Thesis for the degree of Master of Science 106. International Institute for Geo-Information and Earth Observation

Ercanoglu M, Gokceoglu C (2004) Use of fuzzy relations to produce landslide susceptibility map of a landslide prone area (West Black Sea Region, Turkey). Eng Geol 75(3&4):229–250

Ercanoglu M, Temiz FA (2011) Application of logistic regression and fuzzy operators to landslide susceptibility assessment in Azdavay (Kastamonu, Turkey). Environ Earth Sci 64(4):949–964

Erener A, Mutlu A, Düzgün HS (2016) A comparative study for landslide susceptibility mapping using GIS-based multi-criteria decision analysis (MCDA), logistic regression (LR) and association rule mining (ARM). Eng Geol 203:45–55

Goldman RP, Cherniak E (1993) A language construction of belief networks. Pattern Anal Mach Intell 15(3):196–208

Haas TC (1991) A bayesian belief network advisory system for aspen regeneration. For Sci 37(2.), 1991):627–654

Henrion M 1989 Some practical issues in constructing belief networks. In: Kanal LN, Levitt TS, Lemmer JF (eds) Uncertainty in artificial intelligence (3), 161–173.

Kanungo DP, Arora MK, Sarkar S, Gupta RP (2006) A comparative study of conventional, ANN black box, fuzzy and combined neural and fuzzy weighting procedures for landslide susceptibility zonation in Darjeeling Himalayas. Eng Geol 85:347–366

Kanungo DP, Arora MK, Gupta RP, Sarkar S (2008) Landslide risk assessment using concepts of danger pixels and fuzzy set theory in Darjeeling Himalayas. Landslides 5:407–416

Kavzoglu T, Sahin EK, Colkesen I (2014) Landslide susceptibility mapping using GIS based multi-criteria decision analysis, support vector machines, and logistic regression. Landslides 11(3):425–439

Kayastha P, Dhital MR, Smedt FD (2013) Evaluation of the consistency of landslide susceptibility mapping: a case study from the Kankai watershed in east Nepal. Landslides 2013(10):785–799

Keefer DK, Larsen MC (2007) Assessing landslide hazards. Science 316(5828):1136–1138

Mallet FR (1875) On the geology and mineral resources of the Darjeeling District and the Western Duars. Geol Surv India Mem 11:1–50

Mamata D (2006) Identification and mapping of hazard prone areas regarding landslide in the Darjeeling hill areas. Department of Ecology & Environment, Netaji Institute for Asian Studies, 1, Woodburn Park, Kolkata, pp 6–17. Environment 65:167–184

Mondal S, Maiti R (2013) Integrating the analytical hierarchy process (AHP) and the frequency ratio (FR) model in landslide susceptibility mapping of shiv-khola watershed, Darjeeling Himalaya. Int J Disaster Risk Sci 4(4):200–212

Moreiras SM (2005) Landslide susceptibility zonation in the Rio Mendoza Valley, Argentina. Geomorphology 66:345–357

Nefeslioglu HA, Sezer E, Gokceoglu C, Bozkir AS, Duman TY (2010) Assessment of landslide susceptibility by decision trees in the metropolitan area of Istanbul, Turkey. Math Probl Eng 2010:1–15

Osna T, Sezer EA, Akgun A (2014) GeoFIS: an integrated tool for the assessment of landslide susceptibility. Comput Geosci 66:20–30

Ozdemir A, Altural T (2013) A comparative study of frequency ratio, weights of evidence and logistic regression methods for landslide susceptibility mapping: Sultan Mountains, SW Turkey. J Asian Earth Sci 64:180–197

Pareta K, Kumar J, Pareta U (2012) Landslide hazard zonation using quantitative methods in GIS. Int J Geospatial Eng Technol 1(1):1–9

Pearl J, Russell S (1998) Bayesian networks. Computer Science Department, University of California

Ruff M, Czurda K (2008) Landslide susceptibility analysis with a heuristic approach in the Eastern Alps (Vorarlberg, Austria). Geomorphology 94(3):314–324

Russell S, Norvig P, Intelligence A (1995) A modern approach. Artificial Intelligence. Prentice-Hall, Egnlewood Cliffs, 25, 27

Sarkar S, Kanungo DP (2004) An integrated approach for landslide susceptibility mapping using remote sensing and GIS. Photogramm Eng Remote Sens 70(5):617–625

Sarkar S, Roy AK, Martha TR (2013) Landslide susceptibility assessment using information value method in parts of the Darjeeling Himalayas. J Geol Soc India 82(4):351–362

Sezer EA, Pradhan B, Gokceoglu C (2011) Manifestation of an adaptive neuro-fuzzy model on landslide susceptibility mapping: Klang Valley, Malaysia. Expert Syst Appl 38(7):8208–8219

Soja R, Starkel L (2007) Extreme rainfalls in Eastern Himalaya and southern slope of Meghalaya Plateau and their geomorphologic impacts. Geomorphology 84(3–4):170–180

Starkel L, Basu S (eds) (2000) Rains, landslides and floods in the Darjeeling Himalaya. INSA, New Delhi, 168 pp

Upreti BN, Dhital MR (1996) Landslide studies and management in Nepal. ICIMOD, Nepal, p 87

Van Westen CJ, Van Asch TW, Soeters R (2006) Landslide hazard and risk zonation—why is it still so difficult? Bull Eng Geol Environ 65(2):167–184

Van Westen CJ, Castellanos E, Kuriakose SL (2008) Spatial data for landslide susceptibility, hazard, and vulnerability assessment: an overview. Eng Geol 102(3):112–131

Varnes DJ (1984) Landslide hazard zonation: a review of principles and practice. UNESCO, Paris, pp 1–63

Chapter 19
Landslides and Erosion Control Measures by Vetiver System

Chandan Ghosh and Shantanoo Bhattacharya

Abstract Hill roads and allied developmental issues have been in the perennial mode of reconditioning especially after rainy season, which are more aggravated due to climate change-induced flash floods and ill-planned human settlements. Even though technology of stabilized hill slopes has been brought into fruitful practices, bioengineering solutions provide ample scope. Investigation of landslides takes into account of portending scenarios of road widening works, multistory buildings juxtaposed on steep slopes with little or no breathing space, and dwindling lifeline infrastructures in hilly terrain amid unplanned construction activities. As preparedness measures, simple guidelines for crack repairs and treatment, restoration and strengthening methodology, applicability of some sustainable technology in landslide mitigation such as anchored earth and geosynthetics-reinforced walls, and bioengineering methods for slope stabilization are emphasized. This paper explains the application of vetiver grass, which is primarily of Indian origin but increasingly used by more than 100 countries. The intention of trying vetiver as one of the bioengineering measures is to see how it can help replace or complement the conventional engineering measures for slope and erosion problems. Present study illustrated some of the potential applications in the North-Eastern India through methodologies along with some successful and failed examples.

Keywords Vetiver system • Erosion • Landslides • Slope stability • Retaining wall • Geosynthetics • Soil nailing

---

C. Ghosh (✉)
GeoHazards Division, National Institute of Disaster Management, Ministry of Home Affairs, New Delhi, India
e-mail: cghosh24@gmail.com

S. Bhattacharya
Public Works Department (PWD), Guwahati, Assam, India
e-mail: shantanoo.bhattacharya@gmail.com

I. Pal, R. Shaw (eds.), *Disaster Risk Governance in India and Cross Cutting Issues*, Disaster Risk Reduction, DOI 10.1007/978-981-10-3310-0_19

19.1 Introduction

The use of vegetation as a bioengineering tool for erosion and drainage control has been implemented for centuries, but its popularity has increased in the last few decades. Many studies all over the world have shown that vetiver as a hedge is the ideal plant to conserve soil and rehabilitate eroded land. Nature has "designed" vegetation as a means to blanket and stabilize the good earth from erosion and conserve solar energy (Islam 2000). In the tropics or subtropical region, this has evolved into forests comprising big trees, shrubs, and leaf litters covering the organic humus-rich topsoil that offer excellent overall protection. In the light of current awareness and conscientiousness of environmental issues, the preferred option to address the above problems would be to go back and seek the solutions that nature has provided in varied forms. It's very much required to reinstate those areas ravaged by human beings by way of regrowing vegetation, i.e., the "green" or "soft" environment-friendly approach. This is in contrast to the conventional "hard" or "inert" engineering solutions using stones or concrete for protecting slopes (Truong 1999).

The reason why landslides occur in such a high frequency especially along the hill cut roads is to be realized in terms of efficacy of the slope maintenance measures, including drainages. The knowledge and understanding of landslides mechanism as one of the subjects in geology, geomorphology, hydrology, and geotechnical engineering have been phenomenal since the 1960s. However, the technology of landslides controls and mitigation roles played by construction agencies has to be looked into. In Fig. 19.1 rubble masonry wall is constructed along a national high-

Fig. 19.1 Rubble masonry gravity retaining wall along national highway, Himachal Pradesh, India

way in the state of Himachal Pradesh, India, to retain the cut slope. As has been experienced, efficacies of such walls are questionable, and they are not only costly but also unsuitable to the type of soil being tried to retain. Therefore, construction process is to be guided by proper landslide mitigation measures, including bioengineering measures, which will be mainly explained in the chapter.

Some indigenous techniques in controlling soil erosion and landslide problems are exemplified in Fig. 19.2, where bamboo poles are used as check dams. Drainage galleries are provided with concrete steps that facilitate the maintenance of the entire structure. As there is hardly any authentic documentation available for this kind of job, the proliferation of the same for mass-scale usage is limited. Hence, due to extensive skilled manual labor requirement, such practices remained limited application elsewhere. Such practices are quite uncommon these days as there is a wide-scale exploitation of mountain areas for the increased settlement densities, for which modern slope engineering techniques, such as soil nailing, reinforced earth retaining walls, gabion walls, etc., are applied. The efficacy of indigenous method as shown in Fig. 19.2 remains doubtful as they require regular monitoring.

Climate change/extreme weather phenomena vs. impact of deforestation and changes in the flora-fauna of mountains, possibly due to multipurpose hydropower projects and other tourism-driven vulnerable establishments along the towns and riverbanks in Uttarakhand state of India, have exposed badly during the June 2013 flash flood.

Fig. 19.2 Application of bamboo check dams for retaining water and concrete drain as one of the indigenous slope stabilization measures at Kalimpong, West Bengal (Photo: http://savethehills.blogspot.in/)

19.2 India: The Source of Vetiver Grass

Vetiver grass, although known in India centuries earlier and applied in specific locations with indigenous knowledge, only became known worldwide through the initiative of the World Bank in the 1980s, mainly in the agricultural sector. Later as the unique characteristics of vetiver became better known through scientific researches, it has emerged as an ideal plant for bioengineering and phytoremediation. In 1987, that soil erosion is the most serious agricultural problem in the world; Richard Grimshaw and John Greenfield, two agricultural scientists of the World Bank, decided to tackle it on war footing. During their visits to India, they tried the solution in a village near Gundlupet in Karnataka state:

> We learnt from these farmers that they have been successfully growing against soil erosion for centuries. It reduced rainfall runoff by as much as 70%, recharged groundwater (villages that use Vetiver have much higher water levels in their wells), and improved ephemeral stream flow.

It started to gain impressive grounds in other fields during the mid-1990s arising from some breakthrough researches that reveal the unique properties of this grass, *Vetiveria zizanioides*, that lends itself ideally for bioengineering and phytoremediation purposes, hence touted as a miracle grass, wonder grass, or super grass (Greenfield 1990). There are 12 known species and may be hundred different cultivars that exhibit distinctive phenotypic which can be exploited by users depending on need. Widely used varieties are *Vetiveria zizanioides* (Asia subcontinent), *Vetiveria nigratana* (Southern Africa), and *Vetiveria nemoralis* (South East Asia). Among many other functions, Vetiver System prevents soil erosion, preserves rain water, stabilizes earth, beautifies landscapes, purifies wastewater, keeps away pests, and stabilizes slope. Vetiver and its component parts have been widely developed for other miscellaneous uses, i.e., as construction materials, forages for livestock, landscaping and ornamentals, mulch, compost, veneer, fiber board, ash for concrete work, and insecticide. The grass also was brought to get rid of heavy metals from industry sewage, leachate from garbage, and take part in various industrial commercial products. It grows best on deep sandy soils; however, it grows on most soil types ranging from black cracking vertisols through to red alfisols. It also grows on rubble, both acid and alkali, and on both shallow and deep soils. Glasshouse and field studies showed that vetiver grass can produce high biomass (>100 t/ha/year) and highly tolerate extreme climatic variation such as prolonged drought, flood, submergence and temperatures (−15 to 50 °C), soils high in acidity and alkalinity (pH 3–11), high levels of Al (85% saturation percentage), Mn (578 mg/kg), soil salinity (ECse 47.5 dS/m), sodicity (ESP 48%), and a wide range of heavy metals (As, Cd, Cr, Cu, Hg, Ni, Pb, Se, and Zn) (Danh et al. 2009). The longest recorded life period is about 60 years in Zambia and 100 years in Fiji.

19.2.1 Genesis of Vetiver

The plant vetiver belongs to the same group of grass family of rice, maize, sorghum, sugarcane, and lemongrass. It got its generic name *Vetiveria*, from its Tamil name "vetiver" which means root that is dug up. *Zizanioides* means by "the riverside." The fact is that the plant is commonly found along the waterways. Its name was later reclassified as *Chrysopogon zizanioides* (Madhu and Haridas 2011). Vetiver has been known to India since ancient times. It has been considered as a high-class perfume, and copper plate inscriptions list the perfume as one of the articles used by royalty. Two species of *Vetiveria* are found in India, of which *V. zizanioides* is the common source of the well-known oil of vetiver, which is used in medicine and in perfumery. The species of Vetiver System (*Chrysopogon zizanioides*) originates in the state of Tamil Nadu, South India, which is now being promoted in nearly 100 countries. It is sterile and noninvasive and has to be propagated by clump subdivision.

In general, nursery multiplication of bare-rooted plants is the preferred method. The average multiplication rate varies but is normally, in a nursery, about 1:30 after about 3 months, depending on climate and growing conditions. The Vetiver Network International (http://vetiver.org/) is the most thorough and current resource for this plant. Nursery clumps are divided into planting slips of about three tillers each and typically planted 15 cm apart on the contour to create, when mature, a barrier of stiff grass that acts as a buffer and spreader of downslope water flow and a filter to sediment. A good hedge will reduce rainfall runoff by as much as 70% and sediment by as much as 90%. A hedgerow will stay where it is planted, and the sediment that is spread out behind the hedgerow gradually accumulates to form a long-lasting terrace with vetiver protection. It is a very low-cost, labor-intensive technology with very high benefit/cost ratios. When used for civil works protection, its cost is about 1/20 of traditional engineered systems and designs. Engineers attribute the vetiver root to a "living soil nail" with an average tensile strength of one sixth of mild steel.

The Council of Scientific and Industrial Research (CSIR), India, recently launched the "Aroma and Phyto-Pharmaceutical Mission" to boost cultivation of aromatic crops like vetiver, lavender, rosemary, and lemongrass and medicinal plants like ashwagandha and shatavari. The Central Institute of Medicinal and Aromatic Plants (CIMAP) has to lead the movement to promote cultivation of these crops especially in unproductive, marginal waste lands including those affected by water scarcity, drought, salinity, or flood.

19.2.2 Vetiver: Global Scenario

This indigenous knowledge was taken along when Indians migrated overseas and usage resumed in new localities around the globe. Thus, one reads of vetiver usage for slope protection and reinforcement of embankments and cuttings outside of

Table 19.1 Various vernacular names used for vetiver grass across the country

Dialect/ language	Vernacular name
Hindi, Bengali	Khas, Khas-Khas, Khus-Khus, Khus
Gujarati	Valo
Marathi	Vala
Telugu	Kuruveeru, Vettiveellu, Vettiveerum
Tamil	Vattiver
Kannad	Vattiveeru, Laamancha, Kaddu, KaridappasajjeHullu
Malayalam	Ramaccham, Vettiveru
Ayurvedic name	Ushira

farmlands since the early 1900s in the West Indies, South Africa (NRC 1993), Brazil (Grimshaw 1993), Fiji (Truong and Gawander 1996), etc. In 1931, it was on record that vetiver was grown at Serdang, near Kuala Lumpur, Malaysia, where it is used for holding up steep banks. It is well-known to be good for this purpose (World Bank 1995). Vetiver has been used extensively all over the world for erosion and sediment control, such as Brazil, China, the Congo, Australia, Indonesia, Venezuela, Vietnam, Thailand, India, the Philippines, Senegal, Zimbabwe, Nigeria, Colombia, El Salvador, Nicaragua, Guatemala, Costa Rica, and the southern United States. This paper presents some applications of Vetiver System in the erosion and landslide control for the hill states in India. In Western and Northern India, it is popularly known as khus (Table 19.1).

It has been extensively used as a cost-effective stabilization of karst stony slopes in high-altitude region, and the revegetation of barren quarried face by an innovative patented method by vetiver combined with other ancillary works was conceived and implemented in China. Riverbank stabilization has been successfully carried out on a major scale in fresh and brackish water environment in the Mekong Delta in Vietnam subjected to waves caused by motorized boat traffic as well as on the Hanjiang River (a Yangzi River tributary) in China (Islam 2011). Trials on the use of vetiver for beach protection were successfully achieved in Senegal, and slope stabilization of 100 km length of 18 coastal polders by vetiver was attempted in Bangladesh with varying success (Islam 2003). Flume tests were conducted in Australia to throw light on hydraulic characteristics of vetiver in deep flow that will aid in the design of channel stabilization and flood erosion control (Hengchaovanich 1998).

19.2.3 Application Domain

In addition to its very important application in agricultural lands, scientific research conducted in the last three decades has clearly demonstrated that VS is also one of the most effective and low-cost natural methods of environmental and infrastructure protection. Besides, it has great socioeconomic impact on local population and climate change. Historically, the order of development of the five main applications of VS is (Truong 2015):

1. Soil and Water Conservation in Agricultural Land
   (a) In agricultural land, vetiver hedges provided a very effective and low-cost method of soil and water conservation on sloping land and resulted in significant crop yield improvement.
2. Stabilization of Infrastructures
   (a) Its extensive and deep root system provides an ideal tool for erosion control of unconsolidated soil and the stabilization of steep slopes such as road and railway batters, dam wall, river and canal banks, and landslips.
3. Environmental Protection
   (a) Phytoremediation of wastewater: The Vetiver System can dispose and/or treat wastewater by reducing the volume or improving the quality of polluted water.
   (b) Phytoremediation of contaminated lands: Vetiver grass has been used successfully for rehabilitation of mine overburden and phytoremediation of mine tailings.
4. Socioeconomic Impact on Rural Community
   (a) Poverty alleviation: Providing income through supply planting materials and handicraft production
   (b) Rural employment to rural community particularly women and children
5. Positive Impacts on Climate Change
   (a) Disaster mitigation
   (b) Carbon sequestration
   (c) Biofuel
6. Agricultural Applications
   (a) Soil and water conservation
   (b) Erosion control
   (c) Slope stabilization
   (d) Embankment stabilization
   (e) Trapping of agrochemicals and nutrients
7. Nonagricultural Applications
   (a) Bioengineering
      (i) Erosion control
      (ii) Slope stabilization
      (iii) Embankment stabilization

8. Phytoremediation

   (a) Reclamation of Problem Soils

      (i) Reclamation of saline soils
      (ii) Reclamation of sodic soils
      (iii) Reclamation of acid sulfate soils
      (iv) Reclamation of other deteriorated soils

9. Rehabilitation of Contaminated Soils and Water

   (i) Rehabilitation of contaminated soils
   (ii) Treatment of mining spoils
   (iii) Treatment of landfills and garbage dumps
   (iv) Removal of agrochemicals and pesticides
   (v) Absorption of heavy metals

10. Rehabilitation of Contaminated Water

   (i) Water treatment and purification
   (ii) Treatment of eutrophicated water
   (iii) Wetland applications
   (iv) Removal of effluents

19.2.4 Medicinal Use

Vetiver widely cultivated in the tropical regions of the world is a miraculous grass native to India, first developed for soil and water conservation by the World Bank during the mid-1980s. Popularly known as "khus," it is the major source of the well-known oil of vetiver, which is used in medicine, in cosmetics, and in perfumery making agarbattis, soaps, soft drinks, and pan masala. Being a major constituent of "rasayana" in Ayurveda, different parts of the vetiver plant have traditionally been used by the Indian tribes for treating various ailments, diseases, and disorders including boils, burns, epilepsy, fever, scorpion sting, snakebite, sores in the mouth, headache, toothache, weakness, lumbago, sprain, rheumatism, urinary tract infection, malarial fever, and acidity relief and as an antihelminthic. It has also been used in traditional medicine of Asia and Africa; particularly ancient Tamil literature mentions the use of vetiver for medical purposes. The essential oil of vetiver has extensive applications in toiletries and cosmetics, possesses sedative property, and has traditionally been used in aromatherapy for relieving stress, anxiety, nervous tension, and insomnia. Root is also important in traditional medicine as a carminative, stimulant, and diaphoretic. Besides these medicinal properties of the plant, the dried culms of the plant are used as brooms and to thatch roofs. Pulp of the plant is used to prepare straw boards and paper. In India, the roots have been used for making screens, mats, hand fans, and baskets (Luqman 2011).

19.2.5 Disaster Prevention

Growing of vetiver hedges on contours and adopting conservation tillage practices between them have been proven to be an effective method to reduce runoff and soil loss and increase in situ moisture, thereby obtaining higher crop yields. Besides conventional slope stability measures such as Gabion, retaining wall, etc., solving erosion and landslide problems has been synonymous with disaster managers dealing in various other natural and man-made disasters. This paper presents some of the potential application of vetiver grass in India, with some case studies, where it has been applied successfully with potential applications. Soil erosion is a quiet crisis, largely man-made disaster that is unfolding gradually.

More than half of India's cropland is losing productivity because topsoil is being washed or blown away faster than natural forces can replace it. Reducing the topsoil layer means reducing plants' access to essential soil nutrients and water. For at least the next decade environmental issues will dominate the agricultural and natural resource sectors. Already the focus of much attention are the problems of deforestation, increased flooding by major rivers, and reduced dry-season water flows for irrigation and urban and industrial supplies. Not enough attention, however, has been given to the massive problem of soil erosion and, more specifically, to the need to reduce soil and water losses caused by excessive rainfall runoff. Changes in farming practices have made the problem worse in recent years. In response to the growing need for grain for exploding human and livestock populations, farmers switched from traditional rotations and multiple cropping to continuous-row cropping, a practice that encourages rainwater to run off the land at a faster and more destructive pace. As a result, crops are denied the moisture they need for optimum growth (Truong and Baker 1998).

Results of numerous trial and mass applications of vetiver grass in the last 30 years in many countries show that the grass is particularly effective in natural disaster reduction (flood, landslide, road batter failure, riverbank, irrigation canal and coastal erosion, water retaining structure instability, etc.), environmental protection (reduction of land and water contamination, treatment of solid and liquid waste, soil improvement, etc.), and many other uses.

19.3 Vetiver System

Vetiveria zizanioides (Linn.) Nash, a member of the family Poaceae commonly known as the *khas-khas*, *khas*, or *khus* grass in India, is a perennial grass with thick fibrous adventitious roots which are aromatic and highly valued. This tufted grass grows throughout the plains of India ascending up to an elevation of 1200 m. Having wide ecological amplitude, this grass grows in a wide variety of ecological habitats covering all biogeographic provinces of India. Vetiver is most closely related to sorghum but shares many morphological characteristics with other fragrant grasses, such as lemongrass (*Cymbopogon citratus*), citronella (*Cymbopogon nardus*, *C. winterianus*), and palmarosa (*Cymbopogon martinii*). It is a tall, tufted, perennial, scented grass, with a straight stem, long narrow leaves, and a lacework root system (Fig. 19.3) that is abundant, complex, and extensive. It offers an inexpensive yet effective and eco-friendly tool to combat soil erosion. The roots have been used in Asia for centuries for their fragrance and are woven into aromatic matting and screens. The roots of some cultivars and ecotypes possess essential oil that has been utilized as fragrant material since ancient times. The plant used as active ingredients in traditional medicine and also as botanical pesticide.

The plant vetiver belongs to the same group of grass family of rice, maize, sorghum, sugarcane, and lemongrass. It got its generic name *Vetiveria*, from its Tamil name "vetiver" which means root that is dug up. Its specific epithet, *Zizanioides*, means "the riverside." The fact is that the plant is commonly found along the waterways. Its name was later reclassified as *Chrysopogon zizanioides*. There are two types of vetiver; one is seedling and the other is non-seedling. Seedling type is commonly found in North India, whereas non-seedling variety is common in South India. Only the non-seedling variety can be recommended for any purpose connected with vetiver, because the other variety is able to spread unwantedly through seeds. The South Indian variety of vetiver has existed for centuries under cultivation and is widely distributed throughout the continents. This variety can be propagated only through vegetative method.

The plant can be grown over a very wide range of climatic and soil conditions and if planted correctly can be used virtually anywhere under tropical, semitropical, and Mediterranean climates. It has characteristics that in totality are unique to a single species. When vetiver grass is grown in the form of a narrow self-sustaining hedgerow, it exhibits special characteristics that are essential to many of the different applications that comprise the Vetiver System. Vetiver grass can be used for applications that will protect river basins and watersheds against environmental damage, particularly from point source factors relating to sediment flows (often associated with agriculture and infrastructure) and toxic chemical flows resulting from excess nutrients, heavy metals, and pesticides in leachate from agriculture and other industries. It can be established on very acid, sodic, alkaline, or saline soils.

(a) (b)

Fig. 19.3 (**a**) North Indian variety vetiver grass tillers with roots, (**b**) grown vetiver with long roots suitable in any soil with air temperature ranging from −15 to 50 °C at pH 3.0–11.0

Table 19.2 Comparison of metal absorption capacity of vetiver grass (Hengchaovanich and Nilaweera 1996)

	Threshold levels in soil (mg/kg) (available)		Threshold levels in plant (mg/kg)	
Heavy metals	Vetiver	Other plants	Vetiver	Other plants
Arsenic	100–250	20.	21–72	1–10
Cadmium	20–60	1.5	45–48	5–20
Copper	50–100	Not available	13–15	15
Chromium	200–600	Not available	5–18	0.02–0.20
Lead	>1500	Not available	>78	Not available
Mercury	>6	Not available	>0.12	Not available
Nickel	100	7–10	347	10–30
Selenium	>74	2–14	>11	Not available
Zinc	>750	Not available	880	Not available

Vetiver tolerates very high levels of aluminum and manganese and a range of heavy metals in the soil (Table 19.2). Due to its extensive and deep root system, vetiver is very tolerant of drought. It can stand extreme heat (50 °C) and frost (−15 °C) and can be established in areas with an annual rainfall from 450 mm and higher. Vetiver is sensitive to shade, and this will slow growth, especially in young plants.

19.4 Conventional Slope Treatment

Compared to traditional and modern system of slope stabilization (Fig. 19.4) available, VS is a very simple, practical, inexpensive, low-maintenance, and very effective means of soil and water conservation, sediment control, land stabilizations and rehabilitation, and phytoremediation (Islam et al. 2013a, b). Being vegetative, it is also environment-friendly. When planted in single rows, vetiver plants will form a hedge which is very effective in slowing and spreading runoff water, reducing soil erosion, conserving soil moisture, and trapping sediment and farm chemicals on-site. Although many hedges can do this, vetiver grass, due to its extraordinary and unique morphological and physiological characteristics described below, can do it better than all other systems tested. In addition, the extremely deep and massively thick root system of vetiver binds the soil and at the same time makes it very difficult for it to be dislodged under high-velocity water flows. This very deep and fast-growing root system also makes vetiver very drought tolerant and highly suitable for steep slope stabilization.

Typical sediment landslide control measures are hard engineering solutions such as gabions, rock fall netting, and geotextiles. Underappreciated and rarely used are biotechnical methods such as the use of vegetation for both erosion and sediment control, alone or in combination with other structures. The revegetation of slopes can be by means of grassing or leguminous cover crops (for minor surface move-

Fig. 19.4 Vetiver grass and its root system at varied stages of growth signifying root matrix suitable to hold soil from washing out

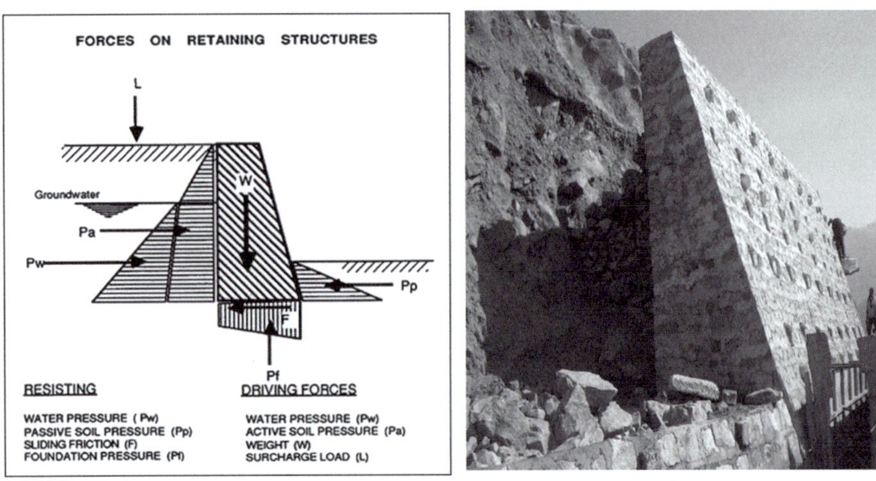

(a) (b)

Fig. 19.5 (**a**) Mechanics of traditional gravity retaining wall, (**b**) constructions of stone masonry walls based on the concept of earth retaining structures

ment) or the use of fast-growing shrubs and trees for the mitigation of deep-seated erosion in the order of 20–150 cm depths. Tree or scenario (Gray 1994) and shrub roots are able to grip and bind the soils needed to prevent the deep-seated surface slips in the event of heavy and prolonged rainstorms, while normal grasses are unable to do so. This is because roots or "inclusions" impart apparent cohesion (c) in similar to "soil nailing" or "soil doweling" in the reinforced soil principle (Fig. 19.5), thus increasing the safety factors of slopes permeated with roots vis-à-vis no-root.

19.5 Special Characteristics of Vetiver Grass

19.5.1 Morphological Characteristics

- Vetiver grass is listed in the Global Compendium of Weeds. Vetiver grass genotypes that produce viable seed exist in certain areas and countries, such as the Caribbean and Australia, and are considered invasive weeds.
- Vetiver grass does not have stolons or rhizomes. Its massive finely structured root system that can grow very fast, in some applications rooting depth can reach 3–4 m in the first year, thus checking erosion of soil. This deep root system makes vetiver plant extremely drought tolerant and difficult to dislodge by strong current.
- It has stiff and erect stems, which can stand up to relatively deep water flow.
- Highly resistant to pests, diseases, and fire.
- A dense hedge is formed when planted close together acting as a very effective sediment filter and water spreader.
- New shoots develop from the underground crown making vetiver resistant to fire, frosts, traffic, and heavy grazing pressure.
- New roots grow from nodes when buried by trapped sediment. Vetiver continues to grow up with the deposited silt eventually forming terraces, if trapped sediment is not removed.
- Trees and shrubs inherently have several drawbacks in that they are too slow to establish to become effective (even with fast-growing species, this process will take about 2–3 years) and the danger of being uprooted, in cases of heavy storms, typhoons, or cyclones. Vetiver does possess several treelike features.

19.5.2 Physiological Characteristics

- Tolerance to extreme climatic variation such as prolonged drought, flood, submergence, and extreme temperature from −15 to +50 °C. The optimal soil temperature for root growth is 25 °C. Root dormancy occurs under a temperature of 5 °C. Under frosty conditions, shoots become dormant and purple, or even die, but the underground growing points survive and can regrow quickly if the conditions improve.
- Ability to regrow very quickly after being affected by drought, frosts, salinity, and adverse conditions after the weather improves or soil ameliorants added.
- Tolerance to wide range of soil pH from 3 to 11 without soil amendment.
- High level of tolerance to herbicides and pesticides.
- Highly efficient in absorbing dissolved nutrients such as N and P and heavy metals in polluted water (Table 19.2).
- Highly tolerant to growing medium high in acidity, alkalinity, salinity, sodicity, and magnesium.
- Highly tolerant to Al, Mn, and heavy metals such as, Cd, Cr, Ni, Pb, Hg, Se, and Zn in the soils.

19.6 Vetiver for Slope Stabilization

In order to stabilize slopes live crib walls, vegetated rock gabions, vegetated rock walls, and joint plantings are some of the soil bioengineering techniques that use porous structures with openings through which vegetative cuttings are inserted and established. The inert structural elements provide immediate resistance to sliding, erosion, and washout. As vegetation becomes established, roots invade and permeate the slope, binding it together into a unified, coherent mass. Over time, the structural elements diminish in importance as the vegetation increases in strength and functionality. Vetiver's unique attributes have been researched, tested, and developed throughout the tropical world, thus ensuring that vetiver is one of the very effective bioengineering tools (Fig. 19.6):

- Although technically a grass, vetiver plants used in land stabilization applications behave more like fast-growing trees or shrubs. Vetiver roots per unit area are found stronger and deeper than tree roots.
- As strong or stronger than those of many hardwood species, vetiver roots have very high tensile strength that has been proven positive for root reinforcement in steep slopes.
- These roots have a mean tested tensile strength of about 75 megapascal (MPa), which is equivalent to one sixth of mild steel reinforcement and a shear strength increment of 39% at a depth of 0.5 m (Table 19.3 and 19.4).
- Slope engineering combines biological elements with engineering design principles practiced in civil-geotechnical discipline. The requirements for both must be considered when planning and designing the measures.

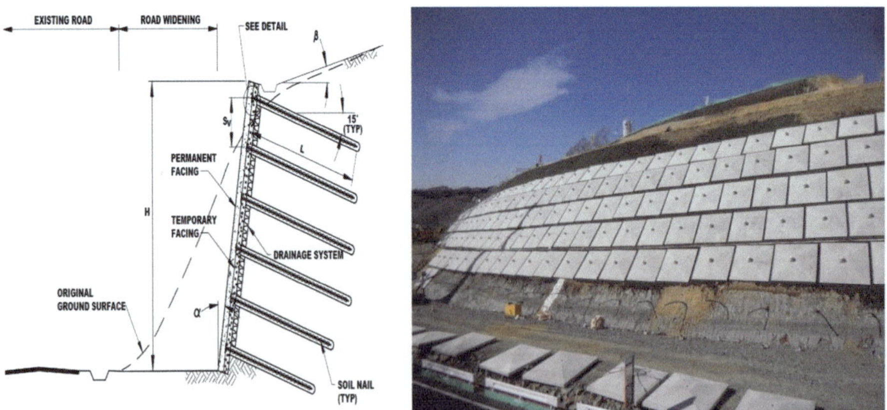

(a) (b)

Fig. 19.6 (**a**) Principles of soil nailing techniques, (**b**) application of the soil nails for slopes in Japan

Table 19.3 Tensile strength of some common grass plant roots (Hengchaovanich and Nilaweera 1996)

Botanical name	Common name	Tensile strength (MPa)
Salix spp.	Willow	9–36
Populus spp.	Poplars	5–38
Alnus spp.	Alders	4–74
Pseudotsuga spp.	Douglas fir	19–61
Acer saccharinum	Silver maple	15–30
Tsuga heterophylla	Western hemlock	27
Vaccinium spp.	Huckleberry	16
Hordeum vulgare	Barley grass	15–31
	Forbs moss	2–20
		2–7 kPa
Chrysopogon zizanioides	Vetiver grass	40–120 (average 75)

Table 19.4 Mean diameter and tensile root strength of various herbs (Hengchaovanich and Nilaweera 1996)

Grass	Mean diameter of roots (mm)	Mean tensile strength (MPa)
Late *Juncellus*	0.38 ± 0.43	24.50 ± 4.2
Dallis grass	0.92 ± 0.28	19.74 ± 3.00
White clover	0.91 ± 0.11	24.64 ± 3.36
Vetiver	0.66 ± 0.32	85.10 ± 31.2
Common centipede grass	0.66 ± 0.05	27.30 ± 1.74
Bahia grass	0.73 ± 0.07	19.23 ± 3.59
Manila grass	0.77 ± 0.67	17.55 ± 2.85
Bermuda grass	0.99 ± 0.17	13.45 ± 2.18

19.6.1 Slope Inspection

19.6.1.1 Topography and Exposure

- Note the degree of slope in stable and unstable areas. Also note the presence or lack of moisture. The likely success of slope treatment can best be determined by observing existing stable slopes in the vicinity of the project site.
- Note the type and density of existing vegetation in areas with and without moisture and on slopes facing different directions. Certain plants grow well on east-facing slopes, but will not survive on south-facing slopes.
- Look for areas of existing vegetation vis-à-vis proposed vetiver grass that may be growing more vigorously than other site vegetation (NRC 1993).

19.6.1.2 Geology and Soils

- Consult geologists about geologic history and types of deposits (colluvium, glacial, alluvium, etc.).
- Note evidence of past sliding. If site evidence exists, determine whether the slide occurred along a deep or shallow failure surface. Leaning or deformed trees may indicate previous slope movement or downhill creep. In addition to site evidence, check aerial photos, which can reveal features that may not be apparent from a site visit.
- Determine soil type and depth. Use the soil survey report, if available, or consult SCS soil scientists.

19.6.1.3 Hydrology

- Determine the drainage area associated with the problem area. Note whether water can be diverted away from the problem area.
- Determine the annual precipitation. Are there concentrated discharges?
- Calculate peak flows or mean discharge through the project area.
- If a seep area was noted, locate the source of the water. Determine whether the water can be intercepted and diverted away from the slope face (NRC 1993).

19.7 Erosion Control by Vetiver

Erosion control at the source is rarely addressed, and resultant sediment control is often underanalyzed, inappropriate, inadequate, or poorly implemented, resulting in frequent slope failures, water body contamination, and road closures. In mountainous terrain, erosion problems are mainly associated with land use changes from native forest to agriculture, grazing, timber harvesting, and road development. The change in land use, including road development, has mainly been done on the municipal and local level. Vetiver planting has been very effective in erosion control or stabilization (Fig. 19.6) in the following conditions:

- It is well adapted to the vegetative barrier practice used to control erosion on farmland because of its strong, compact root system, and numerous stiff stems.
- Slope stabilization along highways and railways especially effective along mountainous rural roads, where retaining structures are either costly or unsuitable (Fig. 19.6).
- Dike and dam batter stabilization, reduction of canal, riverbank and coastal erosion, and protection of hard structures themselves (e.g., rock riprap, concrete retaining walls, gabions, etc.).
- Slope above culvert inlets and outlets (culverts, abutments).
- Interface between cement and rock structures and erodible soil surfaces.

Fig. 19.7 Stabilization of cut slope as of July 2015 by vetiver grass along Guwahati–Shillong National Highway, Meghalaya State, India

- As a filter strip to trap sediment at culvert inlets.
- To reduce energy at culvert outlets.
- To stabilize gully head erosion, when vetiver hedges are planted on contour lines above gully heads.
- To eliminate erosion caused by wave action, by planting a few rows of vetiver on the edge of the high watermark on big farm dam batters or riverbanks (Fig. 19.7a, b).
- In forest plantations, to stabilize the shoulders of access roads on very steep slopes as well as the gullies (logging paths/ways) that develop following harvests.

19.8 Other Applications of Vetiver System

Bioengineering is a relatively new subbranch of civil engineering. It attempts to use live materials, mainly vegetation, on its own or in integration with civil engineering works to address the problems of erosion and slope stabilization. In the late 1980s and the following decade, due to heightened awareness of environmental issues and availability of knowledge and parameters of plants that can aid as well as lend credence to the designs, bioengineering became more well-known and accepted (Xia et al. 1999).

Vetiver hedgerows can reduce sediment/soil loss by 90%; reduce rainfall runoff by as much as 70%; significantly help in the maintenance of soil fertility; provide a source of byproducts that can be used as mulch, forage, and thatch; clean up farm pollution; recharge groundwater; be used as biofuel; and in some cases protect crops from insect damage (Grimshaw 2008):

- Vetiver's extremely deep and massive finely structured root system can extend down to 2 to 3 m in the first year. On fill slope, many experiments show that this grass can reach 3.6 m in 12 months. Vetiver does not penetrate deeply into the groundwater table. Therefore, at sites with a high groundwater level, its root system may not extend as long as in drier soil. Vetiver's extensive and thick root system binds the soil which makes it very difficult to dislodge, and it is found extremely tolerant to drought.
- Vetiver roots can penetrate a compacted soil profile such as hardpan and blocky clay pan common in tropical soils, providing a good anchor for fill and topsoil.
- When planted closely together, vetiver plants form dense hedges that reduce flow velocity, spread and divert runoff water, and create a very effective filter that controls erosion. The hedges slow down the flow and spreads it out, allowing more time for water to soak into the ground.
- Acting as a very effective filter, vetiver hedges help reduce the turbidity of surface runoff. Since new roots develop from nodes when buried by trapped sediment, vetiver continues to rise with the new ground level. Terraces form at the face of the hedges; this sediment should never be removed. The fertile sediment typically contains seeds of local plants, which facilitate their reestablishment (Truong et al. 1996).
- Vetiver grass regrows very quickly following drought, frost, salt, and other adverse soil conditions when the adverse effects are removed.
- Vetiver displays a high level of tolerance to soil acidity, salinity, solidicity, and acid sulfate conditions (Van Du and Truong 2003

19.9 Vetiver Grass Technology

The Vetiver Grass Technology (VGT) is a low-cost and extremely effective system for soil and water conservation, pollution control, wastewater treatment, mitigation and prevention of storm damage, and many other applications. Vetiver can be used in the tropics and semi-tropics, where there are hot summers and winters that do not include permanently frozen soil conditions (Man et al. 2003, 2011). Some of the specific areas where vetiver is beneficial are:

- *A permanent, low-maintenance solution*: Vetiver grass is a perennial plant, which provides a permanent solution with little or no maintenance.
- *Strong anchorage to soil*: The vetiver root mass is very large, and the fibrous roots are very strong (Figs. 19.3 and 19.4).
- *Deep anchorage*: Vetiver roots penetrate several meters into native soil or fill material.
- *A durable surface*: Vetiver foliage is tough; it survives fire and extended flooding.
- *Stops soil erosion*: The dense foliage traps soil particles being washed downhill.

- *Improves water quality*: Sediments are trapped by foliage before entering nearby waterways.
- *Water harvesting, flood mitigation*: Vetiver greatly increases percolation rates.
- *Supports local economies*: Vetiver projects are labor intensive. They employ locals, especially in rural areas.
- *A "Green" solution*: Vetiver is more natural and more attractive than stone or concrete.
- *Side benefits*: Vetiver foliage may be harvested for thatching, fodder, composting, or other purposes where biomass is required.
- *Vetiver is noninvasive*: It sets sterile seed and does not have running stolons.
- *Vetiver is noncompetitive*: Roots grow vertically downward; vetiver does not compete with adjacent plants.
- *Safe from pests and disease*: Vetiver has been shown to have very few pest or disease problems. Vetiver can check weed invasion too. It can block the spread of other grasses including the world's worst creeping grasses.
- *Precedents*: Vetiver has been proven in many projects in many countries around the world.

19.10 Vetiver as Sustainable Living

- *Harvest rainwater* – Vetiver hedges intercept and retain overland flows (storm runoff) and significantly increase soil porosity in the root zone.
- *Protect infrastructure* – Road shoulders/cuttings/banks, causeways, bridges, pathways, canals, and drainage systems.
- *Protect structures* – Stabilizes unconsolidated banks and cuttings and mitigates flood damage.
- *Protect coastlines* – Barrier to windblown sand; grows well in the littoral zone.
- *Protect waterways* – Stabilizes riverbanks and improves water quality (reduces sediment loads) by filtering runoff.
- *Stabilize sloping land* – Permanent bioengineering solution against sheet erosion, gully erosion, and landslides.
- *Protect flood-prone land* – Slows down overland flows and traps sediments.
- *Protect farmland* – Stabilizes slopes, riverbanks, and flood zones and does not compete with adjacent crops; vetiver is noninvasive.
- *Sequester carbon* – Creates a permanent, massive root system comprised mainly of carbon. Estimates of carbon sequestration have been made.
- *Facilitate reforestation and plantation establishment* – Increases survival rates and promotes rapid growth of tree seedlings.
- *Phytoremediation and bioremediation* – Phytoremediation (Greek: phyton, plant; Latin: remediare, remedy) refers to a green technology that uses plants to decontaminate polluted soils and water. It has gained popularity by leaps and bounds during the last few years because of the rediscovery of the vast potential of plants to do very effective jobs at such low costs compared to the "conven-

tional" cleanup solutions, using mechanical or chemical means. Vetiver grass has been shown to enhance the degradation of heavy metals such as aluminum, cadmium, chromium, copper, lead, and nickel and polycyclic aromatic hydrocarbons in the soil. It is used for wastewater treatment and rehabilitation of old mines.

- *Treat liquid wastes* – Nutrient removal via massive, fibrous root system and rapid biomass production, removes other pollutants including some heavy metals in leachates.
- *Bio-nailing of slopes* – Soil bioengineering, in the context of slope protection and erosion reduction, combines mechanical, biological, and ecological concepts to arrest and prevents shallow slope failures and erosion. Root system increases drainage and permeability which results in a decrease of excess pore pressures The root system increases soil shear strength via apparent cohesion and therefore the probability of the occurrence of a landslide is reduced. The tensile strength of the roots hinders crack forming. The tensile strength of the roots hinders pushing off of aggregates of soil. Bio-nailing measures such as live crib walls and brush layering are relatively complex and must be tailored carefully to specific soil and site conditions.

19.11 Suitability of Vetiver System

- It is a perennial plant that does not produce viable seed, stolons, or rhizomes but has a very fibrous and deep root system. It is noninvasive.
- The major advantage of VS over conventional engineering measures is its low cost and longevity. For slope stabilization in China, for example, savings are in the order of 85–90% (Xie 1997; Xia et al. 1999). In Australia, the cost advantage of VS over conventional engineering methods ranges from 64% to 72%, depending on the method used (Bracken and Truong 2000). Its maximum cost is only 30% of the cost of traditional measures. In addition, annual maintenance costs are significantly reduced once vetiver hedgerows are established.
- Vetiver System can be used effectively to control surface erosion and shallow failure of road batter.
- Vetiver System can be used effectively at slope between 30 and 60°.
- Vetiver System could be applied by road authorities to cope with erosion and shallow failure of road slope.
- At road slope >60°, Vetiver Technology is not recommended to be applied solely (must be combined with geotextiles and/or mechanical methods).
- As with other bioengineering technologies, VS is a natural, environment-friendly way to control erosion control and stabilize land that "softens" the harsh look of conventional rigid engineering measures such as concrete and rock structures. This is particularly important in urban and semirural areas where local communities decry the unsightly appearance of infrastructure development.

- Long-term maintenance costs are low. In contrast to conventional engineering structures, green technology improves as the vegetative cover matures. VS requires a planned maintenance program in the first 2 years; however, once established, it is virtually maintenance-free. Therefore, the use of vetiver is particularly well suited to remote areas where maintenance is costly and difficult.
- Vetiver is very effective in poor and highly erodible and dispersible soils.
- VS is particularly well suited to areas with low-cost labor forces.
- Vetiver hedges are a natural, soft bioengineering technique, an eco-friendly alternative to rigid or hard structures.

19.12 Limitation of Vetiver System

- The main disadvantage of VS applications is the vetiver's intolerance to shading, particularly within the establishment phase. Partial shading stunts its growth; significant shading can eliminate it in the long term by reducing its ability to compete with more shade-tolerant species. However, this weakness could be desirable in situations where initial stabilization requires a pioneer to improve the ability of the microenvironment to host the voluntary or planned introduction of native endemic species.
- Constraints on planting times or the availability of the required quantities of suitable plant materials during allowable planting times may limit the usefulness of VS.
- The Vetiver System is effective only when the plants are well established. Effective planning requires an initial establishment period of about 2–3 months in warm weather and 4–6 months in cooler times. This delay can be accommodated by planting early and in the dry season.
- The biotechnical usefulness of vetiver grass would be limited on slopes that are exposed to high-velocity water flow or constant inundation during initial phase.
- Vetiver hedges are fully effective only when plants form closed hedgerows. Gaps between clumps should be timely replanted.
- It is difficult to plant and water vegetation on very high or steep slopes.
- Because of the harsh environment on dune, beaches, and anti-salt dikes, planting bare-root slips should be avoided as the survival rate is slow and the plants are slower to take hold.
- Vetiver requires protection from livestock during its establishment phase.

19.13 Status of Vetiver Technology

Based on the feedback by the vetiver users around the world, the responses are adapted as below (Grimshaw 2008):

(a) *Lack of knowledge and technology dissemination*: This covers a wide range including ignorance of the technology by administrators, policy makers and planners, uninformed technical professionals and lack of profession endorsement, teaching and learning limitations in universities and schools, limited press coverage, absence of mass marketing, lack of publications, language barriers, and not using modern marketing tools.
(b) *Leadership*: New technology introduction requires farsighted leadership with vision and commitment. A committed lead organization is required. Good NGOs and private sector companies can often do this best. Commitment is rarely found in government organizations.
(c) *Low-cost solution and problems thereof*: Not always, but generally VS is seen as a low-cost technology that does not attract high budget allocations, and therefore the opportunities and attractiveness for corrupt practices are much less than for high-cost alternatives.
(d) *Technology*: Majority of solutions have in the past an engineering base. Most engineers have not been trained in bioengineering solutions, particularly those that are low cost. Low-cost biological solutions are often seen as too simple and as such are unattractive. Again applying low-cost solutions results in lower fees for designers and executing contractors. Many higher-cost engineering solutions do not always last long and have to be replaced.
(e) *Specifications*: Specifications and standards should be followed; bad application generally results in failure and detracts potential users. Site specification is important. Often rather general standards are given and followed, and if not properly supervised and fine-tuned, it can lead to failure.
(f) *Multipurpose use*: For some potential user groups such as railway and highway engineers, it is best to have narrowly focused workshops and training on the application at hand. For other users such as farmers and rural planners, there is a need to look at the wider aspects and the multi-benefits that are possible from VS.
(g) *Plant propagation*: Because vetiver has to be vegetatively propagated, an upfront investment and lead time are required. This can be detraction. However, there are plenty of demonstrations showing that small farmer private nurseries can be quickly established if there is a guaranteed market.
(h) *Invasive species and native plant syndrome*: This is more of a problem in developed countries. Sometimes deliberate miscasting of vetiver as an invasive species. Many government projects in the United States will only use native plants. Also entrenched vested interests in other more "profitable" technology work hard to keep VS out, and the "invasive" slur is a handy tool to frighten unaware decision-makers.
(i) *Research*: Some research has been very adequate, but more has to be done for field application.
(j) *Maintenance requirements*: Once vetiver is well established on the selected site, usually within one growing season, it generally becomes self-repairing by regeneration and growth and requires little maintenance. However, a newly installed project will require careful periodic inspections until it is established.

Established vegetation is vulnerable to trampling, drought, grazing, nutrient deficiencies, toxins, and pests and may require special management measures at times.

(k) *Limitation*: Vetiver System (VS) measures should not be viewed as a panacea or solution for all slope failure and surface erosion problems. VS has unique attributes, but is not appropriate for all sites and situations. In certain cases, a conventional vegetative treatment (e.g., grass seeding and hydro-mulching) works satisfactorily at less cost. In other cases, the more appropriate and most effective solution is a structural retaining system alone or in combination with soil bioengineering.

(l) *Vetiver growth*: Vetiver plantation is suitable for protecting slopes in different geographic areas with different soils and climatic conditions. However, the growth of vetiver roots and shoot varies in different areas. These are due to the differences in soil type, nutrient content, salinity, and climatic conditions. Even in high saline zone area, vetiver plantation is a suitable solution to protect the side slopes of shrimp ponds from flood and wave actions.

(m) *Livelihood propagation*: Vetiver grass is traditionally used by communities (Fig. 19.8) for livelihoods, however facilitation of Vetiver products across the country needs motivation and cooperation from government.

As a way forward after the Sixth International Conference on Vetiver (ICV), held in May 2015, Vietnam, it was opined that this VS technology must be used across sectors by communities to resolve a number of climate change issues without having to resort to external funding or assistance thus enabling communities to sustain and improve their quality of life through betterment of their environment (Truong 2015) (Fig. 19.9).

19.14 Conclusions

Vetiver Grass Technology (VGT) has so far been applied as soil conservation technique, and it is being used a bioengineering tool which involves significant engineering design and construction that requires an understanding of biology, soil science, hydraulics, hydrology, and geotechnical principles. The layout design varies with slope gradient, cut or fill slope, soil types, and rainfall of a particular site. It has been playing role in soil and water conservation, infrastructure stabilization, pollution control, wastewater treatment, mitigation and rehabilitation, sediment control, prevention of storm damage, and many other environmental protection applications through bioengineering and phytoremediation. Vetiver, although known as a grass, does possess several treelike features. It therefore becomes an attractive alternative to trees or shrubs when it comes to bioengineering applications. Some of the potential applications of Vetiver System across the 100+ countries demonstrate that this plant, even though originated in India, needs extensive research and development in Indian context. Parting with traditional ways to slope

(a)

(b)

Fig. 19.8 (**a**) Application of geotextile mat (GeoMat) for riverbank erosion control and (**b**) use of Vetiver System at a site in Assam state, India

stabilization, there are several modern earth reinforcing techniques that have been applied to India, which are mostly propitiated by select few industries but not formalized in codal practices in India. Excepting a mere mention in the recent IRC code for the application of vetiver system, many nodal organizations are hesitant to adopt VS. However, a combination of modern techniques with Vetiver System has

Fig. 19.9 Utilization of vetiver grass (khus) after extracting oil for the livelihood regeneration, India

to be emphasized and in order to harness that geo-professional practices needs to evolve technique to apply bioengineering solution in soil erosion and landslide problems. Following considerable research and the successes of the many applications presented elsewhere, it is now established that vetiver, with its many advantages and very few disadvantages, is a very effective, economical, community-based, and environment-friendly sustainable bioengineering tool that protects infrastructure and mitigates natural disasters. Once established, the vetiver plantings will last for decades with little, if any, maintenance. However, it must be stressed that the most important keys to success are good-quality planting material, proper design, and correct planting techniques. By all count vetiver is the new-generation green technology everyone is going to adopt for the future.

Acknowledgment VGT becomes more well-known and accepted in bioengineering through the diligent efforts by the Vetiver Network International (TVNI) with membership exceeding 100 countries, including its regional allied networks, the Royal Development Projects Board (RDPB), through the publication of its newsletters and technical bulletins as well as the dedication of many concerned individuals. The compilation made by the first author contains direct reference to the experts working in this area for more than three decades. The author acknowledges all who have been responsible for the promotion of this miracle grass for the bioengineering tool. Field application of Vetiver System in India has been learned through Mr. P. Haridas, C.K. Ashoke Kumar, Mr. Deepak C. Das, Mr. Agosta Basu, Mr. Gautam Chandola, C.S. Pradeep Kumar, and many others. The author deeply acknowledges their supportive role in the VS applications and continued interests.

References

Bracken N, Truong PN (2000) Application of Vetiver grass technology in the stabilization of road infrastructure in the wet tropical region of Australia. Proc. Second International Vetiver Conf. (ICV-2), Thailand

Danh LT, Truong P, Mammucari R, Tran T, Foster N (2009) Vetiver grass, Vetiveria zizanioides: a choice plant for phytoremediation of heavy metals and organic wastes. Int J Phytoremediat 11(8):664–691. doi:10.1080/ 15226510902787302

Gray DH (1994) Influence of vegetation on stability of slopes. In: Vegetation and slopes. Institution of Civil Engineers, London, pp 2–25

Greenfield JC (1990) Vetiver grass -the hedge against erosion, 3rd edn. The World Bank, Washington, DC

Grimshaw RG (1993) Soil and moisture conservation in central America, vetiver grass technology, Observations from visits to Panama, Costa Rica, Nicaragua, El Salvador, Honduras, and Guatemala. Asia Technical Department, The World Bank, Washington, DC

Grimshaw D (2008) An annual report of Vetiver system. http://vetivernetinternational.blogspot.in/

Hengchaovanich D (1998) Vetiver grass for slope stabilization and erosion control. Tech. Bull. No.1998/2, PRVN / ORDPB, Bangkok, Thailand

Hengchaovanich D, Nilaweera N (1996) An assessment of strength properties of Vetiver grass roots in relation to slope stabilization, Proceedings of the first International Conference on vetiver (ICV-1), Bangkok edn. Office of the Royal Development Projects Board, pp 153–158

Islam MN (2000) Embankment erosion control: towards cheap and simple practical solutions for Bangladesh. Proceedings of the Second International Conference on Vetiver. Office of the Royal Development Projects Board, Bangkok. Pp.307–321

Islam MN (2003) Role of Vetiver in controlling water-borne erosion with particular reference to Bangladesh coastal region. In: Proceedings of the 3rd International Conference on Vetiver (ICV-3), Guangzhou, China, pp 358–367

Islam MN (2011) Riverbank erosion and sustainable protection strategies. J Eng Sci 2., 2011:63–72

Islam MS, Nasrin S, Islam MS, Moury FR (2013a) Use of vegetation and geo-jute in erosion control of slopes in a sub-tropical climate. World Acad Sci Eng Technol 73:1162–1170

Islam MS, Shahriarl BAM, Shahin H (2013b) Study on growth of Vetiver grass in tropical region for slope protection Int. J GEOMATE 5(2):729–734

Luqman S (2011) Investigations on biological activity of Vetiveria zizanioides L. Nash, a palin genesis of some important findings in miracle grass. Proceeding of The Fifth International Conference on Vetiver (ICV-5), Lucknow, India

Madhu AK, Haridas P (2011) Protection of river banks by using Vetiver system technology in Malappurum district of Kerala, Proc. Fifth International Vetiver Conf. (ICV-5), Lucknow, India

Man Tran Van, Elise Pinners, Truong P (2003) Some results of the trial application of Vetiver grass for sand fly, sand flow and river bank erosion control in Central Vietnam. Proc. Third International Vetiver Conf. (ICV-3) China

Man Tran Van, Huynh Van Thang, Truong P (2011) VST in river and canal bank stabilisation in central Vietnam: successes and failures ten years later, Proc. Fifth International Vetiver Conf. 9ICV-5, Lucknow, India

National Research Council (1993) Vetiver grass: a thin green line against erosion. National Academy Press, Washington, DC

Truong P (1999) Vetiver grass technology for land stabilisation, erosion and sediment control in the Asia Pacific region. Proc. First Asia Pacific Conference on Ground and Water Bioengineering for Erosion Control and Slope Stabilization, Manila, the Philippines

Truong P (2015) Summary and highlights of papers Submitted to ICV-6, Source: http://www.vetiver.org/g/conferences.htm, 28 pages

Truong P, Baker D (1998) Vetiver grass system for environmental protection. Technical Bulletin 1998/1. Pacific Rim Vetiver Network, Office of the Royal Development Projects Board, Bangkok, Thailand

Truong P, Gawander JS (1996) Back from the future: do's and don'ts after 50 years of vetiver utilisation in Fiji. Proc. of the First International Conference on vetiver (ICV-1). Office of the Royal Development Projects Board, Bangkok, pp 12–17

Truong P, Gordon I, Baker D (1996) Tolerance of vetiver grass to some adverse soil conditions. Proc. first International vetiver Conf. (ICV-1), Thailand

Le Van Du, Truong P (2003) Vetiver system for erosion control on drainage and irrigation channels on severe acid sulfate soil in Southern Vietnam. Proc. Third International Vetiver Conf. (ICV-3), China

World Bank (1995) Vetiver grass for soil and water conservation. Technical Paper No. 273, RG Grimshaw (ed), Washington DC, USA

Xia HP, Ao HX, Liu SZ, He DQ (1999) Application of the Vetiver grass bio-engineering technology for the prevention of highway slippage in southern China. International Vetiver Workshop, Fuzhou, China, October 1997

Xie FX (1997) Vetiver for highway stabilization in Jian Yang County: Demonstration and Extension. Proceedings International Vetiver Workshop, Fuzhou, China, October 1997

MIX
Papier aus verantwortungsvollen Quellen
Paper from responsible sources
FSC® C105338

If you have any concerns about our products,
you can contact us on
ProductSafety@springernature.com

In case Publisher is established outside the EU,
the EU authorized representative is:
Springer Nature Customer Service Center GmbH
Europaplatz 3, 69115 Heidelberg, Germany

Printed by Libri Plureos GmbH
in Hamburg, Germany